MODERN MATHEMATICAL MODELS OF TIME AND THEIR APPLICATIONS TO PHYSICS AND COSMOLOGY

Proceedings of the International Conference held in Tucson, Arizona, 11–13 April, 1996

Edited by

W.G. TIFFT

Department of Astronomy, University of Arizona,
Tucson, Arizona, U.S.A.

and

W.J. COCKE

Department of Astronomy, University of Arizona,
Tucson, Arizona, U.S.A.

Reprinted from *Astrophysics and Space Science*
Volume 244, Nos. 1–2, 1996

SPRINGER-SCIENCE+BUSINESS MEDIA, B.V.

A C.I.P. Catalogue record for this book is available from the Library of Congress.

ISBN 978-0-7923-4663-0   ISBN 978-94-011-5628-8 (eBook)
DOI 10.1007/978-94-011-5628-8

Printed on acid-free paper

TABLE OF CONTENTS

Session 7. Mathematical Models and Methods

JOINT INTERNATIONAL CONFERENCE ON MATHEMATICAL MODELS OF TIME AND THEIR APPLICATIONS TO PHYSICS AND COSMOLOGY

ORGANIZED BY
Anthony P. Pitucco, William G. Tifft
Carl L. DeVito and William J. Cocke

1. Preface

The following proceedings are the result of a Joint International Conference on Mathematical Models of Time and Their Applications to Physics and Cosmology. The conference was held on April 11 through April 13, 1996, in conjunction with the University of Arizona and Pima Community College West Campus in Tucson, Arizona. The conference was made possible by a grant from Research Corporation, Tucson, Arizona; matching support was provided by the Steward Observatory and the Department of Mathematics at the University of Arizona, and by Pima Community College. The conference brought together a number of researchers in astronomy, astrophysics, cosmology, gravitational physics, particle physics, and mathematics. The purpose was to summarize known and new information relating to the nature of time and discuss how the properties of time might relate to these various areas.

The Idea for such a conference came from a research group consisting of Dr. William G. Tifft, Dr. John Cocke, Department of Astronomy, the University of Arizona; Dr. Carl DeVito, Department of Mathematics, the University of Arizona; and Dr. Anthony Pitucco, Department of Physics, Pima Community College. This group has been examining possible ways to formulate a theory to explain recent findings in physics and astrophysics which appear to relate to properties of time. Over the past 20 years various models have been explored in connection with phenomena such as quantum time, 3-dimensional time, non-Doppler redshifts, and particle properties. It was felt that by bringing together a group of current workers in these fields insight might be provided to assist in formulating a new model relevant to a broad spectrum of researchers.

Astrophysics and Space Science **244**: 1-6, 1996.

The conference was organized in seven sessions consisting of invited reviews and short contributed papers which are reproduced in these proceedings. Questions and panel discussions held at the end of each day were not recorded, but participants have been encouraged to incorporate results of such discussion in their final manuscripts. The papers were presented in in seven sessions distributed over three days. The registered participants for the conference are listed following this introduction. A strong participation by students was encouraged, and a public lecture by Paul Davies served to broaden the scope of the conference beyond the confines of the technical sessions. The seven technical sessions covered the following topics:

1) **The Redshift** Work in two areas, discordant or excess redshifts, and quantized redshifts were reviewed by H. C. Arp and W. G. Tifft. Shorter contributions were given by J. W. Sulentic, W. M. Napier, and Y. Terzian. The association of objects with widely discordant redshifts, and the organization of redshifts into a quantized hierarchy provide primary data which, if correct, must be incorporated into models of cosmology. One way to include such effects involves new models of time.

2) **Critical Properties of the Universe** Reviews by Li-Zhi Fang, S. A. Gregory, and A. L. Peratt, and a short contribution by J. G. Laros were designed to bring out background information and current ideas relating to known properties of the Universe which any model should consider.

3) **Statistical Methods** Reviews given by W. M. Napier, W. I. Newman, and W. J. Cocke and a short contribution by P. A. Sturrock focussed on the testing of results derived from observations which are at variance with classical theory. Since such data are critical to developing new theories it is imperative that statistical testing, by methods such as spectral power analysis, be carefully examined and verified by direct simulations.

4) **New Approaches to Cosmology** Reviews by G. Burbidge, J. V. Narlikar, and W. G. Tifft examined the general need for, and possible models for, alternative cosmologies which can incorporate some types of data at variance with classical models. This includes Machian models with time dependence of mass and 3-dimensional time where quantized redshifts and properties of fundamental particles and forces may be determined.

5) **Gravitation and Time in General Relativity** Reviews by W. J. Cocke, and R. Penrose discussed various extensions of relativity theory relating to aspects of time. Aspects of Machian models were discussed by D. F. Roscoe, and properties of the gravitational force were reviewed by T. Van Flandern. A contribution relating to the role of the cosmological constant was given by P. C. W. Davies, and L. Halpern contributed a paper discussing views on the limits of physical laws. Two papers read in absentia

are included here, a review by M. Sachs and a contribution by M. Saniga. Gravitation and its connection to time and cosmology stands out as an area where extensive theory exists but where major critical unsolved problems may require important changes.

6) **Nuclear and Particle Physics** A review by V. Icke examined the problem which arises when one attempts to relate the current standard model in particle physics with general relativity on the large scale. A. Lehto reviewed concepts which use 3-d time to construct fundamental particles and describe some of their properties, such as magnetic moments, and C. Wolf reviewed quantum processes which might arise out of conditions in the early universe. Contributions by A. C. Elitzur, L. W. Morrow, and M. Kokus expand upon experimental and theoretical details. These diverse papers touch in various ways on the uncertainties which underlie our understanding of the fundamental structure of matter and how these relate to large scale space-time physics and cosmology. These papers serve in part to define the scope of the problem; possible connections between the redshift and particle properties suggest that understanding the nature of time will be critical to relating these extremes.

7) **Mathematical Models and Methods** Reviews by C. L. DeVito and A. P. Pitucco began the development of a general mathematical structure for looking at multi-dimensional time without restrictions directly related to spatial concepts. DeVito gives an axiomatic treatment of a non-linear model of time. Pitucco uses a geometric construction from classical mechanics to find temporal analogies for some of the concepts used in the construction. Contributions by W. M. Stuckey and B. R. Frieden examined other approaches related to the study of time. Time in conventional physics is viewed as a simple one-dimensional parameter. This restricted viewpoint may obscure some very fundamental new possibilities in physics.

The conference brings together a range of subject matter and controvertial ideas which are sometimes individually mentioned, but rarely looked at in a comprehensive manner. The common thread which ties the subjects together seems to point toward a need for a better understanding of the nature of time. There are serious limitations in the great existing theories and critical measurements, especially at the cosmological scale, which suggest that new concepts of multi-dimensional time may be important. This conference has attempted to define the problem more clearly, and perhaps begin a serious effort to look at alternative models of time.

There are a number of individuals whose efforts were critical to the success of the conference. The conference grant was administered through the Pima College Foundation. We would particularly like to thank Joseph Nevin, executive director, and Joyce Gee, who oversaw the registration and

Figure 1.   The conference organizers, from left to right, A. Pitucco, W. J. Cocke, W. G. Tifft and C. L. Devito. Photo by Claire Cocke.

business matters. Buddy Powell and Alan Newell, at the University of Arizona, and Robert Jensen, Chancellor, Carol Gorsuch, Vice Chancellor, and Colin Campbell and Chuck Wachsman, Associate Deans, at Pima Community College, provided additional funds. We wish to express our gratitude to G. Burbidge, P. Sturrock, M. Stuckey, P. Strittmatter, L. Halpern, L. Fagg, and S. Barr, who chaired the technical sessions. H. Arp provided a final summary and S. Barr presented the paper of M. Sachs in absentia. At the University of Arizona, the administrative staffs of Steward Observatory and the Department of Mathematics were extremely helpful, as were those the Pima Community College West Campus. The staff at the Rodeway Inn North Tucson, particularly Steffenie Lehman and Karilyn Daniels, provided essential help in arrangements at the host hotel. Jim Dotzler assisted with transportation. Claire Cocke acted as photographer, and Janet Tifft provided the artistic renditions which introduce the daily sessions in these proceedings.

The organizing committee consisted of Anthony P. Pitucco of Pima Community College and Willian G. Tifft, Carl L. DeVito, and William J. Cocke at the University of Arizona. These proceedings were assembled and edited by William G. Tifft and William J. Cocke. They were prepared in part through the facilities at the National Radio Astronomy Observatory at Green Bank, West Virginia.

2. Participants

Halton Arp, Max Planck Institut fur Astrophysics, Garching, Germany
Antonio Aurilia, California State Poly. Univ., Pomona, Calif.
Stuart Barr, Pima Community College, Tucson, Ariz.
Geoffery Burbidge, Univ. California at San Diego, La Jolla, Calif.
William Cocke, Univ. Arizona, Tucson, Ariz.
Paul Davies, Univ. Adelaide, Adelaide, Australia
Harry Dart, Tucson, Arizona
Carl DeVito, Univ. Arizona, Tucson, Ariz.
Jim Dotzler, Univ. Arizona, Tucson, Ariz.
Avshalom Elitzur, Tel Aviv Univ., Tel Aviv, Israel
Lawrence Fagg, Catholic Univ. Amer., Washington, D.C.
Li-Zi Fang, Univ. Arizona, Tucson, Ariz.
Roy Frieden, Univ. Arizona, Tucson, Ariz.
Chris Fulton, Main Sequence Systems, Canyon Country, Calif.
Robert Gentry, Earth Science Assoc., Knoxville, Tenn.
Steven Gregory, Univ. New Mexico, Albuquerque, New Mex.
Leopold Halpern, Florida State Univ., Talahasse, Florida
Kieth Hege, Univ. Arizona, Tucson, Ariz.
Shandelle Henson, Univ. Arizona, Tucson, Ariz.
Vincent Icke, Sterrewacht Leiden, Leiden, Netherlands
Kurt Just, Univ. Arizona, Tucson, Ariz.
Roy Keys, Montreal, Canada
Martin Kokus, Hopewell, Penn.
John Laros, Univ. Arizona, Tucson, Ariz.
Flectcher Lance, Free Lance Academy, Jersey City, New Jersey
Ari Lehto, Univ. Helsinki, Helsinki, Finland
Jim Liebert, Univ. Arizona, Tucson, Ariz.
Roger Lynds, Natonal Optical Astron. Obs., Tucson, Ariz.
Clifford Matthews, Wilmette, Illinois
Thomas Miller, Columbia, Univ., Cooperstown, New York
LeRoy Morrow, Pittsburgh, Penn.
William Napier, Armagh Obs., Armagh, Northern Ireland
Jayant Narlikar, Inter-Univ. Centre for Astron. Astroph., Pune, India
William Newman, Univ. California at Los Angeles, Los Angeles, Calif.
Goetz Oertel, A.U.R.A., Washington, D.C. (in absentia)
Roger Penrose, Oxford Univ., Oxford, England
Anthony Peratt, Los Alamos National Lab., Los Alamos, New Mex.
Peter Phillips, Washington Univ., St Louis, Mo.
Anthony Pitucco, Pima Community College, Tucson, Ariz.
Y. Pinelis, Univ. Arizona, Tucson, Ariz.

David Roscoe, Sheffield Univ., Sheffield, England
George Rieke, Univ. Arizona, Tucson, Ariz.
Mendel Sachs, State Univ. New York at Buffalo, Buffalo, New York (in absentia)
Metod Saniga, Slovak Acad. Sci., Slovak Republic (in absentia)
Nigel Sharp, Natonal Optical Astron. Obs., Tucson, Ariz.
Shannon Smith, Tampa, Fla.
Peter Strittmatter, Univ. Arizona, Tucson, Ariz.
Mark Stuckey, Elizabethtown College, Elizabethtown, Penn.
Peter Sturrock, Stanford Univ., Stanford, Calif.
Jack Sulentic, Univ. Alabama, Tuscaloosa, Ala.
Yervant Terzian, Cornell Univ., Ithica, New York
William Tifft, Univ. Arizona, Tucson, Ariz.
Tom Van Flandern, Meta Res., Washington, D.C.
Kenneth Wallace, Paradise Valley, Arizona
Carl Wolf, North Adams State College, North Adams, Mass.
Xiang-Ping Wu, Univ. Arizona, Tucson, Ariz.

3. Student Participants

S. Agut, V. Alexandrov, D. Anderson, P. Benjamin, J. Christensen, J. Cuny, M. Elowitz, N. Goodman, S. Gottily, J. Glasscock, K. Green, B. Helgason, S. Hoell, C. Hoey, P. Kolumbus, R. Lorentz, M. Murphey, K. Pentland, M. Ramirez, M. Rippa, J. Savredo, W. Sarlls, B. Smith, D. Souza, S. Taunton, J. Tollakson, J. Williams, C. Yarbrough

THE REDSHIFT
CRITICAL PROPERTIES OF THE UNIVERSE
STATISTICAL METHODS

DAY 1
Rodeway Inn, North

Figure 1. Sketch art courtesy Janet A. Tifft

THE PAIR OF X-RAY SOURCES ACROSS NGC 4258: ITS RELATION TO INTRINSIC REDSHIFTS, EJECTION AND QUANTIZATION

HALTON ARP

Max–Planck–Institut für Astrophysik
85740 Garching, Germany

Abstract. The chance that the pair of X-ray sources observed across NGC 4258 is accidental can be calculated as 5×10^{-6}. The recent confirmation as quasars, and determination of the redshifts of the pair, at $z = 0.40$ and 0.65 by E.M. Burbidge enables the final accidental probability of the configuration to be calculated as $< 4 \times 10^{-7}$. In addition there are a number of observations which indicate the central Seyfert galaxy is ejecting material from its active nucleus.

The NGC 4258 association is compared to four other examples of close association of pairs of X-ray quasars with low redshift galaxies. It is concluded that in each of these five cases the chance of accidental association is less than one in a million. The ejection speed calculated from the redshift differences of the X-ray quasars is 0.12c. This agrees with the ejection velocity of 0.1c calculated in 1968 from radio quasars associated with low redshift galaxies. When corrected for ejection velocities the observed redshift peaks become narrower – simultaneously strengthening the ejection origin for quasars and the quantization property of their redshift.

1. Introduction

The first associations of high redshift quasars with low redshift galaxies was made more than 30 years ago (Arp 1966b, 1967, 1968). The most recent, striking evidence has come from X-ray sources, paired across Seyfert galaxies, which have turned out to be quasars of considerably higher redshift than the galaxy (Radecke 1996, Arp 1996). Evidence for smaller intrinsic redshifts of galaxies has also accumulated (Arp 1994b). The evidence

Astrophysics and Space Science **244**:9-22,1996.
© 1996 *Kluwer Academic Publishers.*

demonstrates that part of the redshift of these extragalactic objects must be intrinsic (non-velocity).

The pairs are particularly important in that they allow us to estimate the ejection velocities necessary to get the quasars out of their parent galaxies. In the cases available, if the ejection velocities are subtracted from the component quasar redshifts, the two members have more closely the same redshift – as if material of the same intrinsic redshift was ejected in opposite directions. After correction for the velocity component, the redshifts also fall closer to the well marked peaks in the redshift distribution (Arp *et. al.* 1990). Therefore quantization of the redshifts becomes more clearly established. The fact of quantization independently reinforces the earlier result that the quasar redshifts are not primarily due to velocity but to an intrinsic property of matter. Perhaps most important of all, existence of quantization represents one of the strongest empirical clues to the physical reason for the intrinsic redshifts of these recently ejected, compact, energetic objects.

2. Ejection

It has long been accepted that radio galaxies eject material out in roughly opposite directions to approximately equal distances from their active nuclei. It has even been possible to study optical emission from material within these radiolobes (egs. see Fosbury 1984 and Morganti, Robinson and Fosbury 1984). More recently, as X-ray observations began to accumulate, it appeared that material which emits high energy X-rays also accompanies these radio ejections. Some well known examples are X-ray jets within the strong radio ejections from Virgo A and Cen A (egs. Feigelson *et. al.* 1981), the X-ray hot spots within the lobes of Cyg A and the X-ray material extending far out along the ejection direction in Cen A (Arp 1994a).

It seems, therefore, that both radio emission and X-ray emission are characteristic of the material ejected from galaxies. It should then perhaps not be surprising when the phenomenon of X-ray sources paired across active galaxies starts to turn up in the same way that pairs of radio sources were initially discovered across (what later turned out to be) galaxies with active nuclei. The study of these X-ray pairs and associations will have to proceed empirically as did the early association of radio pairs. Now, as then, the first step is to test the statistical significance of associations and build up a list of secure associations in which to study their empirical characteristics.

The identification of secure associations must utilize the *a priori* criterion of tendency toward: 1) opposite ejection 2) equal separation and 3) similarity of ejected sources. All these are demonstrated properties of accepted ejections and rule out any question of *a posteriori* probability calculations. Since in many of the X-ray cases (as in a number of previous radio and optical associations) objects of differing redshifts are identified, one cannot interpret them on the basis of a present theory or understanding of ejection mechanisms (if indeed it is solely an ejection phenomenon). A number of cases must be accumulated, studied and a working theoretical explanation suggested from an inductive analysis.

In order to make the *a priori* statistical criteria specific we refer to the early data of pairing of radio sources across active and disturbed galaxies: The Atlas of Peculiar Galaxies (Arp 1966a) showed numerous cases of radio sources paired across galaxies, particularly in that section containing galaxies with morphological evidence of ejection. The improbability of these paired radio sources being accidental resided principally in the closeness of the sources at their brightness and secondarily in the tendency for the sources to be aligned across the central galaxy and tertiarily to be equally spaced across the galaxy (Arp 1967). Although some sources were aligned to within $\pm 1°$, the average over the 26 original associations was $12°.7$, cf the order of alignment of radio lobes and knots which are conventionally believed to have been ejected from active galaxies.

A follow-up analysis of pairs of radio sources in the Parkes Survey (Arp 1968) showed those pairs which had galaxies located between them had similar properties, demonstrating physical association. Of these radio sources in pairs, 16 were identified as quasars and disturbances in some of the central galaxies indicated the time since ejection. This enabled calculation of ejection speeds for the quasars of 0.1c. In an important result of the present analysis, just this predicted velocity is now calculated directly from the measured redshifts of the pairs of X-ray quasars across such galaxies as NGC 4258 as well as in previous associations of pairs of X-ray quasars.

3. The NGC 4258 Configuration

ROSAT X-ray measures of a $20'$ radius field around the active Seyfert II galaxy, NGC 4258 revealed a striking pair of X-ray sources aligned across the center of the galaxy (Pietsch *et. al.* 1994 and Fig.1 here). The authors commented: "If the connection of these sources with the galaxy is real they may be bipolar ejecta from the nucleus".

Using the parameters listed in Tables 1 and 2 of Pietsch *et. al.* we construct Table 1 here showing the properties of the two sources. The fluxes (F_x) are computed in two steps: First the 0.4 to 2.4 keV band counts = B are obtained from $B = \frac{\text{SASS}}{2}(1+\text{HR1})$. Then the count-to-energy conversion factor of 1.4×10^{-11}erg em^{-2} cts^{-1} (see Pietsch in Arp 1994a) is used to compute F_x. This F_x is close to the system of Hasinger *et. al.* (1993) and we

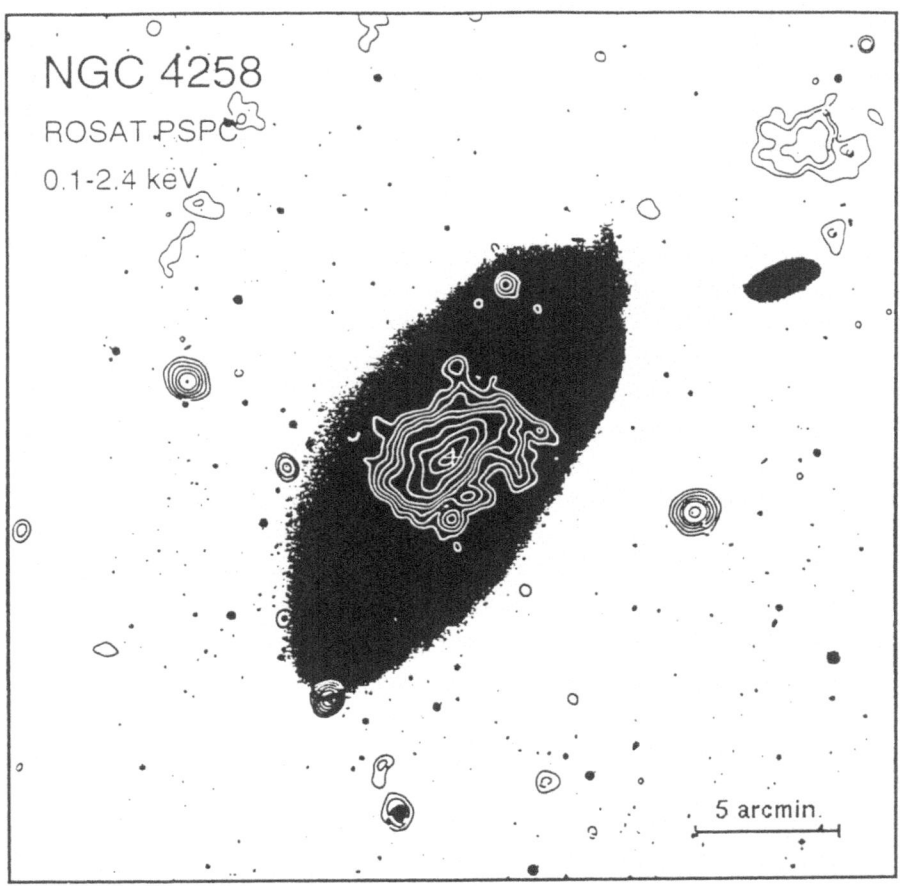

Figure 1.   The pair of X-ray sources across NGC 4258 as discovered by Pietsch *et. al.*
(1994). #26, with $z = .65$, is to the NE and #8, with $z = .40$, is to the SW. The blue
stellar objects are seen at the center of each source.

can therefore use their $\log N(> S) - \log(S)$ curve to compute the density
of sources of the strength of #26 and #8 in an average field. That density
comes out to be about 5 and 2 per deg.^{2} respectively.

TABLE 1. Parameters of X-ray Pair

	r arc min	p.a. deg	F_x (0.4 to 2.4 keV) $\times 10^{-13}$(cgs)	HR1	ct rate $\times 10^{-3} s^{-1}$
Source #26 (NE)	9.66	73.3	0.8	−.4	17.9
Source #8 (SN)	8.57	256.6	1.4	−.2	25.1

Therefore the chance of finding two such background sources acciden-tally within their measured distances of NGC 4258 is only $p_1 = 0.052$. Further the alignment is within 3.3 degrees giving an additional factor of $p_2 = 0.018$. The spacing across the nucleus differs by only 1.09 arc min. Considering a posssible range for the spacing of 20 arc min., an additional factor of $2.18/20 = p_3 = .109$ is required. The total probability of two such bright sources pairing so closely across an arbitrary point in the sky is then $p_1 \times p_2 \times p_3 = 1.0 \times 10^{-4}$.

It is necessary, however, to consider the similarity in hardness ratio and count rate of the two sources. This enables us to compute the probability that the two sources are not just randomly drawn from an average field of X-ray sources. Using Table 2 of Pietsch *et. al.* (1994) one can compute that the median range about zero hardness ratio (HR1) is 1.03. The chance therefore of #8 and #26 falling within ±.2 of each other is $p_4 = 0.2$. Similarly the ratio of count rates for other sources in the field show a median range of 1.53. The chance of #8 and #26 falling within ±40% of each other is then $p_5 = 0.26$. The combined probability of the two sources being only accidentally so similar in intensity and spectrum is then $p_4 \times p_5 = .052$. Therefore the total probability of the pair being accidental, just from the X-ray properties, is $p_{\text{tot}} = 5 \times 10^{-6}$.

This formal calculation quantifies from the X-ray properties alone what the eye and qualitative judgement of the viewer takes in at a glance, namely that there is a very small chance that this is not a pair of X-ray sources physically associated with NGC 4258.

4. Evidence For Ejection

As early as 1961 gaseous emission filaments emanating from the nucleus marked NGC 4258 as ejecting material (Courtes and Cruvellier 1961). E. M. Burbidge, G. R. Burbidge and Prendergast (1963) as well as Chincarini and Walker (1967) showed that large deviations from circular motion occur in this galaxy, indicating large scale eruptive activity. Later van der Kruit, Oort and Mathewson (1972) from radio measures suggested the emission filaments and radio arms were caused by "...clouds expelled from the nu-cleus in two opposite directions in the equatorial plane about 18 million years ago, at velocities ranging from about 800-1600 km s^{-1}."

Of course the X-ray pair is aligned within 13 and 17 degrees of the position angle of 60° (van Albada 1980) of the minor axis of NGC 4258, a direction in which one empirically expects ejection activity. In addition, however, is the fact that if one looks closely at the X-ray isophotes around source #26 it is clear that they are all elongated, both inner and outer isophotes, generally in the direction back towards the nucleus of NGC 4258.

The isophotes on the SW side of #26 are elongated at p.a. 237°, i.e. only 16° from a line back to the NGC 4258 nucleus. It can be ascertained from inspection of similar sources in representative ROSAT fields that this is not likely to be an instrumental effect.

Evidence from completely different wavelengths comes from the unusual water maser observations reported by Miyoshi *et. al.* (1995). They observe, within 8 mas of the center of NGC 4258, two small clumps of emission on either side of the nucleus. The one in the direction of the NE X-ray source has a redshift relative to the galaxy of $\Delta cz = +960$ to $+760$ km s^{-1} and the one in the direction of the SW X-ray source has $\Delta cz = -940$ to -860 km s^{-1}. The authors place the major axis of a supposed disk at p.a. $= 86°$ and interpret the redshift as due to Keplerian rotation around a black hole 40 times more massive than any previous candidate. Since black holes are commonly modeled to have accretion disks which have bipolar ejection, this would be an argument for unusually strong ejection activity in NGC 4258. The model would place such an accretion disk, very small, in the nucleus and presently oriented with its minor axis almost 90° to the supposed quasar ejection line. However, there are 5 different unpredicted aspects of this model for which the authors must invoke probable or possible explanations. But in simple essence, this observation merely consists of some points with a relative redshift of $+900$ km s^{-1} aligned in a direction only 13° different from the direction to X-ray source #26 and some points with a redshift of -900 km s^{-1} in a direction only 11° different from source no. 8. If #26 is identified as having an appreciably higher redshift than #8 this might be interpreted as evidence for #26 to have been ejected away, and #8 towards the observer from the nucleus of NGC 4258.

The degree of alignment found for the X-ray pair across NGC 4258 (3°), their alignment with the minor axis (13° and 17°) and their alignment with the water maser redshift anomalies (11° and 13°) are all then within the tolerances of normally accepted radio ejection phenomena.

5. Relation to Other Examples of Ejection from Active Galaxies

5.1. RADIO QUASARS ASSOCIATED WITH LOW REDSHIFT GALAXIES.

After radio source pairs were found across disturbed central galaxies in the Atlas of Peculiar Galaxies (Arp 1967) the analysis was turned around and pairs of radio sources were searched for in a region of the sky which was covered by the then new Parkes Survey at a frequency of 408 Mhz. In the region between R.A. $= 22^h$ to 4^h and Dec $= +20°$ to $-30°$, thirteen conspicuous, bright radio pairs on the sky were found which, in addition, had bright galaxies between them (Arp 1968). Between 10 and 12 of the radio sources in these pairs were quasars. Using the disturbances in the

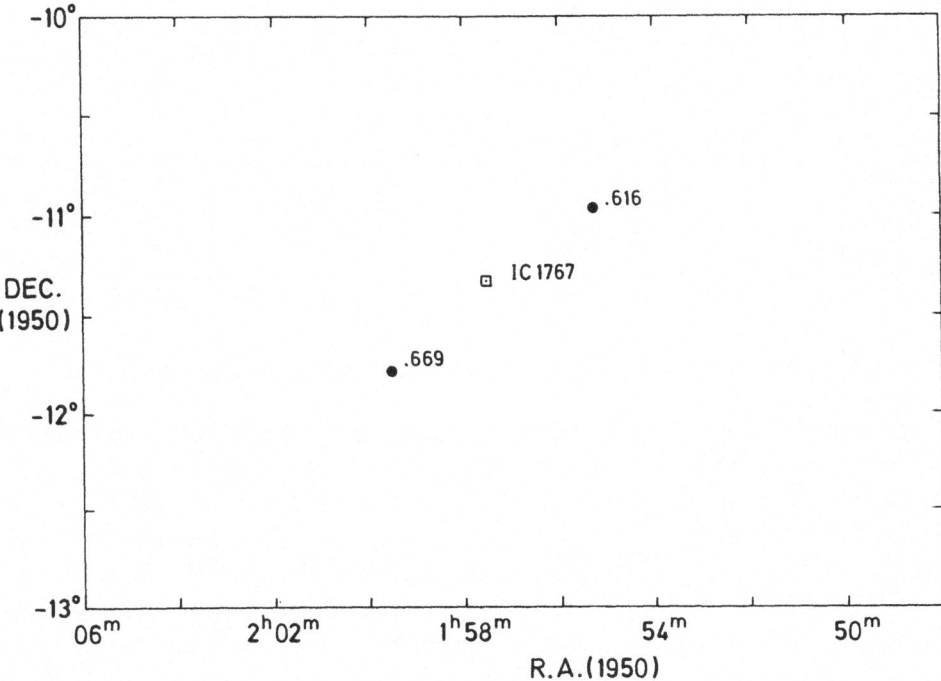

Figure 2. The radio quasars PKS 0155-10 and 0159-11 are shown paired across the disturbed spiral galaxy IC 1767. Their redshifts are similar to within $\Delta z = 0.05$. All radio sources > 0.5 f.u. at 20 cm within the area are plotted. The $z = .616$ quasar has $s = 2.0$ f.u. while the $z = .669$ quasar has $s = 2.9$ f.u. (Arp 1968).

central galaxies to estimate a time since ejection of the order of 10^7 years, ejection speeds of the order of 0.1c were calculated. It is very impressive to now see in Table 2 of the present paper just from the difference in redshift of the pairs of X-ray quasars, an average of about 0.15c projected ejection velocity.

In addition Figure 2 shows the best case of aligned radio sources across a central galaxy found in 1968. It was known at that time that both were quasars but it went unremarked in the following years when their redshifts were measured, that they came out so extremely similar at $z = 0.62$ and 0.67. In fact, as we shall see now from examining the more recent pairings of X-ray quasars across low redshift galaxies, as significant as the new cases are, the best case of all as shown in Figure 2, has been known for the order of 27 years.

TABLE 2. Some X-ray Pairs Across Active Galaxies

Galaxy	z_G	$r'_1 ; r'_2$	$\Delta\theta^\circ$	$F_{x,1}$ $\times 10^{-13}$	$F_{x,2}$ $\times 10^{-13}$	$z_1 ; z_2$	p_1	p_{tot}
NGC 4258	0.002	8.6;9.7	3	1.4cgs	0.8cgs	0.40;0.65	$5 \cdot 10^{-2}$	$< 4 \cdot 10^{-7}$
Mark 205	0.07	13.8;15.7	44	2.3	2.7	0.64;0.46	$2 \cdot 10^{-2}$	connected
PG1211+143	0.085	2.6;5.5	8	0.2	1.4	1.28;1.02	$1 \cdot 10^{-2}$	$< 10^{-6}$
NGC 3842	0.02	1.0;1.2	33	1.0	0.3	0.95;0.34	$7 \cdot 10^{-5}$	$6 \cdot 10^{-8}$
NGC 4472	0.003	4.4;6.0	1	~ 3000	~ 800	0.004;0.16	$2 \cdot 10^{-4}$	$< 10^{-6}$

5.2. X-RAY QUASARS ASSOCIATED WITH LOW REDSHIFT GALAXIES.

If we restrict ourselves to just X-ray pairs across galaxies, there have been a number of notable cases discovered during the relatively short time that X-ray observations have been accessible. Some of these are listed here in Table 2. When all properties of each pairing are taken into account the chances of any of these five cases being accidental is $\leq 10^{-6}$. We see then that NGC 4258 is merely the latest confirmation in a series of compelling examples of X-ray sources, mostly identified as quasars, which have been physically associated with low redshift galaxies in the same way that radio sources were in the past.

It is of interest to comment individually on each of the five cases listed in Table 2. In the table the subscript 1 designates the nearest source. The F_x values are estimated for the 0.4-2.4 keV band, except for the last entry which refers to M87 and 3C273 and uses the HEAO 1, 2-10 keV band. p_1 designates the accidental probability of finding the sources of listed F_x at r_1 and r_2. $1 - p_{tot}$ gives the estimated probability of physical association.

NGC 4258   Although the pair of X-ray sources across NGC 4258 was identified prior to September 1993, it was not until February 1995 that spectra were obtained by E.M. Burbidge (1995) which showed that the two BSO, X-ray identifications were quasars $z = 0.398$ and 0.653. It was calculated earlier that the chance of having two X-ray sources brighter than the observed flux within the observed distance was $p_1 = .052$. When the alignment,spacing and similarity of the sources is taken into account the chance of accidental pairing drops to 5×10^{-6}. But this does not take into account the fact that both are now confirmed quasars with similarities in their optical magnitudes and redshifts. It is very unusual to observe two such similar redshifts for adjacent quasars. For example, of 26 quasars of $m_v = 19$ to 20 mag. observed by Arp, Wolstencraft and He (1984), only two fell in the interval $.1 < z < .8$. This would reduce this improbability of accidental association to 4×10^{-7}. Still further we should take into account the tendency for the alignment of the pair to be in the direction of NGC

4258's minor axis and also within $11° - 13°$ of the alignment of anomalous redshift water maser lines. (The high redshift water maser sources lie in the direction of the $z = 0.65$ quasar and the low redshift sources lie in the direction of the 0.39 quasar.) Finally there is the extension of the outer X-ray isophotes of the $z = 0.65$ quasar back toward the nucleus of NGC 4258. Overall we would have to say the chances of this being not a physical association are much less than one in a million.

Mark 205   It is striking how similar the Mark 205 association (Arp 1995a) is to that of the just discussed NGC 4258. In Mark 205 the two bright X-ray sources are somewhat more separated but the quasars they are identified with are optically and X-ray brighter than in NGC 4258 as if the whole system were a factor of 1.5 closer. It is particularly striking to note how similar the quasar redshifts are in the two systems. In fact if we propose that in NGC 4258 a pair of quasars of intrinsic $z = 0.53$ were ejected towards and away from us with a velocity of $.13c$ then we could say that a pair of $z = .55$ quasars were ejected with $.09c$ from Mark 205.

Of course the two Mark 205 quasars are not at all well aligned in their observed positions. But the luminous X-ray connection from the $z = .46$ quasar curves back toward Mark 205 (Arp 1995a) in such a way that it must enter Mark 205 in a more northerly direction — in a direction initially more aligned with the $z = .64$ quasar. Such a situation would be analogous to ejected radio lobes when one side is bent or curved. There is also the possibility of a three way ejection that would conserve momentum.

But regardless of details of possible models, just the fact of finding two such optically bright quasars this close to an arbitrary point in the sky is of the order of 10^{-3}. As for the overall probability, the X-ray filament to the SSW contains two imbedded quasars, the $z = .64$ and a $z = 1.259$ quasar as well. (Arp1995a,d). If they are physically connected to the low redshift Seyfert, of course there is essentially zero probability of being accidental.

PG1211+143   Although classified as a quasar, this central galaxy has a Seyfert spectrum and properties very similar to Mark 205. It's associated X-ray sources, however, are fainter and closer to the central galaxy than either of the two preceding cases. The alignment and similarity of the two quasars makes the chance probability $\sim 10^{-6}$ (Arp 1995b). But of course, if the apparent alignment of radio sources so far observed (Kellerman *et. al.* 1994) is confirmed, its coincidence with the X-ray alignment would not only make certain the physical association but also the liklihood of an ejection origin for the quasars.

A recent spectroscopic observation by courtesy of IUCAA in Pune India and the Beijing Astronomical Observatory enables the redshift of the second quasar to be determined as $z = 1.015$. This is particularly important

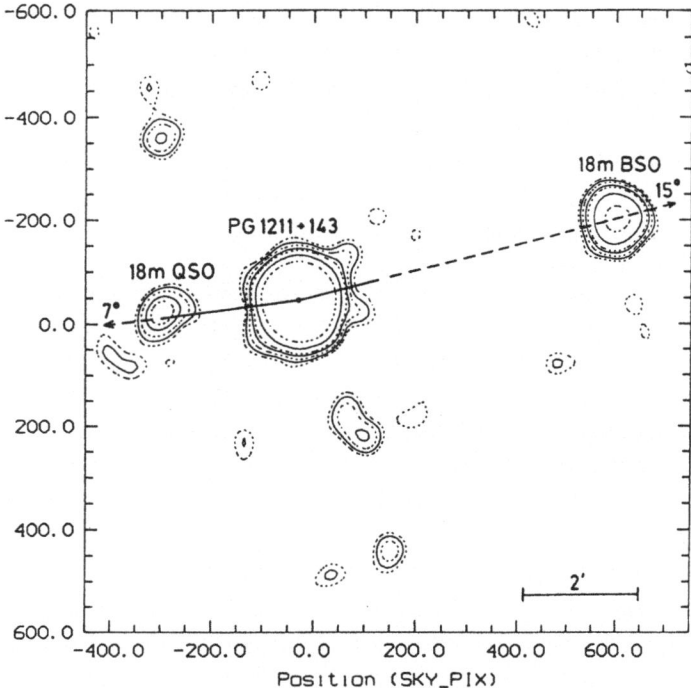

Figure 3.   The Seyfert-like object is PG 1211+143, flanked by quasars of $z = 1.28$ and $z = 1.02$. The full line represents the direction and extent of a line of radio sources which appears to coincide with the line of X-ray sources (Arp 1995b).

because it now gives Δz for the first three pairs in Table 2 of 0.25, 0.18 and 0.26. If one wishes to interpret these pairs on an ejection hypothesis it, for the first time, gives a numerical estimate for the projected ejection velocity as about $0.12c$ — very close to the ejection velocity computed in 1968 from radio quasars. (Note: In Fig. 2 if the two quasars were ejected with $\sim 0.1c$ nearly perpendicular to our line of sight, and if IC 1767 were ~ 2 times closer than Mark 205, the configuration would be similar in scale.)

NGC 3842   The close spacing of this X-ray pair across the central galaxy accounts for most of the improbability of its being accidental (the X-ray fluxes are from Bechtold *et. al.* , 1983). It is important to note, however, that there is a third quasar associated with NGC 3842 which makes an approximate equilateral triangle with the first two. This not only lowers the improbability of chance association to 6×10^{-8} (Arp 1987, p13), but it also offers a natural explanation for why, in a three way ejection event, the quasars would not need to be closely aligned. In view of the enormous significance of this association it would seem to be the one to which the properties of others would be compared for purposes of confirming the nature of the association. The most obvious property, which led to this

Figure 4. The E galaxy NGC 3842, brightest in a cluster, is seen flanked by two X-ray sources which turned out to be quasars (QSO1 and QSO2). QSO3 is a third quasar discovered as a radio source (Arp 1987).

discovery, was the existence of two point X-ray sources very close to a galaxy, as in NGC 4258.

NGC 4472   The X-ray fluxes of the two sources flanking NGC 4472 (M49) are so strong that HEAO1 measures in the 2-10 keV band are tabulated in Table 2. The exact brightness is not of import here. The point rather is that M87 is one of the strongest X-ray galaxies in the sky and 3C 273 is the strongest X-ray quasar in the sky. Assuming we have $\sim 35,000$ deg^2 unobscured of the 41,253 deg^2 in the sky and that M87 is one of the three brightest X-ray sources and that 3C 273 is one of the first 10 we can compute $p_1 \approx 2 \times 10^{-4}$ that such bright sources fall so close to an arbitrary point in the sky. We see from Table 2 that this proximity of X-ray sources

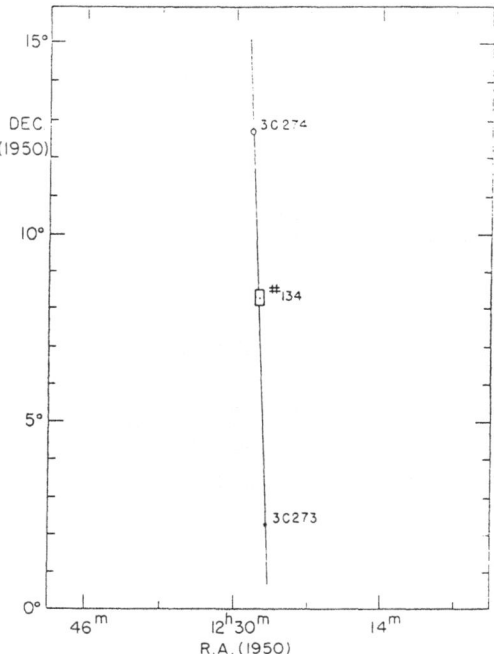

Figure 5.   The brightest galaxy in the Virgo Cluster is Atlas of Peculiar Galaxies #134
= NGC 4472 = M49. 3C 274 is one of the brightest radio galaxies, M87, and 3C 273 is
the brightest apparent magnitude quasar in the sky at $z = .16$ (Arp 1967). The latter
pair of sources are among the brightest X-ray sources in the sky.

is more improbable than NGC 4258 and better aligned.

Of course, the total improbability of the association being accidental
was computed (Arp 1967) as $\sim 10^{-6}$. But perhaps even more convincing
was the qualitative question: Is it significant that the brightest radio, X-ray
galaxy in the dominant galaxy cluster in our sky, and the brightest radio,
X-ray quasar in the sky are near to and almost exactly aligned across the
brightest galaxy in the center of that galaxy cluster? Most recently it has
been shown that X-ray emitting material continuously connects M49 in one
direction to M87 and in the other direction to 3C 273. (Arp 1995c) It would
seem that the best evidence of all for linking higher redshift active objects
to a lower redshift central galaxy had been already found in 1966.

6. Summary

It has been shown that, in agreement with visual impression, the alignment
of the recently observed X-ray quasars across the nucleus of the Seyfert
galaxy NGC 4258 is extremely significant. Further it is shown that NGC

4258 is not an isolated case but there are a number of other associations of X-ray quasars with active galaxies that are of comparable or greater significance.In addition to the cases discussed here there are cases of X-ray jets pointing from active galaxies toward nearby higher redshift quasars (Arp 1995a). Undoubtedly it is the high energy wavelengths of the X-ray observations which emphasize the young, active objects like quasars, make easy their identification and render conspicious their relation to the active central objects. There is considerable evidence to support an ejection origin for these quasars in analogy with the characteristic ejection of radio lobes and radio sources from active galaxies. Of the five cases discussed in Table 2, four have central galaxies which are active and two show strong evidence for ejection from the active nucleus. It would be an obvious prediction that more cases like the ones discussed here would be available for study after carrying out systematic X-ray searches in the vicinity of active galaxies.

The quantized redshift peaks in the X-ray quasar range are: $z = 0.061$, 0.30. 0.60, 0.96 and 1.41 (Arp *et. al.* 1990). Table 3 shows the most recent pairs (Arp 1996), which when the ejection velocity components are averaged out, gives an intrinsic $\bar{z}$ of 0.58. It is apparent that the value falls very close to the $z = 0.60$ quantization peak, and drastically reduces the dispersion around the peak of the individual redshift values.

TABLE 3.   Quasar Redshifts in X-Ray Pairs

Galaxy	Redshifts
NCC 5548	$z = 0.56$ and 0.67
NCC 4258	$z = 0.40$ and 0.65
Mark 205	$z = 0.46$ and 0.64
IC 1767	$z = 0.62$ and 0.67
Average z	$\bar{z} = 0.58$

Another pair recently measured across NGC 2639 (E. M. Burbidge, H. Arp, and H. D. Radecki in preparation) have redshifts of $z = 0.307$ and 0.325. In addition BL Lac objects of $z = 0.308$ and 0.615 have been shown to be associated with NGC 1365 and NGC 4151 respectively (Arp 1996). In sum, the evidence for quantized redshifts in quasars is becoming stronger as more observations become available.

References

Arp, H.: 1966a, *ApJS*, **16**, 1
Arp, H.: 1966b, *Science*, **151**, 1214

Arp, H.: 1967, *ApJ*, **148**, 321
Arp, H.: 1968, *Astrofiska* (Armenian Acad. Sci), **4**, 59
Arp, H.: 1987, *Quasars, Redshifts and Controversies* (Interstellar Media)
Arp, H.: 1994a, *A&A*, **296**, 738
Arp, H.: 1994b, *ApJ*, **430**, 74
Arp, H.: 1995a, in *IAU Symp. 168* (Kluwer) in press
Arp, H.: 1995b, *A&A*, **296**, L5
Arp, H.: 1995c, *Physics Letters A*, **203**, 61
Arp, H.: 1995d, *A&A*, **294**, L45
Arp, H.: 1996, *A&A* in press
Arp, H., Bi, H. G., Chu, Y., Zhu, X.: 1990, *A&A*, **239**, 33
Arp, H., Wolstencraft, R. D., He, X.T.: 1984, *ApJ*, **285**, 44
Bechtold, J., Forman, W., Giacconi, R., Jones, C., Schwarz, J., Tucker, W., Van Spey-
    brock, L.: 1983, *ApJ*, **265**, 26
Burbidge, E. M.: 1995, *A&A*, **298**, L1
Burbidge, E. M., Burbidge, G. R. and Prendergast, K. H.: 1963, *ApJ*, **138**, 375
Chincarini, G., Walker, M. F.: 1967, *ApJ*, **149**, 487
Courtès, G. and Cruvellier, P.: 1961, *C.R. Academy Sci.*, Paris **253**, 218
Feigelson, F. D., Schreier, E. J., Delaville, J. P., Giacconi, R., Grindlay, J. E., Lightman,
    A. P.: 1981, *ApJ*, **251**, 1981
Fosbury, R. A. E.: 1989, ESO Workshop *Extranuclear Activity in Galaxies*, p. 169
Hasinger, G., Burg, R., Giacconi, R., Hartner, G., Schmidt, M., Trümper, J., Zamorani,
    G.: 1993, *A&A*, **275**, 1
Kellerman, K. I., Sramek, R. A., Schmidt, M., Green, R. F., Schaffer, D. S.: 1994, *A.J.*
    108, 1163
Miyoshi, M., Moran, J., Herrerstein, J., Greenhill, L., Nakai, N., Diamond, P. Inoue, M.:
    1995, Nature **373**, 127
Morganti, R., Robinson, A. Fosbury, R. A. E. 1989:, ESO Workshop *Extranuclear Activity
    in Galaxies*, p. 433
Piccinotti, G., Mushotzky, R. F., Boldt, E. A., Holt, S. S., Marshall, F. E., Serlemitsos,
    P. J., Shafer, R. A.: 1982, *ApJ* **253**, 485
Pietsch, W., Vogler, A., Kahabka, P., Jain, A., Klein, U.; 1994, *A&A* **284**, 386
Radecki, H. D.: 1996, *A&A* in press
van Albada, G. D.: 1980, *A&A* **90**, 123
van der Kruit, P. C., Oort, J. H., Mathewson, D. S.: 1972, *A&A* **21**, 169

A FRESH LOOK AT DISCORDANT REDSHIFT GALAXIES IN COMPACT GROUPS

JACK W. SULENTIC AND J. BRETT SMITH
Department of Physics and Astronomy
University of Alabama
Tuscaloosa, Alabama 35487

Abstract. We reexamine the statistics of discordant redshift galaxies in compact groups. We find that 43 out of 100 groups in the Hickson catalog contain at least one discordant redshift galaxy. We show that, despite the prevailing impression, all previous attempts have failed to explain this large number of discordant redshift galaxies. The order of magnitude excess survives all of our attempts to refine the sample.

1. Introduction

Compact groups are aggregates of four or more galaxies with surface density enhancements 10^2–10^3 times their local surroundings. In addition to the challenges that such groups present to ideas about galaxy interactions, they also contain a large number of discordant redshift components. In this context, a galaxy redshift (expressed in velocity units where V= cz) is considered discordant if it differs from the median group value by $\Delta V \geq 1000$ km s^{-1}. The statistics of discordant components are insensitive to an increase of this limit by several times 10^3 km s^{-1}. The median velocity dispersion for accordant redshift groups is 200 km s^{-1}(Hickson et al. 1992). The standard paradigm requires that all of the discordant galaxies are chance projections of foreground or background interlopers.

The main problem with the chance projection hypothesis involves the rarity of physically dense and interacting compact groups. That makes the probability of such a chance projection quite small. Of course such an *a posteriori* probability estimate has little value. One needs a reasonably complete sample of compact groups in order to make meaningful estimates of interloper contamination. The belief in 1960 was that the vast majority of

Astrophysics and Space Science 244:23-28,1996.

compact groups, when finally cataloged, would show accordant redshifts.
The first discovered discordant system, Stephan's Quintet, would then be
an example of a rare compact quartet made more prominent by the superposition of a bright discordant galaxy. The level of surprise grew considerably
when redshifts were measured a decade later for two other famous compact
groups (Seyfert's Sextet and VV172: Burbidge and Sargent 1970). Both
quintets showed single discordant components ($\Delta V \sim$ 11000 and 21000 km
s^{-1}respectively).

2. Statistics of Discordant Redshifts in Compact Groups

The statistics of compact groups have improved considerably in the past
15 years. A reasonably complete catalog of compact groups has been published (Hickson 1982) and the groups have been studied extensively. Near
completion of the redshift measures reveals that 43 out of 100 groups have
at least one discordant redshift component. This result would have caused
considerable discussion had it occurred 25 years ago but, coming in recent
years, it passed with very little comment. The result can be contrasted with
the 51/602 discordant redshift binary galaxies (8-9%) found in a reasonably
complete and similarly compiled catalog. If pairs and compact groups fall
in regions with similar galaxy surface density then we expect similar levels
of interloper contamination (allowing for differences in surface area).

How can we reconcile the discordant redshift galaxies in compact groups
without challenging the redshift-distance relation upon which modern cosmology is constructed? There have been several attempts to explain the
large number of discordant systems as chance projections. The efforts have
focussed on estimating the number of discordant quintets that are expected
given the observed population of accordant quartets. This approach evolved
from the original discovery of three discordant quintets (4+1 systems) and
because quartets comprise the largest subsample in the Hickson catalog.
The arguments are identical for the next largest discordant population involving triplets with one or more interlopers (3+n systems where n= 1, 2 or
3). The number of discordant quintets (n_5), for example, can be estimated
from the expression:

$$n_5 = n_4 \times \sigma \times A$$

where σ is the field galaxy surface density (in galaxies deg^{-2}) and A is
the surface area subtended by the groups (in deg^2). We have three possible
variables that can be considered in attempting to explain the discordant
quintets: we need a large value(s) for n_4, σ and/or A. All three possibilities
have been considered and are briefly summarized below.

2.1. MANY QUARTETS (N_4)

A few years after discovery of the three famous quintets, and several years before the publication of the Hickson (1982) catalog, a survey of galaxy quartets was published (Rose 1977). This survey suggested that 400-500 suitably bright quartets existed on the sky. When combined with a reasonable extimate for σA, it was concluded that this quartet population could explain the (exactly three) discordant quintets known at that time. The most surprising thing about this claim was that redshifts were then known for approximately 12 of this vast population of quartets. Nottale and Moles (1978) showed that the probability of drawing the three expected discordant groups from such a small part of the purportedly large quartet sample was vanishingly small. Another quartet survey was made in the early eighties (Sulentic 1983) and it was demonstrated that the Rose (1976) estimate for the number of quartets was about an order of magnitude too high. This reanalysis was independently confirmed by the publication of the Hickson (1982) catalog which listed 100 compact groups north of declination -30° (including the three famous discordant quintets and approximately 35 accordant quartets). The number of discordant quintets in the Hickson catalog has now increased to seven plus three additional quartets with two or three discordant companions. Even allowing for incompleteness in the numbers, it is clear that one can not account for discordant redshifts using any reasonable value for n_4.

2.2. MANY POTENTIAL INTERLOPERS (σ)

There have been three attempts to estimate the accordant/discordant interloper population from local galaxy counts (Sulentic 1987; Rood and Williams 1989 and Kindl 1990: see also Palumbo et al. 1995). Local surface densities were derived for each group using a range of magnitude limits and search radii with (generally) similar results being obtained. Typical surface densities are small with less than 10% of the Hickson groups found in rich group or cluster environments. Many others are associated with loose groups that are part of the large scale structure in the local universe. Much has been made of this fact in the past few years (Vennik et al. 1993; Ramella et al. 1994; Rood and Struble 1994) but it is not surprising to find that compact groups are associated with large scale structure. If groups avoid clusters and if no underlying "continuum" of field galaxies exists, what else could they belong to? What is surprising is how often compact groups are found in regions of very low galaxy surface density. If one takes the product of individual estimates of sigma and the sky areas subtended by each group ($\sum A\sigma$) one obtains expectations of 2-5 discordant redshift systems compared with the observed value of 43. One cannot explain the discordant redshift population by arguing that compact groups lie in regions with higher than average galaxy surface density.

2.3. ADJUSTING SURFACE AREA (A)

The most recent attempts to reconcile the discordant excess involve the argument that one must not use the observed surface areas of the groups when calculating the interloper expectation (Hickson et al. 1988; Mendes de Oliviera 1995). Instead it is argued that many of the groups would satisfy the Hickson selection criteria even if another galaxy was located much farther from the accordant members. Hickson et al. (1988) reported Monte Carlo simulations of hypothetical groups with random projections of an interloper population. They show that one can reproduce the observed number of discordant systems by using the maximum possible group areas in the expectation calculation. Actually their procedure is equivalent to calculating the largest radius that each observed compact group could have (e.g. the largest radial distance where an additional galaxy could fall) without violating the imposed isolation and surface brightness criteria used in assembling the Hickson Catalog. The increase in surface area is quite dramatic in some cases ($250\times$ in the case of Seyfert's Sextet) since a doubling of A results from a modest (0.4) increase in group radius. Some cataloged groups are indeed so bright, compact and isolated that the addition of another galaxy many group diameters distant would still result in a compact group satisfying the formal definition.

We do not believe that this approach provides a solution to the problem. The observed population of discordant galaxies tend to fall near the accordant galaxies in each group. This is either a) an indication that they are physical (or lensed?) members of the groups or b) an indication of a selection bias in the group catalog. The chance projection hypothesis indeed predicts that many interlopers will fall at the outer edge of any selected search radius. The observed number of interlopers should be proportional to surface area (R^2) so half of any complete interloper population will fall outside of 0.7R where R is the normalized group radius.

Figure 1 shows the radial distribution of the discordant redshift members in units of the maximum possible group radii that were employed by Hickson et al. (1988). We find a large excess of discordants within 0.4R when we expect half of the sample to fall outside of 0.7R (which is approximately the result we obtain using the actual Hickson radii for the groups). A two bin $\chi^2 = 13.8$ suggests that the observed distribution of discordants is significantly different from the expectation (indicated by the dots in Figure 1). If the Hickson et al. (1988) maximum radii are the correct ones to use in estimating the interloper frequency then Figure 1 shows that the Hickson

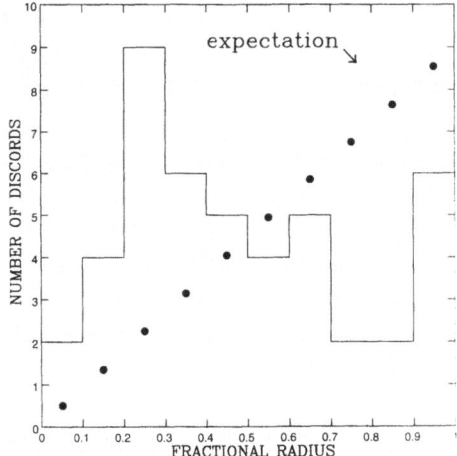

Figure 1. The radial distribution of discordant redshift galaxies in compact groups. Units are normalized maximum possible group radius. Dots represent expected distribution under a random projection hypothesis.

catalog is seriously incomplete. However use of the maximum radii resulted in an interloper expectation that approaches the currently observed value. Completion of the compact group catalog will add a large number of additional discordant systems. The result is that the problem does not go away.

3. A Refined Sample of Compact Groups

Having failed to account for the excess in any of the three ways discussed above we are left with the possibility that the Hickson catalog is somehow seriously incomplete and/or biased in a way that favors finding the discordant groups much more efficiently than accordant ones. This includes the possibility that many discordant groups are near limits, or in violation, of the stated selection criteria (see Hickson 1992). We reanalyzed the sample in order to verify that at least four members of each group satisfy all of the selection criteria. We found that 18 groups either 1) do not satisfy the stated isolation criterion or 2) show a larger scatter in (R band since Hickson searched the E prints of the Palomar Sky Survey in assembling his catalog) apparent magnitude than the stated maximum dispersion criterion ($\Delta m \leq 3.0$). We are left with 82 groups out of which 35 contain at least one discordant member. It is clear that the order of magnitude excess discordant population does not go away with sample refinement.

This is as far as we can go without considering redshifts. Of course we have already used the redshifts to identify discordant groups Further interpretation requires some conventional assumptions: 1) that redshift is proportional to distance and 2) that true interlopers exist in the sample. We can then identify two populations of compact groups: 1) 22 false groups containing $n \leq 3$ accordant members and 2) 60 groups that represent "physical" ($n \geq 4$) multiplets in redshift space. The latter population contains 47 completely accordant multiplets and 13 discordant redshift (4+n) systems. If we play by the standard rules we can only treat the second population in a statistically meaningful way. We are unable to consider the large number of false groups (mostly 3+n systems) without information on the statistics of accordant triplets on the sky. The small number of accordant triplets found in the only published survey (Karachentseva et al. 1979; see also Sulentic 1983)) suggests that the situation for 3+n systems closely parallels that for 4+n groups.

The 13 discordant groups containing $n \geq 4$ accordant multiplets are susceptible to statistical test. These systems come from an optimally defined sample including $n_4 = 43$ groups with 33 pure quartets plus 10 quartets with one or more discordant members. Summing the surface areas and σ values for this sample yields an expectation of 0.98 interlopers compared with the observed number n= 10: the order of magnitude excess persists. An additional fact makes this result even more robust. The majority (9/10) of the accordant groups would have satisfied the Hickson selection criteria even without the interloper. In other words, the addition of the interloper did not push an otherwise too faint aggregate over the selection threshold. The underlying groups satisfy the Hickson criteria independently of the discordant component. The excess discordant redshift population in compact groups remain a paradox.

References

Burbidge, E. M. & Sargent, W.: 1970, In *Nuclei of Galaxies*, ed. D. O'Connell, (North Holland), p. 351

Hickson, P.: 1982, *ApJ*, **255**, 382

Hickson, P. Kindl, E. & Huchra, J.: 1988, *ApJ Lett.*, **329**, L65

Hickson, P, Mendes de Oliveira, C., Huchra, J. & Palumbo, G.: 1992, *ApJ*, **399**, 353

Karachentseva, V. et al.: 1979, *Astroph. Issled.*, **11**, 3

Kindl, E.: 1990, Ph.D. Thesis, University of British Columbia

Mendes de Oliveira, C.: 1995, *MNRAS*, **273**, 139

Nottale, L. & Moles, M.: 1978, *A& A*, **66**, 355

Palumbo, G., Saracco, P., Hickson, P. & Mendes de Oliveira, C.: 1995, *AJ*, **109**, 1476

Ramella, M., Diaferio, A., Geller, M. & Huchra, J.: 1994, *AJ*, **107**, 1623

Rood, H. & Williams, B.: 1989, *ApJ*, **339**, 772

Rood, H. & Struble, M.: 1994, *PASP*, **106**, 413

Rose, J.: 1977, *ApJ*, **211**, 311

Sulentic, J. W.: 1983, *ApJ*, **270**, 417

Sulentic, J. W.: 1987, *ApJ*, **322**, 605

Sulentic, J. W.: 1993, In *Progress in New Cosmologies: Beyond the Big Bang*, ed. H. Arp, R. Keys, K. Rudnicki, (Plenum), p. 49

Vennik, J., Richter, G. & Longo, G.: 1993, *Astron. Nach.*, **314** 393

EVIDENCE FOR QUANTIZED AND VARIABLE REDSHIFTS IN THE COSMIC BACKGROUND REST FRAME

W. G. TIFFT
Steward Observatory
University of Arizona
Tucson, Arizona 85721

Abstract. Evidence is presented for redshift quantization and variability as detected in global studies done in the rest frame of the cosmic background radiation. Quantization is strong and consistent with predictions derived from concepts associated with multidimensional time. Nine families of periods are possible but not equally likely. The most basic family contains previously known periods of 73 and 36 km s^{-1} and shorter harmonics at 18.3 and 9.15 km s^{-1}. Several approaches to evaluating the significance of quantization are employed and the dependence on redshift, the width and shape of 21 cm profiles and morphology is discussed. Common properties between samples define several basic classes of galaxies. Quantization is consistently optimized for a transformation vertex very close to the vertex of the cosmic background dipole. Relationships between cosmocentric and galactocentric rest frames are discussed.

1. Introduction

This paper presents evidence for global redshift quantization and examines its properties. By global quantization we mean that redshifts of homogeneous classes of galaxies from all over the sky contain specific periods when viewed in an appropriate rest frame; the redshift is not a continuous variable as conventionally expected. Work prior to 1992 is summarized eleswhere (Tifft 1995a). Early discussions of cosmology are contained in Tifft (1995b) and Tifft, Cocke & DeVito (1996). A current empirical model is discussed elsewhere (Tifft 1996a).

Two advances in quantization work occurred in 1992 and early 1993. The 3 degree cosmic background radiation rest frame was recognized to be the

Astrophysics and Space Science **244**:29-56,1996.

primary reference frame for global quantization (Cocke & Tifft 1996), and a possible model (Lehto, 1990) was identified which predicts redshift periods in terms of the Planck energy. The model, henceforth the 'temporal model', views time as three dimensional and connects the structure of matter to redshift effects and cosmology.

Early work on global redshift quantization used a galactic center rest frame (Tifft 1978a,b Tifft & Cocke 1984). Quantization was found for galaxies with wide and narrow 21 cm profiles, but not for intermediate objects. In the CBR rest frame it is now possible to work with all types of galaxies. The cosmic frame appears to be fundamental. The effect can induce large non-random effects in other rest frames, especially the galactic center where Guthrie and Napier (1991, 1996) independently confirm periodicities.

Before 1993 redshift periods were empirical and related by simple factors to a period near 72 km s^{-1}. One important period is near 36 km s^{-1}. Precise predicted periods given by the 3-d temporal model now remove any ambiguity introduced by period uncertainty. One aspect of this paper is to show how accurately these periods fit observations. The first work with the CBR reference frame used the old periods. By combining the CBR rest frame and the 3-d temporal concepts we can simultaneously demonstrate the period match and the presence of a consistent CBR vertex. See Tifft (1996b) for details.

From the earliest studies it has been apparent that redshift periods and phasings depend upon properties of the galaxies involved; it is essential to use accurate homogeneous data sets or periods and transformations are masked or distorted. Accuracy largely limits studies to 21 cm redshifts where measures can achieve sub km s^{-1} precision (Tifft & Cocke 1988, Tifft 1990). With 21 cm data, homogeneity is improved by sorting according redshift, 21 cm profile width, W, sometimes profile shape or asymmetry, A, and standard morphology, the t index. Low quality data is rejectd using signal-to-noise, S/N, or flux-to-width, F/W, ratios. F/W also separates samples by luminosity in deep redshift surveys. Studies of 21 cm redshift precision suggests that variability is present; redshifts seem to shift within the periodic pattern. Because of this, redshift sources are rarely combined, and samples are limited to defined time intervals.

The most recent galactic center transformation is $(232.2, -36.6, 0.9)$ km s^{-1}. The numbers are the transverse, radial, and perpendicular components in galactic coordinates. This transformation is now usually applied relativistically. Older values rarely differ by more than 1 or 2 km s^{-1}. The most recent CBR transformation is $(-243, -31, 275)$ km s^{-1}. It is usually applied as a Galilean transformation sequential with a relativistic galactic center transformation. A direct Galilean CBR transformation of $(-241.7, -30.8, 275.1)$ km s^{-1} has also been used. Differences are negligible except at very short periods. It seems likely that a relativistic transformation to the CBR is not correct but further studies of vertices and the form of the transformation are needed. The present uncertainties do not affect any important conclusions.

The association of redshift quantization with the CBR reference frame provides one link with cosmology. A second link comes from nonlinearity in the periodicity with lookback time. Before redshifts are analyzed a correction is applied according to

$$V_{corr} = 4c[(1 + z)^{1/4} - 1] + \cdots. \tag{1}$$

Higher terms are a function of q_0, and cancel for $q_0 = 1/2$. See Tifft (1991) for a derivation. A demonstration that $q_0 = 1/2$ is appropriate is given in Tifft (1996b). The correction is important only for short periods and high redshifts.

2. The Periodicity Rule

Lehto (1990) describes properties of matter by assuming a minimum unit of time, the Planck time, and expanding it into observable time intervals by a doubling process. The Planck time is.

$$t_0 = \frac{1}{\nu_0} = \sqrt{\frac{hG}{c^5}} = 1.3506 \times 10^{-43} \text{ s}, \tag{2}$$

where h is Planck's constant, G the gravitational constant, and c the speed of light. ν_0 is a corresponding maximum frequency which defines a maximum unit energy and mass.

$$E_0 = h\nu_0 = \frac{h}{t_0} = 4.905 \times 10^{16} \text{ erg}, \tag{3}$$

$$m_0 = \frac{E_0}{c^2} = 5.458 \times 10^{-5} \text{ gm}. \tag{4}$$

These units alone are sufficient to model the properties of fundamental particles and redshift periodicities. Only time and energy, in energy or mass form, is involved. This is all that is assumed in the temporal model.

Lehto (1990) also includes a minimum spatial unit, the Planck distance

$$r_0 = ct_0 = 4.049 \times 10^{-33} \text{ cm}. \tag{5}$$

Recent work in quantum gravity suggests that such minimum spatial intervals could be a property of space. In a discrete space-time lattice structure the basic unit of velocity is

$$v_0 = \frac{r_0}{t_0} = c. \tag{6}$$

As noted above spatial quantization is not required to quantize redshifts. The present temporal model assumes that space is continuous.

Given basic units, Lehto assumed a period doubling process, known to operate in some chaotic decay processes, to extend the Planck units into the observable domain. The simplest doubling process relates observable values to the fundamental by a factor $2^{\pm D}$, where D is the number of doublings. Such a scheme requires that fundamental masses, energies, etc be related by integer powers of 2. What Lehto found was that exponents seemed to concentrate at $1/3$ integer values. He interpreted this in terms of three-dimensional time. One dimensional perceived time could be related to a three dimensional volume in temporal 3-space. Volume doubling is reduced to one dimension by taking cube roots; perceived time is a scalar with sign and magnitude only. Lehto wrote perceived times as

$$t = t_0 2^{\frac{L}{3}} = t_0 2^{\frac{N_x + N_y + N_z}{3}} = t_0 2^{\frac{3D+M}{3}}. \tag{7}$$

N values are individual axial doublings, D is the net doubling and M is a temporal fraction, 0, 1 or 2. The D, M notation distinguishes three doubling families; values with constant M are related by powers of two.

Equation (7), combined with (3) and (4), permits calculation of particle masses and particle pair energies. Lehto found that the mass corresponding to $L = 227$ is equivalent to the electron mass within the uncertainty in the physical constants defining t_0. Recent extensions model most of the basic stable particles and forces (Tifft & Lehto 1996).

Equation (7) assumes that volumes follow a doubling rule where all three axes scale by $2^{1/3}$ simultaneously. Removing this restriction allows growth in steps of $2^{1/9}$

$$t = t_0 2^{\frac{9D+T}{9}}. \tag{8}$$

The integer T ranges from 0 to 8 and defines 9 period doubling series.

Possible redshifts, in velocity units, are given by

$$v = P = c2^{-\frac{9D+T}{9}}. \tag{9}$$

This equation also represents all first order velocity differences, hence possible periods, P. This is the periodicity rule which is found to fit observed redshift periods very well. The basic $T = 0$ sequence contains periods of 73.2 and 36.6 km s^{-1}, which closely match the empirical periods discovered in the 70s. Although the 3-d temporal model has proven to be quite fruitful, this paper is intended primarily to show how well equation (9) fits redshift data without regard to any interpretation of the equation.

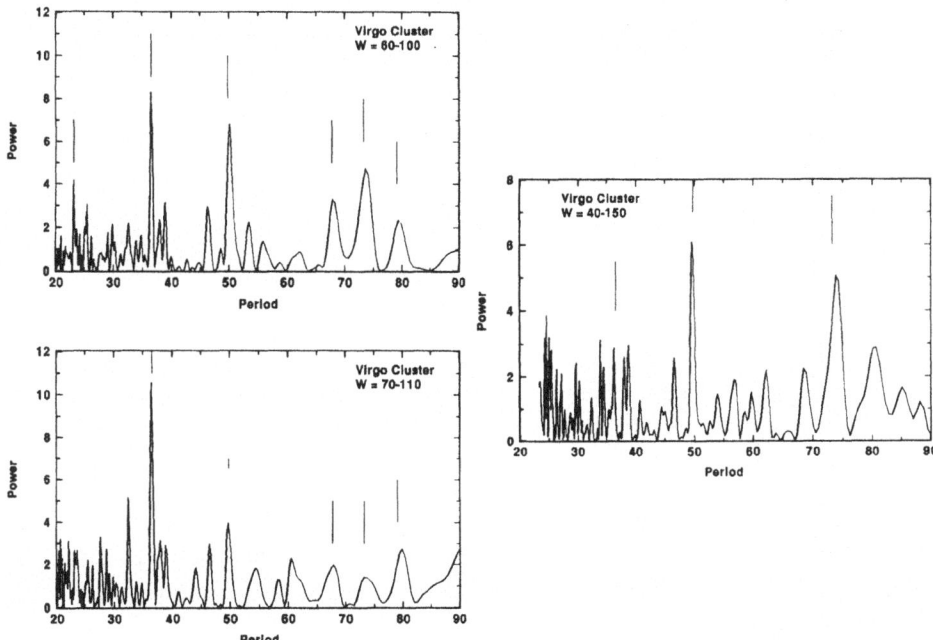

Figure 1. Power spectra of the Virgo region sample subdivided into 21 cm profile width intervals, W, in km s^{-1}. The abscissa is redshift period in km s^{-1}. Vertical lines mark predicted periods. The power at a given period depends upon the profile width interval utilized but the locations of the peaks do not.

3. First Tests, Virgo Dwarfs

Certain classes of local galaxies show periodic redshift patterns in the galactic center rest frame; the patterns are not seen in 21 cm data on Virgo cluster dwarf galaxies by Hoffmann et al (1987). Early tests of the CBR association also used primarily local galaxies (Cocke & Tifft 1996). The Virgo galaxies provided the first test combining the CBR association with periodicity predictions of equation (9). Clear periodicities matching the predictions are present; details are given in Tifft (1996b). Significance has been examined several ways, but the simplest way to show the periodicities is by spectral power analysis. As used here the average power and power dispersion for random data should be close to unity. The probability of finding a given power at one specified frequency is approximately inversely exponential in the power. See Cocke, DeVito & Pitucco (1996).

Figure 1 contains power spectra for galaxies in overlapping intervals of 21 cm profile width, W. Three periods stand out; two match the original 36 and 72 km s^{-1} periods. Essentially all significant peaks match predicted periods as shown with vertical lines. Table 1 summarizes the three main

TABLE 1. Spectral Power Data for Virgo

Width km s^{-1}	Pk Pow P=73.2	Pk/Per	Pk Pow P=36.6	Pk/Per	Pk Pow P=49.8	Pk/Per
50-90	5.6	1.0004	< 4	⋯	7.7	1.0084
60-100	4.2	1.0029	8.3	0.9966	7.1	1.0033
70-110	< 4	⋯	9.8	0.9947	4.1	0.9980
100-250	⋯	⋯	⋯	⋯	4.8	0.9912
60-125	< 4	⋯	5.0	0.9926	7.1	0.9979
40-150	5.1	1.0111	< 4	⋯	6.3	0.9975
All	4.0	1.0126	< 4	⋯	6.1	0.9970

periodicities. The peak power and the ratio of the period at peak power to the predicted period (the pk/per ratio) are given as a function of profile width. The following statements summarize findings and some results to be brought out later:

1) As W is varied, power may shift between periods, especially harmonics within one T family, but the peaks closely track the predicted periods. Width adjustment varies the power but does *not* 'tune' periods.

1a) Distinct phase shifts within the same period occur near certain profile widths. This is not apparent in the dwarf-dominated Virgo sample but will be shown later. Certain periods or T values tend to associate with particular morphology and profile width intervals.

1b) Redshifts generally concentrate in absolute phase around simple common fractions of the periods. Concentrations are not randomly spread in phase. This is again most easily shown with later samples.

2) The pk/per ratio is a measure of the quality of fit to the set of predicted periods. The ratio between adjacent predicted values is 1.0801, the ninth root of two. Power peaks concentrate strongly around 1.00.

3) Periods are not distributed randomly in T; the basic $T = 0$ family is usually dominant along with the $T = 6$ cube-root family. Some ninth-root families tend to occur with the dominant families, $T = 1$, 5 and 7 being the most important. The 49.8 km s^{-1} Virgo period has $T = 5$.

Figure 2 shows the mapping of power, at the 36.6 km s^{-1} period, around expected values of the transverse and radial transformation components appropriate to the CBR and galactic center rest frames. The small box is the

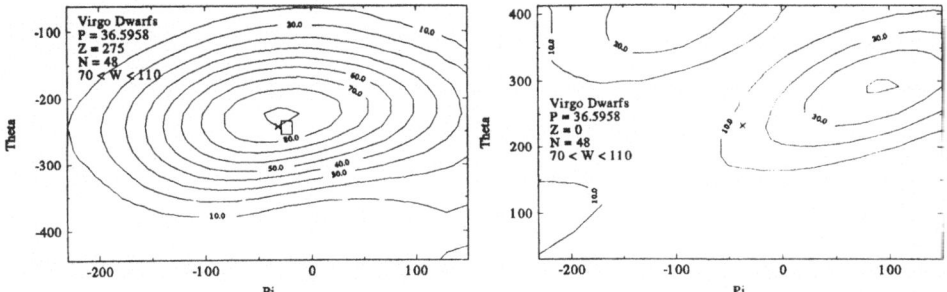

Figure 2. Contour diagrams of power, at the 36.6 km s^{-1} period, for a sample of Virgo cluster galaxies. W is profile width in km s^{-1}. Power is shown scaled by a factor of 10. The axes are the θ and π components, in km s^{-1}, of the vector used to transform to the CBR rest frame (left) or galactic center (right). Each diagram refers to a constant Z component as indicated. The box is the COBE error box for the CBR dipole vertex. The x in each frame is the approximate location of the rest frame vertex based upon previous quantization studies. The Virgo periodicities associate with the CBR vertex, but not with the galactic center.

COBE error box for the CBR vertex. X symbols mark adopted quantization vertices. The Virgo galaxies, not widely spread on the sky, show a broad power concentration associated with the CBR but not the galactic center.

4) Power at predicted periods is maximized when redshifts are transformed to a rest frame close to the COBE CBR vertex. The radial component is typically slightly more negative than the COBE value.

The significance of the periodicities and the match to equation (9) has been evaluated three different ways. The fact that the most prominent power peak matches the previously known period near 36 km s^{-1} with a power of 10 is significant by itself. About 20 predicted periods fall in the range studied; a match at any period at power 10 has a likelihood of accidental occurrence below the 0.001 level. Since at least three periods are matched above power 6 and no high power peaks fail to match periods, we conclude that equation (9) well describes the periodicities present.

Since questions have been raised about using extreme power values to estimate likelihoods we have used a binomial test to evaluate the degree to which all peaks above power 4 fit predicted periods. The power spectra for all the profile width intervals in the 20 to 100 km s^{-1} period range contains 14 independent power peaks greater than 4.0. The fits vary slightly but 8 of the 14 peaks lie near or within 0.005 of unity in the pk/per fitting index. The index should be uniformly populated between 1.00 ± 0.04 for a random distribution, hence the probability of falling within 0.005 of unity is

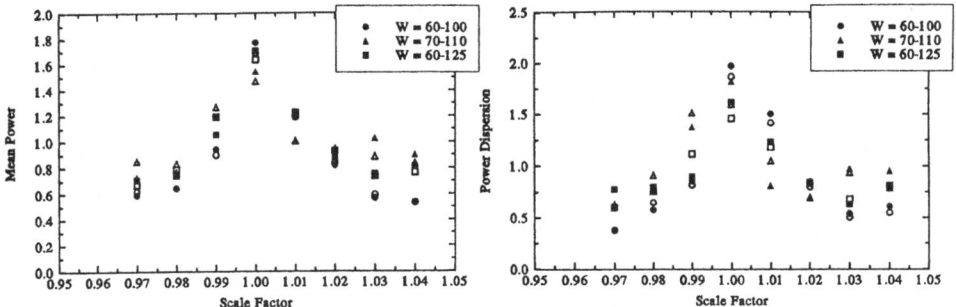

Figure 3. Mean power and power dispersion for 25 periods between 23 and 146 km s^{-1}. Symbols distinguish profile width subsets of Virgo galaxies; W is in km s^{-1}. Samples with open symbols cover the complete redshift range; samples with filled symbols omit redshifts below 500 km s^{-1}. Power is measured at periods scaled from the predicted periods by the scale factor on the abscissa. Power and dispersion peak at the predicted periods.

about 1/8. The likelihood of 8 or more accidental fits in 14 trials within this tolerance is 8×10^{-5}. The likelihood of finding more than 4 fits is < 0.05.

A third test uses the unit mean and dispersion in power expected for a random distribution. Eight sets of periods were generated by scaling the predicted periods by factors 0.97, 0.98, $\cdots$, 1.04 to cover the range between predicted periods. The mean and dispersion in power was then found for the 25 periods between 23 and 150 km s^{-1} using three data sets in the 60 to 100 km s^{-1} width range. Figure 3 shows the result as a function of the scaling factor. The distributions peak sharply at 1.00, the predicted period set. Comparing the mean and dispersion at 1.00 with the values midway between the predicted periods using a Student's t test yields t = 4.1, or about a 10^{-4} likelihood that they arise from the same parent population. Using power levels, peak locations, or the power distributions we find consistent significant results. The Virgo dwarf galaxies contain periodicities uniquely consistent with equation (9) when viewed from the CBR rest frame.

The third test confirms that adjusting profile width intervals does not arbitrarily tune periods. If this were not the case we should have found many intermediate periods. Real periods will shift only slightly and show power variation according to their actual profile width dependence as is observed. To test for tuning of spurious periods one must use periods which are not predicted, but such periods are not found in significant numbers. Arbitrary periods are rare since the redshift distribution is not random, it contains stable predicted periods.

4. A Second Sample, The Perseus Supercluster

A second sample uses redshifts from the deep Arecibo survey of the Perseus supercluster from Giovanelli and Haynes (1985, 1989). There is a special reason for this choice. A common criticism of periodicity studies is that parameters are adjusted to optimize power and period fits; in the first samples discussed here all parameters were set prior to the discovery of the ninth-root rule. See Tifft (1996b) for details. The samples were defined in older studies. In 1984 Tifft & Cocke (1984) found that local galaxies with 21 cm profiles wider than about 420 km s^{-1} contained a 36 km s^{-1} galactocentric periodicity. Martin Croasdale (1989) generally verified this using independent data, some of which came from the 1985 Perseus survey. To examine and extend Croasdale's work the 154 galaxies with unblended 21 cm profiles wider than 400 km s^{-1} were compiled from both Perseus region studies. Only 22 of the galaxies are in common with Croasdale.

This ready-made sample was used in 1992 in an early study of the CBR association, prior deriving equation (9). 73 galaxies with S/N < 4 were set aside as lower in quality. Using the 81 best redshifts it was noted that galaxies with small flux-to-width ratios contained a 36.05 km s^{-1} period which appeared to associate with the CBR rest frame. The F/W ratio provides a rough luminosity distinction. Using this distinction the 81 galaxies were divided into 53 with F/W < 0.01 and 28 with F/W > 0.01 to roughly optimize the 36.05 km s^{-1} periodicity; this period is *not* one of the periods predicted later by equation (9). By defining the samples using this period we introduce some sample homogeneity but no bias relating to the periods predicted later.

Figure 4 shows redshift phase, plotted in a double cycle to show periodic clumping, for the 81 galaxies with S/N > 4. The predicted 36.5859 $T = 0$ period is used; The abscissa is F/W. The right hand frame shows part of the 53 point power spectrum. Table 2 summarizes period fits giving the peak power and the (peak period)/(predicted period) ratios. The dominant peak fits the $T = 0$ family at the 18.3 km s^{-1} harmonic. There is no significant change if galaxies in common with Croasdale (1989) are deleted. The 18.3 km s^{-1} period is present. This period along with the 36.6 km s^{-1} and 9.15 km s^{-1} harmonics is often dominant. The period is aligned on phase 0.0, simply phased with the CBR rest frame. The Perseus data also illustrate the general preference for cube-root families; the $T = 3$ family is distinct. The ninth-root families flanking $T = 6$ are present and match the detection in Virgo. T values do not occur randomly; there are similarities between samples. The low S/N data are also consistent. When the low quality data are combined with the 28 point sample the longer period $T = 3$ periods are reinforced. Low quality data destroy short periods but preserve long ones.

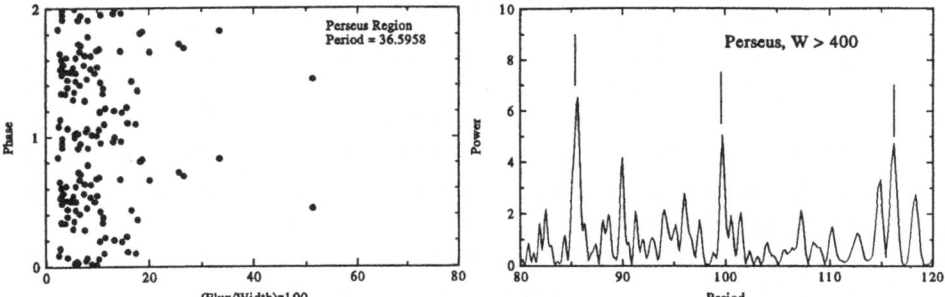

Figure 4.   (left) Phase diagram for Perseus galaxies with wide 21 cm profiles. Points are plotted twice in a double phase cycle at the 36.6 km s^{-1} period. The abscissa is the flux-to-width ratio scaled by 100. There is a regular 18.3 km s^{-1} pattern aligned at phase 0.0 and 0.5. (right) A section of the power spectrum for the 53 point subsample of wide profile Perseus galaxies; W is in km s^{-1}. The abscissa gives periods in km s^{-1}. Three predicted periods are shown with vertical lines. A disproportionate fraction of peaks in excess of power 4 associate closely with predicted periods.

To evaluate significance we found all power peaks greater than 4.0 in the period range 17 to 250 km s^{-1}. The limits are set by the breadth of power peaks at long periods and excess noise at short periods. We then count the matches within a limiting pk/per range and apply binomial statistics. The 53 point sample has 9 fits within 0.004 of unity (1/10 the possible range) and 27 power peaks above 4. The probability of 9 or more accidental hits in 27 trials is only 0.0009. The 28 point sample has 4 hits out of 9, unlikely at the 0.008 level. If we raise the cutoff period to 36 km s^{-1} to reduce noise, the 53 point sample returns 6 hits out of 10, unlikely at the 0.0002 level. Table 3 summarizes some of the binomial test results for both the Virgo and Perseus investigations. We again note that there was no optimization for the wide profile Perseus samples; they were defined before equation (9) was found.

The Giovanelli and Haynes (1989) data were next examined using galaxies with narrower 21 cm profiles. Findings are similar; Some of the results are given in Table 3. One subsample of the 472 galaxies available used 179 galaxies with S/N > 4 and 200 < W < 400 km s^{-1}. A period matching study found 10 of 21 power peaks above 4 falling within 1/5 of the range around predicted periods. This returns a random likelihood of 0.004. Of special interest is the continuity of the 18.3 km s^{-1} period at W = 400 km s^{-1}. Table 2 (lower part) shows that the pk/per ratio match is within 1% on both sides, 1.0003 and 1.0002; the power values are 6.9 and 7.1. Such continuity between width intervals, here with a phase shift, is extremely unlikely by accident but is not considered in evaluating significance. Periods

TABLE 2. Spectral Power Data for Perseus

Period km s^{-1}	T	Power N=53	Pk/Per	Power N=28	Pk/Per	Power 28+73	Pk/Per
36.5958	0	4.4	0.9990	...	...	...	...
18.2979	0	7.1	1.0003	...	...	...	...
232.3689	3	...	...	5.0	0.9968	...	...
116.1845	3	4.8	0.9999	...	...	...	...
58.0922	3	...	...	...	...	7.6	0.9984
29.0461	3	4.9	0.9977	4.4	1.0029	...	...
99.5984	5	4.9	1.0016	...	...	...	...
85.3802	7	6.8	1.0029	...	...	...	...

TABLE 2. Spectral Power Data for Perseus
Comparison Above and Below W = 400

V km s^{-1}	W km s^{-1}	S/N (F/W)	N	Pow(Pk/Per) P=18.2979
4500-17500	> 400	(< 0.01)	53	7.1(1.0003)
0-20000	300-400	> 4	76	6.9(1.0002)
0-20000	200-400	> 4	179	6.0(1.0003)

are intrinsic to the data, power but not periods can be tuned by selecting profile width intervals.

The samples on either side of the $W = 400$ km s^{-1} boundary also show the stability of the CBR association. Figure 5 contains power maps, at P = 18.3 km s^{-1}, for a range of the tangential and radial transformation components near the COBE vertex as shown earlier for Virgo. The independent maps agree closely.

5. A Third Sample, The Cancer Supercluster

A deep Arecibo survey of the Cancer region by Bicay & Giovanelli (1986ab, 1987) provides additional information. The 643 galaxy study provides a good illustration of the common $T = 0$ periods. For detail see Tifft (1996b). The familiar 36 km s^{-1} periodicity is strong in low redshift foreground galaxies viewed from the CBR rest frame. The top part of Figure 6 show

TABLE 3. Binomial Tests of the Distribution of Power Peaks

Sample	N	V km/s	W km/s	P km/s	Fit	Pks	Hit	Prob
Virgo	137	All	All	20-100	1/8	14	8	0.00008
					1/4	14	9	0.002
Per-W	53	All	> 400*	17-250	1/10	27	9	0.0009
S/N>4	*			36-250	1/10	10	6	0.0002
	28 **	All	> 400*	17-250	1/10	9	4	0.008
Per-565	179	All	200-400	16-250	1/5	21	10	0.004
	31	All	225-250	16-250	1/4	10	7	0.004
	56	6-8000	50-250	10-100	1/16	8	4	0.0009
	34	6-8000	200-350	10-100	1/16	9	3	0.02

* Predetermined F/W<0.01, ** F/W>0.01

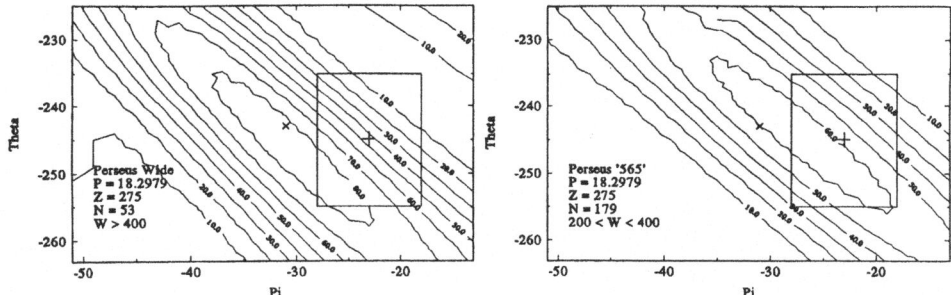

Figure 5. Contour diagrams of power, at the 18.3 km s^{-1} period, for samples of Perseus galaxies with 21 cm profiles wider and narrower than 400 km s^{-1}. Power is shown scaled by a factor of 10. The axes are the θ and π components, in km s^{-1}, of the vector used to transform to the CBR rest frame. The diagrams refer to a Z component of 275 km s^{-1}. The box is the COBE error box for the CBR dipole vertex; the x is the approximate location of the vertex from previous quantization studies. Samples on opposite sides of a phase shift, which occurs near $W = 400$ km s^{-1}, define the same period and vertex.

the 36.6 km s^{-1} phase-width pattern and power spectrum for the 58 galaxies with redshifts below 2000 km s^{-1}. The narrow profiles and 36 km s^{-1} period resemble the Virgo galaxies. The power spectrum sharpens and peaks very near the predicted period for the 33 galaxies with $90 < W < 190$ km s^{-1}. The lower right panel shows the CBR association for the 33 objects. The

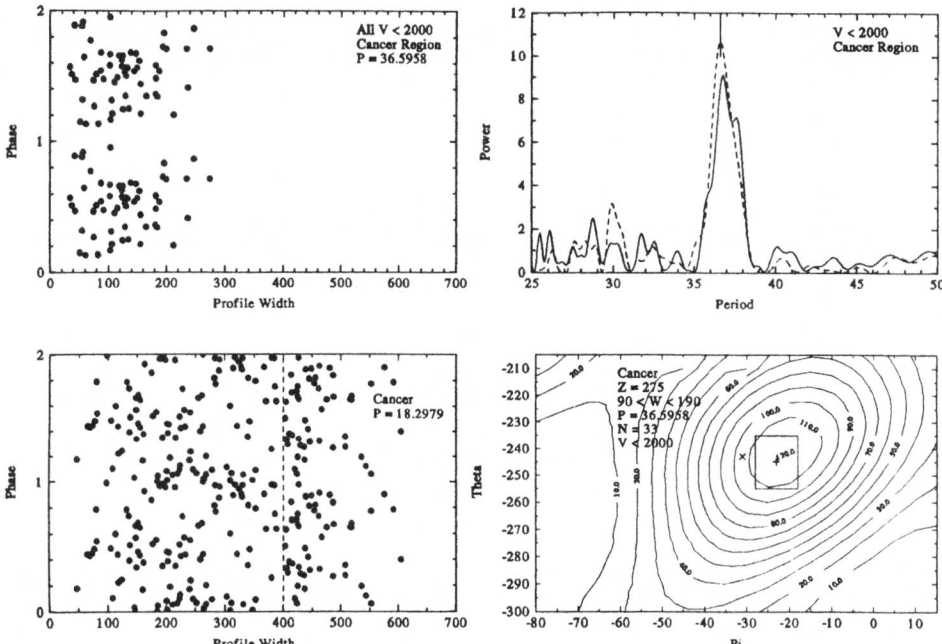

Figure 6. The left panels show phase-profile width diagrams for Cancer galaxy samples. Phase is plotted in a double cycle. Periods, P, profile widths, W, and redshift, V, are in km s^{-1}. The upper left panel shows the 36.6 km s^{-1} periodicity for low redshift galaxies. The lower left panel shows an alternating periodicity pattern found for higher redshift galaxies when $P = 18.3$ km s^{-1}. The selection criterion, discussed in the text, changes at $W = 400$ km s^{-1} where a phase shift occurs. The power spectrum (upper right) refers to the low redshift sample. The solid line is for all 58 low redshift galaxies; 33 galaxies with $90 < W < 190$ km s^{-1} produce the dashed spectrum. A line marks the predicted period. The lower right frame shows the power concentration, for the 33 galaxy set, associated with the CBR dipole vertex. See Figure 5 for a description of axes and symbols.

first part of Table 4 summarizes the foreground analysis. The lower redshift end of the Cancer complex, including the Cancer cluster, shows the 36.6 km s^{-1} periodicity among the wider profile galaxies. The 18.3 km s^{-1} period dominates when lower luminosity sources are included and can be traced through nearly the entire sample.

The lower left panel of Figure 6 and the last part of Table 4 trace the $T = 0$ periods through the higher redshift data. The 9.15 and 18.30 harmonics alternate as a function of profile width. F/W and S/N levels are set high to show the shorter period clearly. Periods track precisely, usually within 1% of the pk/per range about predictions, through successive independent width intervals while remaining aligned at 0.0 and 0.5 in phase on the 18.3 km s^{-1} scale. Above $W = 400$ km s^{-1} an expanded sample shows the phase break which occurs near $W = 400$ km s^{-1}. As found for

TABLE 4. Spectral Power Data for Cancer

$V/10^3$ km/s	W km/s	S/N	N	Pow(Pk/Per) P=9.1490	P=18.2979	P=36.5958
0-2	All	All	58	$\cdots$	$\cdots$	9.1(1.0043)
	90-190		33	$\cdots$	$\cdots$	10.7(0.9998)
3.5-7	> 250	> 8	128	$\cdots$	4.7(1.0023)	7.8(0.9907)
	All		184	$\cdots$	6.9(1.0027)	$\cdots$
2-10	All		562	$\cdots$	6.6(1.0022)	$\cdots$
5-10	0-175	*	30	$\cdots$	4.2(1.0032)	$\cdots$
	175-275		40	6.7(1.0002)	$\cdots$	$\cdots$
	275-400		31	$\cdots$	7.3(1.0000)	$\cdots$
	0-400		100	8.4(1.0003)	$\cdots$	$\cdots$
0-10	> 400	> 8	49	4.3(0.9999)	$\cdots$	$\cdots$

* F/W > 0.015

Perseus data there is no change in period but there is a phase shift. These Cancer data give a clear picture of how phase, harmonics, and width can be interrelated.

6. Short Periods and q_0 Determinations

Studies of short periods over wide redshift intervals are sensitive to the nonlinearity in z from equation (1). This equation was derived as a Taylor expansion about q_0 of 1/2 (Tifft 1991); higher order terms permit a determination of q_0. Short periods in the basic $T = 6$ cube-root family are strong in Perseus, Cancer and local redshift data (Tifft 1996b). Precise period matches occur when q_0 is equal to 1/2. Table 5 contains examples. The peak power location, near power 10 in these examples, shifts slightly as q_0 is varied. The pk/per ratio passes through 1.00000 when q_0 approaches 1/2. This result, found for several independent samples, gives considerable confidence in the significance of both equations (1) and (9). A classical interpretation of q_0 is unlikely in the temporal model.

The determination of q_0 is quite insensitive to the CBR vertex assumed. Figure 7 shows peak power maps as a function of the transverse and radial transformation components. Two large independent samples from the Cancer data are shown; one includes 89 narrow, the other 128 wide profile galaxies. Peak power exceeds 10 close to the standard vertex we have assumed. At this high resolution (P = 2.88 km s^{-1}) only the edge of the

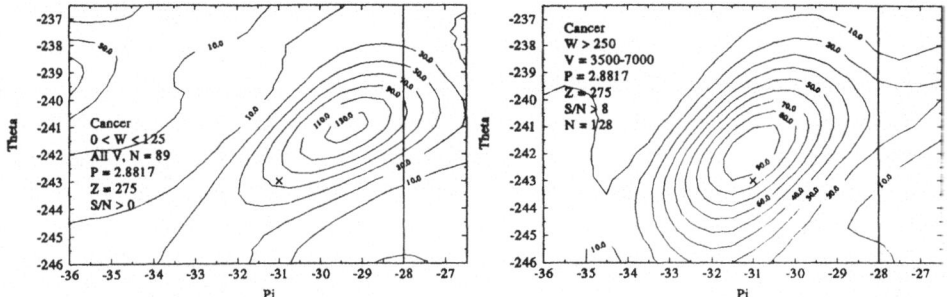

Figure 7. The power distribution, in the vicinity of the CBR dipole vertex, for two subsets of Cancer region galaxies. W and V give profile width and redshift ranges in km s^{-1}. S/N refers to signal-to-noise limits and N gives the sample size. At the short period shown the power peaks are sensitive to q_0, and match predicted periods when $q_0 = 0.5$. The independent samples conform closely to the same vertex. See Figure 5 for description of axes and symbols. Only the edge of the COBE error box, the line at right, falls within the frame for such short periods.

TABLE 5. Estimation of q_0

Period km/s	q_0	Peak km/s	Pk/Per	N	V km/s	W km/s	F/W
5.76348	0.52	5.7626	0.99985	88	4300-17900	>450	<0.015
	0.51	5.7633	0.99997		$(-241.5, -24.2, 275.0)$		
	0.50	5.7641	1.00011				
	0.49	5.7650	1.00026				
	0.48	5.7659	1.00042				
	0.46	5.7678	1.00075				
2.88174	0.51	2.8815	0.99992	128	3500-7000	>250	S/N>8
	0.50	2.8818	1.00004		$(-241.7, -30.8, 275.1)$		
	0.49	2.8821	1.00012				
	0.47	2.8826	1.00030				

COBE CBR dipole error box is visible at the right.

7. Local Data

The samples so far discussed involve single epoch studies of non-local galaxies, including ones in external superclusters. We now turn to multi-epoch data for local galaxies, drawing primarily on 21 cm data from the Fisher-

Tully (1981) survey from the 70s and data from Tifft & Cocke (1988) and Tifft (1990) from the 80s. Using different epochs we can investigate variability. The new data also include quantitative measures of 21 profile shape, the A = asymmetry index. Some material here is from Tifft (1996b); details about variability will be discussed in Tifft (1997).

Quantization effects change near certain profile widths. One such change occurs near W = 400 km s^{-1}; a second occurs near 200 km s^{-1}. The top frames of Figure 8 compare an original study of the CBR association using Fisher-Tully galaxies with the lower left frame from Cancer work in Figure 6. Below $W \approx 175$ km s^{-1} the Fisher-Tully data are periodic. The periodicity blurs but returns, with a phase shift, by $W \approx 300$ km s^{-1}. We see the same shift in Cancer at the 18.3 km s^{-1} $T = 0$ period, and recognize the shift as a transition through the 9.15 km s^{-1} harmonic. local galaxy work focussed on the $100 < W < 300$ km s^{-1} interval to verify the dominance of $T = 0$ periods and investigate the transition region. Detail are revealed by the way 'deviations', redshift differences between epochs, relate to phase and asymmetry. 'Phase-deviation' diagram are used. Deviation usually refers to the redshift difference Tifft & Cocke (1988) minus Fisher-Tully (1981), written TC−FT. The TCF sample contains 454 galaxies for which this difference is available.

Figure 9 contains phase-deviation diagrams for TCF galaxies. The upper left frame includes all 249 galaxies with $100 < W < 300$ km s^{-1}. The peak power is above 9 in the CBR rest frame as shown at upper right. The pk/per ratio is 1.0010; the peak falls within the central 2.5% of the range between predicted periods. The systematic shift between modern redshifts and Fisher-Tully values is apparent. The lower panels show enhanced periodicity when profile asymmetry is restricted to less than 10%, and morphology to t < 9. The lower right frame shrinks the width interval. The transition region contains a finer harmonic structure and is the same in Cancer and locally. S/N has little effect; scatter in phase is not due to observational uncertainty, it is due to finer structure. Scatter, from modern redshifts, is not much larger than the point size. Scatter in Fisher-Tully data affects deviations only.

To see finer structure we must introduce redshift variability. Figure 10 repeats a variant of the lower right frame of Figure 9. The vector shows that if a redshift were to decrease by one period between epochs a galaxy could shift from one end of the deviation pattern to the other and remain in phase. This is what seems to occur, occasional rapid transitions retain a periodic phasing and generate periodic deviations. A secular downward drift seems to occur from the high redshift end of levels. Intermediate steps may or may not be seen. Data from galaxies with wider profiles (lower right) show a staggered pattern which would be generated if intermediate levels

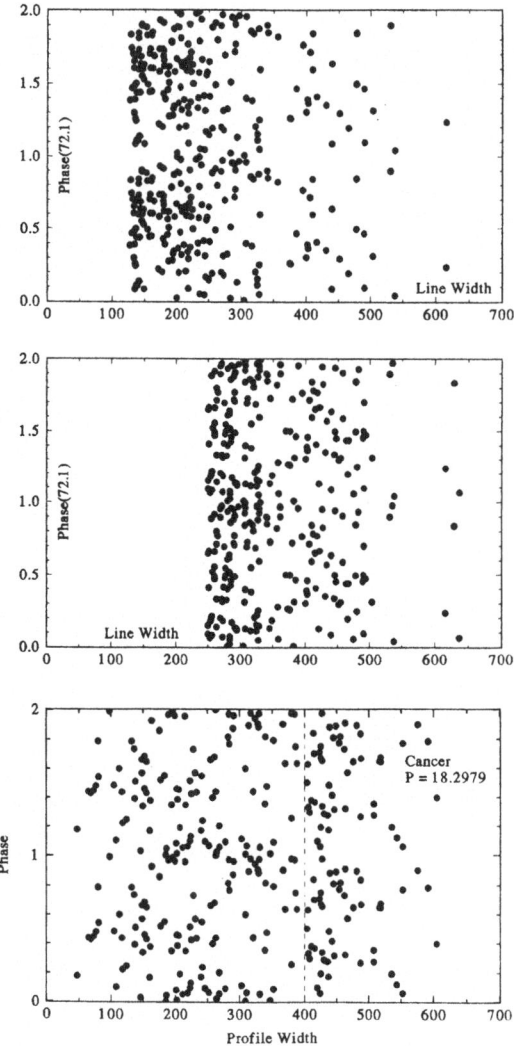

Figure 8.   The upper two frames are phase-width diagrams for local galaxies using Fisher-Tully data; W is in km s^{-1}. The diagrams are from a study of the association of redshift quantization with the CBR rest frame before precise periods were predicted; an empirical 72.1 km s^{-1} period was used. There are distinct periodicities, with a large phase shift, on either side of a transition region near $W = 200$ km s^{-1}. The lower frame repeats a Cancer region phase-width diagram from Fig. 6 to show that the same type of transition, involving subharmonics of the period, occurs at this width. The phase shift effect at $W = 400$ km s^{-1} can also be seen in both samples.

occur. Periodicity in such cases cannot be recognized without deviation information. A Student's t test comparing deviations in the half phase intervals dividing at .0 and .5 easily shows the periodicity.

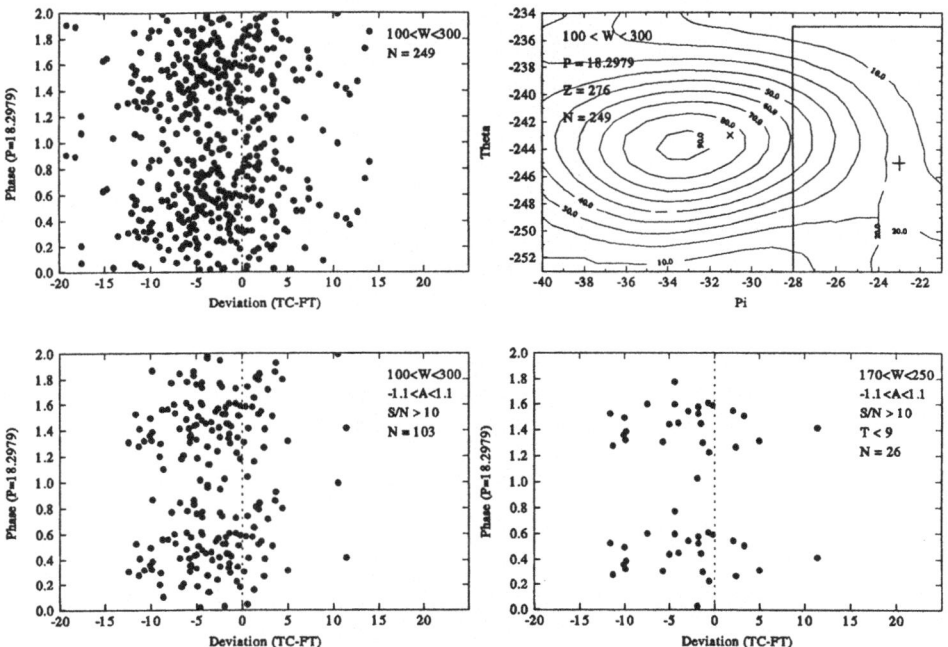

Figure 9. Phase-deviation diagrams, and a cosmocentric power map, for local galaxies. The 18.3 km s^{-1} predicted period is used. Deviation is the redshift difference, in km s^{-1}, Tifft-Cocke minus Fisher-Tully (a 10± year interval). The upper left diagram shows a complete set of 249 galaxies with profile widths, $100 < W < 300$ km s^{-1}; points concentrate around phase 0.5. At upper right the association of power with the CBR rest frame is shown. Refer to Fig. 5 for a description of axes and symbols. More restricted samples are shown in the lower frames. Restrictions, in W, profile asymmetry (A index), signal-to-noise, and morphology (t index), improve homogeneity and enhance power, but have no significant affect on the period or vertex location (compare Fig. 10).

The upper right panel of Figure 10 contains the power spectrum of the sample at upper left. The dashed line shows the spectrum without an asymmetry restriction. As with width adjustments, narrowing a parameter range can affect the power, but does not significantly affect periods. The periods are intrinsic to the galaxies. The lower left frame shows the CBR association of the restricted sample. As with periods, restricting the sample does not significantly affect the vertex location. The second peak in the power spectrum matches the $T = 1$ period at 16.94 km s^{-1}. The adjacent ninth root families seem to appear when the cube-root families show evidence of recent or current change. Near the beginning or end of a doubling process one axis may be out of synchronization with the others.

Asymmetry restrictions often improve the resolution of periodicities. If phase scatter is due to intermediate states, asymmetry may discriminate.

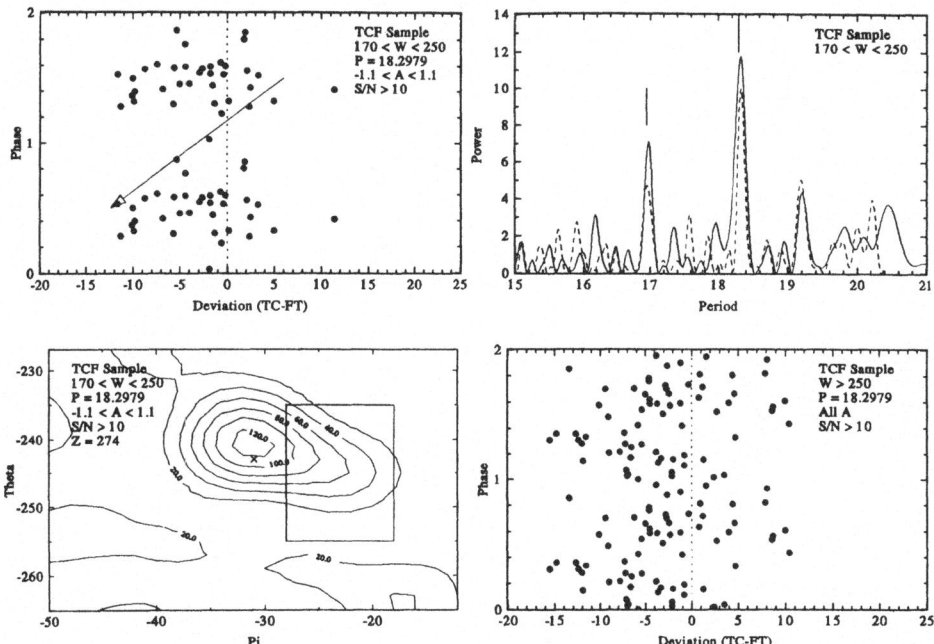

Figure 10. Redshift variability effects in a restricted local sample (see Fig. 9). The vector in the P = 18.3 km s^{-1} phase-deviation diagram (upper left) shows how a shift of one cycle between epochs will shift a point, produce a related deviation, and retain the periodicity. Intermediate levels seem to occur in other width intervals, (lower right), yielding a characteristic staggered pattern. The power spectrum (upper right) of the 36-point restricted sample shows a precise period match; predictions are marked with lines. Removing S/N and asymmetry, (A), restrictions generates a 92-point sample. Power drops slightly (dashed spectrum) but does not affect the period or CBR association (lower left, see Fig. 5 for a description). The power spectrum contains the next shorter predicted period; such periods seem to occur when recent or current variation is suspected.

Figure 11 shows positively asymmetric galaxies; they concentrate on the wide side of the 200 km s^{-1} transition, populate the 9.149 km s^{-1} $T = 0$ harmonic and show a sharply staggered pattern consistent with stepwise redshift decay. Objects with $A < 0$ tend to favor the narrow side of the 200 km s^{-1} transition with a shift in phase. In the 0 to -10 km s^{-1} deviation range the $A > 0$ objects show a power, at 9.149 km s^{-1}, in excess of 10. They map into the standard CBR rest frame as shown at upper right. This negative wing is expanded at lower right where points seem to clump laterally at still higher $T = 0$ harmonics. Redshifts seem to change in discrete steps, cascading between relatively stable levels in ways associated with the width and shape of the 21 cm profiles. The *velocity distribution function* appears to determine transition likelihoods. Multiple epoch data and detailed profile shape information are essential for the study or even

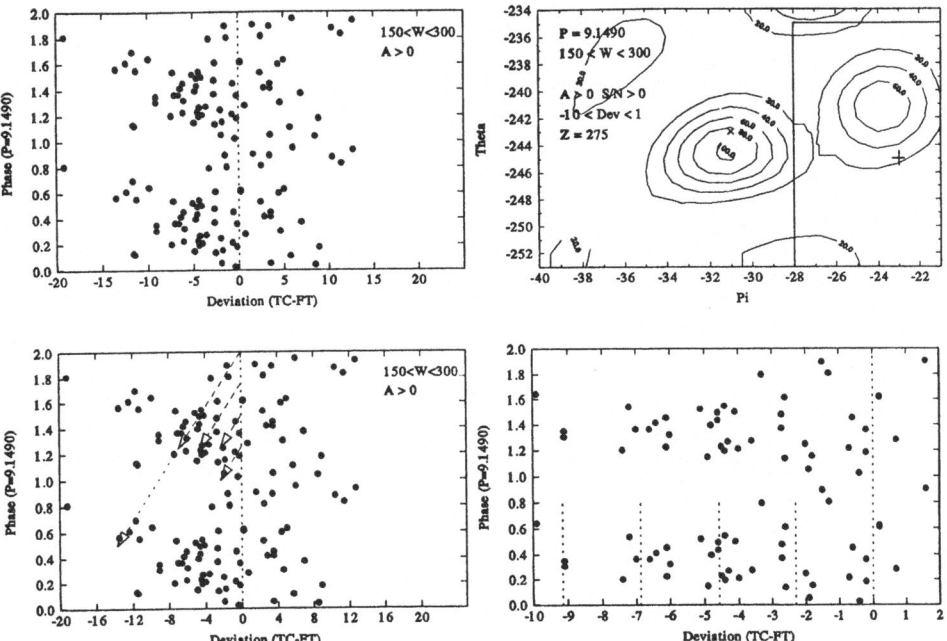

Figure 11. Phase-deviation diagrams and a power contour map for local galaxies with positive asymmetry and profile widths in the upper portion of the $100 < W < 300$ km s^{-1} transition region. A strong negative deviation is present and periodic at 9.15 km s^{-1}. The pattern is consistent with discrete changes (lower left) involving a substructure at still higher harmonics (lower right). Power contours for galaxies with $-10 < \text{dev} < -1$ km s^{-1} (upper right, see Fig. 5 for a description of axes and symbols) show power in excess of 10 very close to the adopted CBR dipole vertex.

the detection of such variability. Lookback time may cause differences as a function of distance so samples in different redshift ranges should not be casually combined.

The basic $T = 0$ doubling family dominates intermediate width profile galaxies consisting of relatively normal spirals. There seem to be three other categories, two classes of dwarf galaxies with narrow 21 cm profiles, and a general class of wide profile objects. Extreme dwarf galaxies, morphology t = 9 or 10 with $W < 75$ km s^{-1}, are locally periodic in the $T = 7$ family with a clear asymmetry dependence. This is shown in Figure 12 for the extreme dwarfs in the TCF sample. Objects with negative asymmetry or symmetric profiles show strong negative deviations periodic at the 10.6726 km s^{-1} $T = 7$ period. Galaxies with positive asymmetry are sharply phase-shifted. The sample, independent of the $T = 0$ sample, shows a clear CBR association and hypothetical transition patterns tuned in absolute phase.

The 10.67 km s^{-1} period and its 5.33 km s^{-1} harmonic were detected

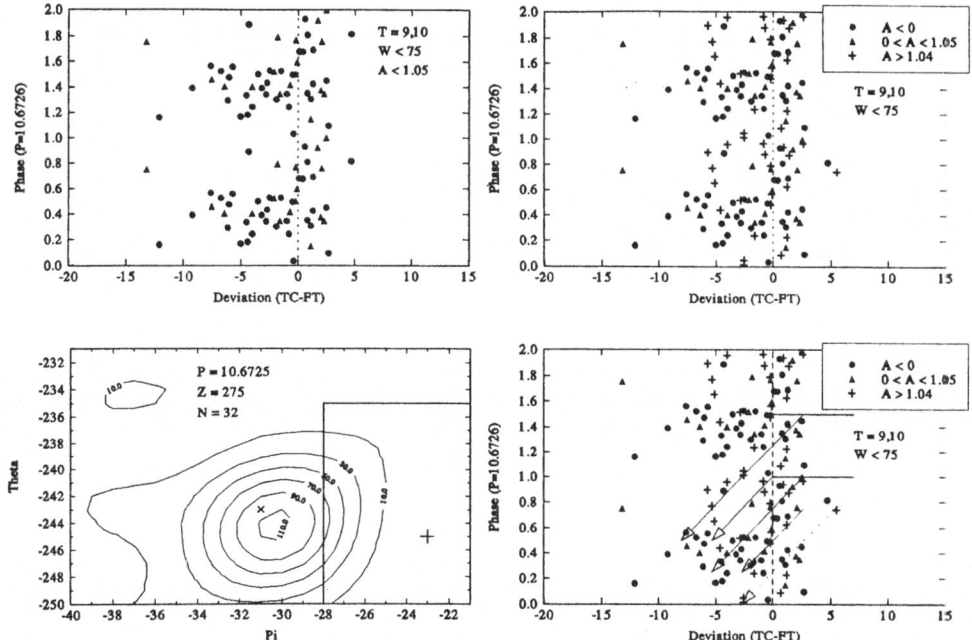

Figure 12. Phase-deviation diagrams and power contours for local galaxies with extreme dwarf characteristics, $t = 9, 10$ and $W < 75$ km s^{-1}. A strong negative deviation is present for galaxies with symmetric and negatively asymmetric 21 cm profiles (upper left). The period is 10.67 km s^{-1}, shifted one ninth-root cycle from the common $T = 6$ family of periods. Galaxies with positively asymmetric profiles, are phase shifted (upper right). The pattern is consistent with discrete changes (lower right) where asymmetry distinguishes between stages. Power contours for the negative wing of galaxies (upper left) show power in excess of 11 very close to the adopted CBR dipole vertex (lower left, see Fig. 5 for axes and symbols)

prior to the derivation of equation (9) (Tifft 1991). Their precise fit into the pattern of predicted periods helped to focus attention on equation (9). The $T = 7$ ninth-root family bears the same shifted relationship to $T = 6$ as the $T = 1$ periods have with $T = 0$. The shifted ninth-root patterns may characterize regions undergoing changes. The $W = 75$ km s^{-1} boundary for this group is not arbitrary; it is the narrow cutoff width for dwarfs found in the original global redshift studies (Tifft & Cocke 1984).

Between the extreme dwarfs and normal spirals of $t = 8$ or earlier, there is a large class of objects. They can be characterized by $t = 9$ or 10 with $75 < W < 250$. These common local galaxies show no overt periodicity until deviations are examined. The $T = 6$ cube-root family is then apparent. Figure 13 shows the 46.1078 km s^{-1} period where a characteristic staggered phase-deviation pattern appears. A Student's t test, comparing

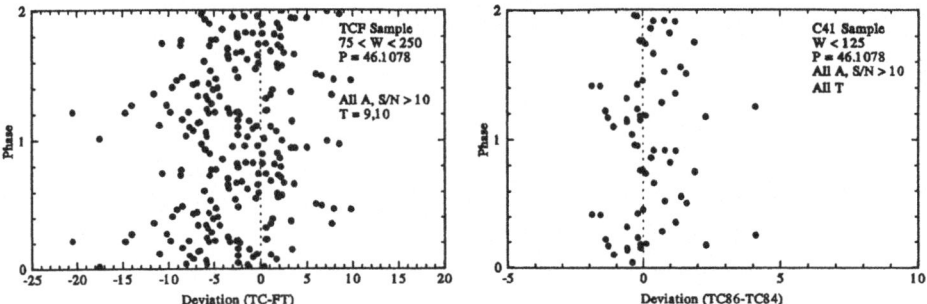

Figure 13. Phase-deviation diagrams for local dwarf galaxies; a staggered deviation pattern, associated with variability, occurs at the 46.1 km s^{-1} period. The comparison between modern redshifts and Fisher-Tully data is shown at left for galaxies with $t = 9, 10$ and $75 < W < 250$ km s^{-1}. The pattern continued to develop between 1984 and 1986 as shown for dwarf galaxies with $W < 125$ km s^{-1} at right. The deviation scale, in km s^{-1}, is expanded by a factor of 3 in the right frame.

mean deviations in half phase intervals dividing .at .0 and .5, indicates that the segments have only an 0.001 chance of being from the same deviation distribution. One interval shows no deviation; the other deviation is consistent with the 5.7635 km s^{-1} $T = 6$ period, a subharmonic of the 46 km s^{-1} period. This class of objects is interesting since recent observations alone show changes. The right panel of Figure 13 compares galaxies of the same general type using 1984 and 1986 observations. The same periodic wave is present with a low amplitude consistent with the short time interval involved; the change was still in progress in 1984. Evidence for variability is not limited to Fisher-Tully data. Consistent shifts for spirals can also be found in some of the oldest available 21 cm data.

The wide profile galaxies show a strong 5.7635 km s^{-1} $T = 6$ periodicity. This is illustrated in Figure 14, first for a restricted class of symmetric profile galaxies with $W > 250$ km s^{-1}, then for an unrestricted sample of objects with $W > 200$ km s^{-1}. A staggered pattern is present with the negative wing aligned on phase 0.0. The power spectrum of galaxies in the deviation interval from -2 to -8 km s^{-1} peaks at power 17 for P $= 5.7631$ for the 36 symmetrical profiles, and at power 16.3 for P $= 5.7624$ for all 64 objects. The period is matched within one-half of one percent of the Pk/Per range in phase with the CBR rest frame.

Sample adjustments have no significant effect on power, period or the CBR association shown at lower right. This pattern is clear in the wide profile Perseus galaxies, and its harmonic in the wide-profile Cancer galaxies, where it was used in q_0 determinations. The short $T = 6$ periods seem

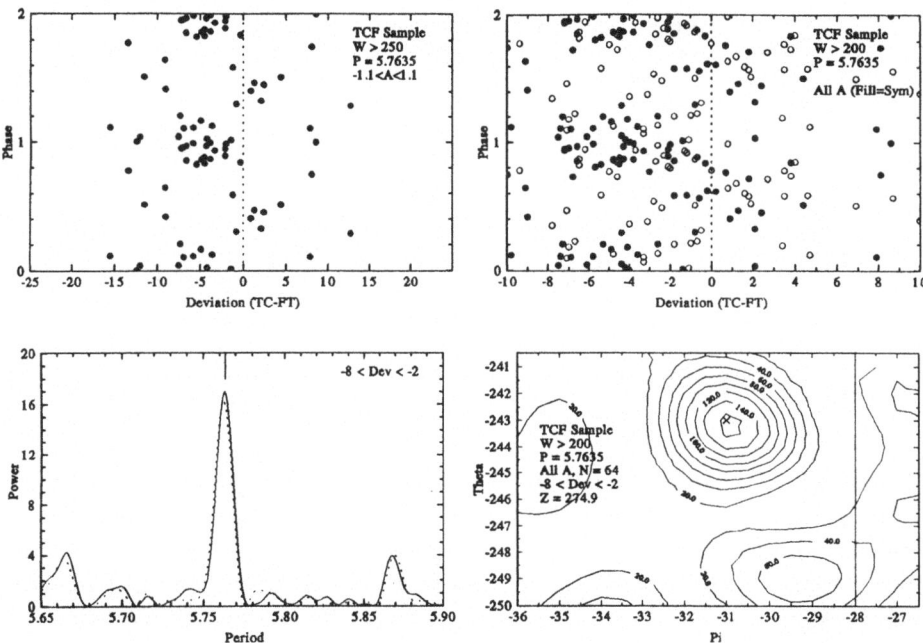

Figure 14. Phase-deviation diagrams, power spectra and power contours for local super-cluster spirals with wide 21 cm profiles. Galaxies with $W > 250$ km s^{-1} and symmetric profiles (upper left) show a staggered pattern of deviations at the 5.76 km s^{-1} period This period is common in galaxies with wide 21 cm profiles. Deviations between -8 and -2 km s^{-1} are aligned on phase 0.0 and achieve a power of 17 (lower left) precisely at the predicted period. When asymmetry restrictions are removed and the W range lowered to 200 km s^{-1} (upper right) there is no significant change of power or period (dashed line in spectrum). The power peak occurs at the adopted CBR dipole vertex (lower right, see Fig. 5 for axes and symbols)

to be quite common among the wide profile galaxies. The $T = 0$ family dominates in galaxies with intermediate profile widths, then longer period $T = 6$ periods return among the dwarfs. The extreme dwarfs shift to the adjacent $T = 7$ family. Specific classes of galaxies associate with different T values and dominant period ranges; the $T = 0$ and $T = 6$ cube-root families stand out. Periods are not randomly distributed in T. Table 6 summarizes the major sets of periods.

8. Relationship Between the CBR and Galactocentric Frames

There seem to be two significant quantization rest frames. To understand this we believe it is necessary to abandon the idea that galaxies move with respect to one another. Such motion for our Galaxy can explain the CBR dipole observation, but combined with similar random motions for other

TABLE 6. Selected Redshift Periods $P = c2^{-\frac{9D+T}{9}}$

$D\backslash T$	7	6	5	1	0
17	⋯	⋯	⋯	⋯	2.2872
16	2.6681	2.8817	⋯	⋯	4.5745
15	5.3363	5.7635	⋯	⋯	9.1490
14	10.6725	11.5270	⋯	16.9416	18.2979
13	⋯	⋯	⋯	33.8831	36.5958
12	⋯	46.1078	49.7992	⋯	73.1916
11	⋯	92.2157	99.5984	⋯	146.3833
10	⋯	184.4313	⋯	⋯	⋯

galaxies would destroy global quantization. If we accept quantization we require a non-velocity explanation of the CBR transformation. In the temporal model, described in a separate review, galaxies are quantized structures dispersed, and evolving, in three-dimensional time. The CBR transformation places the observer in the quantized frame associated with 3-d temporal space. We see quantized patterns which depend only on lookback time and the character of the type of galaxies involved.

At the same time we cannot overlook the fact that within our Galaxy we have real spatial motion with respect to the galactic center. When we apply the galactocentric transformation we remove this motion, but this is not sufficient to place us in the quantized temporal frame. As a 3-d temporal object the Galaxy has certain properties; the CBR transformation removes these along with the spatial solar motion. The galactocentric transformation removes only the spatial part. This seems to be sufficient to induce local resonant patterns which appear as strong fluctuations rather than a regular global pattern.

The two transformations appear to be linked; within scatter in local random stellar motions the tangential components are equal and opposite (-243 and $+232$ km s^{-1}) while the radial terms are the same, (-31 and -36 km s^{-1}). In the temporal model the dynamics of a galaxy may be associated with the temporal structure. The CBR vertex is nearly opposite the galactocentric one in longitude; transformation terms in longitude simply change sign. Given strong tuning in the CBR frame one might expect resonances opposite the longitude of the CBR vertex using only the spatial correction. This is especially true if one examines galaxies with similar characteristics.

Figure 15 (left) takes the sample of local spirals, $V < 1000$ km s^{-1}, used by Guthrie and Napier (1991), and refers them to the galactic center

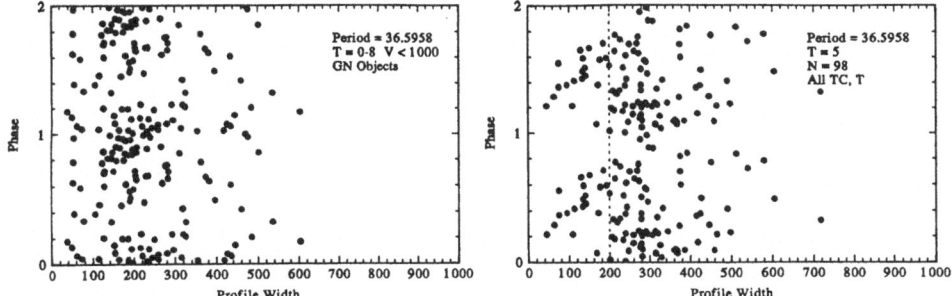

Figure 15. Phase-profile width diagrams, in the galactocentric rest frame, for spiral galaxies. The basic 36.6 km s^{-1} period is used; W is in km s^{-1}. The left frame contains a 106 point Guthrie-Napier sample of nearby spirals. These very local objects are periodic but are heavily dwarf dominated. The right frame contains 98 classical Sc ($t = 5$) galaxies. The periodicity is again detectable in specific width intervals, but is not phased with other types and shows a distinct change near $W = 200$ km s^{-1}. Such periodicities, much more regular in the CBR rest frame, may induce strong periodic fluctuations in local galaxy samples observed in the galactocentric frame.

for the predicted 36.6 km s^{-1} period. The galaxies are mostly dwarf object but similar enough internally that just the removal of dynamical motion reveals the periodicity. This is also true (right) for all Sc galaxies ($t = 5$) with recent Tifft-Cocke or Tifft redshifts. These extend further in redshift, show the basic periodicities at 36.6 and 18.3 km s^{-1}, and the 200 km s^{-1} profile width transition. They are not in phase with the local dwarf objects, however. Given such periodic clumping any local sample can be far from random in its redshift distribution. A search near the galactic center, or the anti-longitude of the CBR vertex, should detect fluctuations.

A search was carried out using a recent 104 point set of spirals defined by Guthrie and Napier. Figure 16 shows the locations of strong power peaks for an 18.8 km s^{-1} period related to the Guthrie and Napier findings. A pattern of power peaks is concentrated around the anti-longitude of the CBR dipole vertex (shown as a cross). Our standard galactocentric vertex is shown with an X. Triangles mark strong 37 km s^{-1} periodicities mapped by Guthrie and Napier (1996). The close proximity of the galactocentric vertex and the CBR anti-longitude point makes it difficult to separate galactic and CBR correlations. All or most galactocentric findings may be traced back to the cosmocentric effect through connections between the transformations.

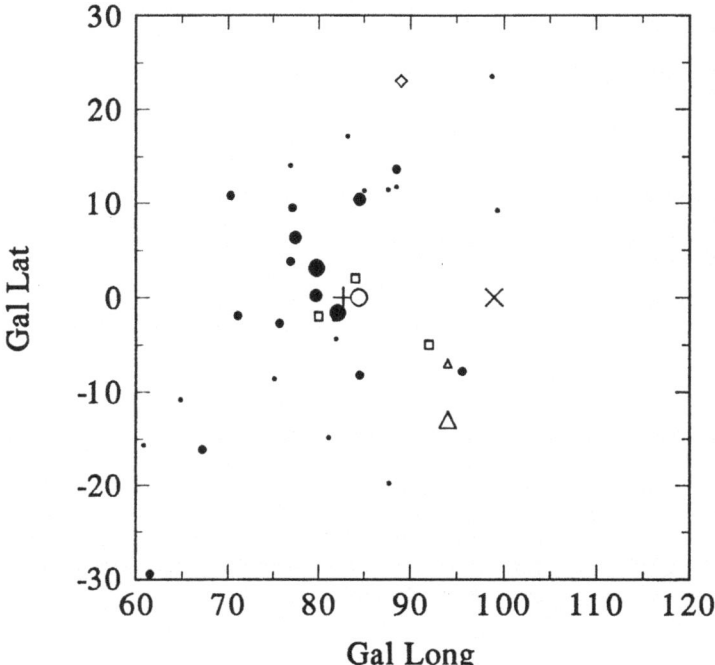

Figure 16. Locations of power peaks, in galactic coordinates, for 104 local supercluster galaxies examined recently by Guthrie and Napier. An X marks the standard galactocentric quantization vertex; the cross and open circle mark points, in the galactic plane, opposite the longitude of the cosmocentric quantization and COBE vertices. Large power fluctuations (filled symbols) at P = 18.8 km s^{-1}, half of the 37.5 km s^{-1} period studied by Guthrie and Napier, associate with the anti-CBR point. Power varies from 9 for the smallest symbols to 17 for the largest. They were found in a 10^6 point search through a velocity cube using a 2 km s^{-1} resolution. Open symbols identify other periodicities found by Guthrie and Napier. These features occur at frequencies greatly in excess of random expectations and may be local fluctuations induced by a basic cosmocentric pattern.

9. Summary

In 1992 and 1993 redshift quantization was associated with the CBR rest frame, and a model involving 3-d time was developed which permits predictions of global periodicities. Two equations now exist, one linearizes periodicities in z, and the other defines periods. They are:

$$V_{corr} = 4c[(1+z)^{1/4} - 1] + \cdots \qquad \text{and} \qquad P = c2^{-\frac{9D+T}{9}}. \qquad (10)$$

The equations are consistent with a model combining 3-d time with 3-d space within which fundamental particle properties at one extreme, and cosmological observations at the other, may be related. The following statements summarize and expand on characteristics introduced in Section 3.

1) Width, redshift range, and profile asymmetry adjustments influence spectral power but do not tune periods. As parameters are varied, power may shift between periods, especially harmonics within one T family, but peaks closely track predicted periods.

1a) Distinct phase shifts within the same period occur near certain profile widths. The shifts involve steps through subharmonics which suggest that redshift transitions occur from time to time. Patterns of offset redshift deviations between different epochs are consistent with such changes.

1b) Redshifts concentrate in absolute phase around simple fractions of the periods. Concentrations are not randomly spread in phase.

2) The pk/per ratio is a measure of the quality of fit to the set of predicted periods. Power peaks concentrate strongly around 1.00 and avoid values near 1.04 midway between predictions. Short periods at high redshifts indicate that $q_0 = 1/2$.

3) Certain periods or T values tend to associate with particular morphology and profile width intervals. Four major classes have been identified in local redshift data.

3a) Periods are not distributed randomly in T; two cube root families, $T = 0$ and 6, dominate. Ninth-root families often associate with the dominant families, $T = 1$, 5 and 7 being the most important. Table 6 summarizes common periods associated with classes of galaxies.

4) Power at predicted periods is maximized when redshifts are transformed to a rest frame close to the COBE CBR vertex. The radial component is slightly more negative than the COBE value. The galactocentric rest frame seems to be intimately related to the CBR frame.

10. Acknowledgements

The author is indebted to John Cocke who was instrumental in the derivation of equation (1) and the pursuit of the CBR connection, and to Ari Lehto for his original work with 3-d temporal concepts, particularly those relating to fundamental particles. Carl DeVito and Anthony Pitucco also contributed ideas which assisted in developing the current 3-d temporal model.

References

Bicay, M. D. and Giovanelli, R.: 1986a, *AJ*, **91**, 705
Bicay, M. D. and Giovanelli, R.: 1986b, *AJ*, **91**, 732
Bicay, M. D. and Giovanelli, R.: 1987, *AJ*, **93**, 1326
Cocke, W. J. and Tifft, W. G.: 1996, *Astroph. & Space Sci.*, in press

Cocke, W. J., DeVito, C. and Pitucco, A.: 1996, this conference
Croasdale, M. R.: 1989, *ApJ*, **345**, 72
Fisher, J. R. and Tully, R. B.: 1981, *ApJS*, **47**, 139
Giovanelli, R. and Haynes, M. P.: 1985, *AJ*, **90**, 2445
Giovanelli, R. and Haynes, M. P.: 1989, *AJ*, **97**, 633
Guthrie, B. N. G. and Napier, W. N.: 1991, *MNRAS*, **253**, 533
Guthrie, B. N. G. and Napier, W. N.: 1996, *A&A*, in press
Hoffmann, G. L.,Helou, G.,Salpeter, E. E., Glosson, J., Sandage, A.: 1987, *ApJS*, **63**, 247
Lehto, A.: 1990, *Chinese J. Phys.*, **28**, 215
Tifft, W. G.: 1978a, *ApJ*, **221**, 449
Tifft, W. G.: 1978b, *ApJ*, **221**, 756
Tifft, W. G.: 1990, *ApJS*, **73**, 603
Tifft, W. G.: 1991, *ApJ*, **382**, 396
Tifft, W. G.: 1995a, *Astroph. & Space Sci.*, **227**, 25
Tifft, W. G.: 1995b, *Mercury*, **24**, 12
Tifft, W. G.: 1996a, this conference
Tifft, W. G.: 1996b, *ApJ*, in press (Sept. 10)
Tifft, W. G.: 1997, in preparation
Tifft, W. G.,& Cocke, W. J.: 1984, *ApJ*, **287**, 492
Tifft, W. G.,& Cocke, W. J.: 1988, *ApJS*, **67**, 1
Tifft, W. G., Cocke, W. J., & DeVito, C.: 1996, *Astroph. & Space Sci.*, in press
Tifft, W. G.,& Lehto, A.: 1996, in preparation

TESTING FOR QUANTIZED REDSHIFTS. I. THE PROJECT

W. M. NAPIER
Armagh Observatory
College Hill, Armagh BT61 9DG, Northern Ireland

AND

B. N. G. GUTHRIE
5 Arden Street
Edinburgh EH9 1BR, Scotland

Abstract. A project intended to examine the long-standing claims that extragalactic redshifts are periodic or 'quantized' was initiated some years ago at the Royal Observatory, Edinburgh. The approach taken is outlined. and the main conclusions to date are summarized. The existence of a galactocentric redshift quantization is confirmed at a high confidence level.

1. Introduction

Persistent claims have been made over the last 25 years or so that at leas: some extragalactic redshifts are non–cosmological in origin. Perhaps the least credible of these claims is that the redshifts of galaxies are periodic or 'quantized', tending to occur at intervals of $\sim$72 km s^{-1} within binaries, groups and clusters (Tifft 1976, 1977, 1980; Arp & Sulentic 1985; Arp 1987), with a related global redshift periodicity of $\sim$24 or $\sim$36 km s^{-1} for field galaxies when a suitable correction for the solar motion is made (Tifft & Cocke 1984, hereinafter TC). The quantization claim is extraordinary, and if confirmed would have profound repercussions for cosmology. Given the perceived success of standard paradigms, a correspondingly high standard of proof would be required before the alleged periodicity could be accepted (say at the level where a cosmological model which failed to incorporate it would lack credibility). Testing for the quantization is however a 'clean', well–posed statistical problem, while new high–precision 21 cm redshifts are now available in adequate numbers for confirmation or otherwise of the

Astrophysics and Space Science **244**:57-63,1996.

claim to be possible. A series of research programmes was therefore initiated at the Royal Observatory, Edinburgh to investigate the issue. Rigorous statistical analysis, utilising power spectrum analysis (PSA), was employed throughout: the pitfalls in the latter, and our means of avoiding them, are described in the companion paper (Paper II). Two pilot studies, involving 48 and 40 high-precision redshifts respectively, yielded positive results, and so were followed by a major analysis involving over 200 spiral galaxies in the Local Supercluster. We summarize herein the progress of this work.

2. The Virgo Cluster

We first examined the distribution of the most accurately measured HI redshifts of galaxies in the region of the nearby Virgo cluster, which had not previously been used in formulating the quantization hypothesis (Guthrie & Napier 1990). We compiled two samples of galaxies within $10°$ of the central galaxy M87, comprising 112 bright spirals and 77 dwarf irregulars. Their heliocentric redshifts cz are $<3\,000$ km s^{-1} (the upper limit for the cluster) and have stated accuracies of ± 10 km s^{-1} or better.

We first tested each sample for the existence of a redshift periodicity somewhere in the range 70–75 km s^{-1}, in accordance with the original claim made by Tifft (1976). No significant periodicity in this range was found for either sample of heliocentric redshifts. However, when the individual redshifts were corrected for the estimated solar motion with respect to the centroid of the Local Group [$V_\odot = 252$ km s^{-1} towards $(l_\odot, b_\odot) = (100°, 0°)$], a possible periodicity of ~ 71.3 km s^{-1} emerged for the sample of 112 spirals. The periodicity appeared to be stronger for the 56 outer spirals at $5°–10°$ from M87. Accordingly, a sub-sample of 48 spirals in low-density regions of the cluster was compiled from a chart of bright galaxies in the region, the criterion for low density being adjusted to maximize the periodicity signal. Taking account of the number of independent trials involved in testing the period range 70–75 km s^{-1} and the number of trials used in selecting the optimum criterion for low density, we found that the periodicity (71.1 km s^{-1}) was significant at a confidence level $0.997 \lesssim C \lesssim 0.999$.

Since the Virgo cluster covers only a small area of sky, the differential correction for the solar motion is small and the exact choice of solar vector is not critical. When the apex was varied over the whole sky, it was found that the periodicity appeared most strongly for correcting vectors $(l_\odot, b_\odot) = (98°, 60°)$ and $(101°, -30°)$; the previously adopted apex $(100°, 0°)$ lies on a north–south ridge encompassing these twin peaks (see figure 8 in Guthrie & Napier 1990). The significance of the peaks was assessed by comparison with 60 whole-sky maps constructed for sets of 48 synthetic, random redshifts with the same overall distribution in space and redshift as the real data. These non-periodic datasets failed to reproduce the observed power contours, and periodicity was preferred over chance at a confidence level $0.996 \lesssim C \lesssim 0.999$. For the real data, whole-sky maps for other solar speeds between 150 and 300 km s^{-1} yielded broadly similar results, the twin peaks and ridge (and therefore the underlying periodicity which generates them) being significantly stronger for $V_\odot \simeq 200$ km s^{-1}.

3. Nearby Field Galaxies

For our second pilot study (Guthrie & Napier 1991) we compiled samples of nearby field galaxies (corrected redshifts <1000 km s^{-1}) to test the TC hypothesis of a global periodicity of ~ 24 or ~ 36 km s^{-1}. Since these periods are small, high redshift accuracy is very important. Redshifts with listed standard errors $\sigma_{cz} \lesssim 4$ km s^{-1} were taken from the extragalactic HI database compiled by Bottinelli et al. (1990). Excluding galaxies in the region of the Virgo cluster, we had 106 spirals and 62 irregulars. Eliminating also the galaxies previously used by TC, we obtained an independent sample of 89 spirals of which 40 had redshift errors $\sigma_{cz} \leq 3$ km s^{-1}.

The heliocentric redshifts of the 89 spirals were individually corrected for the solar vector found by TC (233.6 km s^{-1}, 98.°6, 0.°2), and a preliminary PSA was applied to the corrected redshifts to search for periodicities in the range 20–200 km s^{-1}. A prominent peak was found at $P=37.1$ km s^{-1}, close to the periodicity of 36.3 km s^{-1} claimed by TC for galaxies with broad HI line profiles, but there was no evidence for the periodicity of 24.2 km s^{-1} claimed for narrow-line galaxies. The significance of the peak at 37.1 km s^{-1} was assessed by Monte Carlo trials: synthetic datasets were constructed by adding to each of the 89 redshifts a random displacement in the range 0–60 km s^{-1} as well as the correction for the TC solar vector. Thus all the essential features of the real dataset were preserved except for the local redshifts. PSA of 3 000 synthetic datasets showed that the probability of obtaining a periodicity of the observed strength in the range 70–75 km s^{-1} or one of the submultiple ranges 23.3–25 and 35–37.5 km s^{-1} by chance in a single trial is ~ 0.003. Thus the TC hypothesis of redshift periodicity is preferred over the null hypothesis of a random redshift distribution at a confidence level $C \sim 0.997$. No significant periodicity was found for the uncorrected, heliocentric redshifts.

There is inevitably some uncertainty in the solar vector found by TC by maximizing the periodicity signal, but it is fairly close to estimates of the solar motion around the Galactic centre. We therefore tested a wide range of vectors around the solar motions relative to the Galactic centre and the centroid of the Local Group, and found two strong peaks at $P = 37.2$ and 37.5 km s^{-1} respectively, both close to the adopted galactocentric solar motion. Taking account of the proximity of these peaks to this latter vector, and noting that the periodicities were stronger for the 40 spirals with $\sigma_{cz} \leq 3$ km s^{-1} than for the other 49 'less accurate' spirals (as would be expected for a real phenomenon), the overall probabilities of chance coincidence were found to be $\sim 3 \times 10^{-5}$ for the 37.2 km s^{-1} periodicity and $\sim 1 \times 10^{-4}$ for the 37.5 km s^{-1} one. On the other hand, no significant periodicities were found for the sample of 62 irregular galaxies.

4. The Local Supercluster

Our two pilot studies raised a number of questions. In particular the existence of multiple peaks clearly complicated the matter of determining the unique solar vector (if such existed) for which the periodicity was seen most strongly. Trials with synthetic data revealed that a single redshift periodicity, for a specific solar vector $V_{\odot}$, yielded a plethora of ghost peaks or side lobes for many $V_{\odot}$, some far from the genuine one. Thus while the existence of a periodicity might readily be inferred, it was not always easy to discriminate the 'true' peak from the 'ghosts' in a velocity map. Further, our derived probabilities were obtained in part from the proximity (in velocity space) of individual high peaks to the galactocentric solar vector derived from Galactic HI data. They were thus an example of extreme value statistics, and sensitive to uncertainties in the true $V_{\odot}$.

On the other hand, these preliminary analyses also yielded positive results: periodicities were observed in the independent datasets, at reasonably high confidence levels, close to the $(P, V_{\odot})$ values previously claimed. We therefore undertook a much more extensive analysis involving over 200 galaxies out to the edge of the Local Supercluster (Guthrie & Napier 1996). We employed a more robust statistical procedure, for example using the overall power in a volume of $(P, V_{\odot})$ space as a statistic rather than the height of individual high peaks; we used pattern–matching of peaks obtained from synthetic datasets to find the 'true' vector corresponding to the periodicity; and we compared with the recent estimate of the galactocentric solar vector due to Merrifield (1992). Various tests for robustness (partitioning of data, sub–division by radio telescope and so on) were also applied. This study is reported elsewhere in these proceedings and is only summarised here.

First, we compiled a list of 97 spirals with corrected redshifts $<2\,600$ km s^{-1} and $\sigma_{cz} \leq 3$ km s^{-1}, not previously used by TC. The overall power distribution in $(P, V_{\odot})$ space was found and compared with distributions obtained from properly constructed synthetic datasets. We found that, when corrected for related vectors close to recent estimates of the Sun's galactocentric motion, the redshifts are strongly periodic ($P \simeq 37.6$ km s^{-1}).

Thus the basic hypothesis of redshift quantization is confirmed at an extremely high formal confidence level. However we also found evidence that the *intrinsic* strength of the global periodicity slowly weakens with distance from the Sun, while remaining strong for galaxies linked by group membership. Thus while the above trials establish that the hypothesis 'redshifts are periodic' is strongly preferred over the hypothesis 'they are not', the evidence may also indicate that the periodicity is strongest for adjacent galaxies but weakens as their separation increases (as would happen, for example, with an imperfectly coherent wave). This of course represents a significant modification of the original hypothesis, and in accordance with orthodox statistical procedure it was tested against a further sample. The latter comprised 117 spirals for which HI profiles with signal-to-noise ratios >10 had been obtained with the 300-foot Green Bank telescope; these galaxies have a higher mean redshift than the 97 and are more widely separated. The new sample indeed provided only weak evidence for a periodicity consistent with that found for the 97 spirals; the additional group-linked galaxies in the combined samples, on the other hand, were found to have a very strong galactocentric redshift periodicity of $\sim$38 km s^{-1}. The differential redshifts of the group–linked galaxies in the combined samples exhibit an extremely strong periodicity which is virtually impossible to ascribe to chance (Fig. 1). For both the Virgo cluster and the field samples, the vector with respect to which the periodicity holds was found to lie within the error box of the solar galactocentric motion. The hypothesis we tested was, however, limited and specific, and we cannot exclude the possibility that other vectors (such as the COBE one) and periodicities may exist for other samples, as claimed by Tifft (1996) in recent work.

5. Discussion and Conclusions

In physics, observation and reproducibility are as important as formal statistical verification. In treating the quantized redshift question purely as an exercise in statistics, there is a risk of obscuring the fact that the periodicity is easily seen by eye. This is particularly so in the case of high-precision differential redshifts within groups (Fig. 1). In general, the more precisely the redshifts are measured, the stronger the effect is seen to be; conversely, it vanishes rapidly as the precision of the redshifts employed degrades. Likewise as the size of the samples increases, so also does the strength of the imbedded signal, while the correcting vector holds with remarkable stability. Simulations reveal this behaviour to be consistent with those of a real periodicity at a quantitative level. The galactocentric nature of the periodicity in the samples studied so far is difficult to reconcile with an artefact, which would of course create a spurious periodicity in the frame of reference

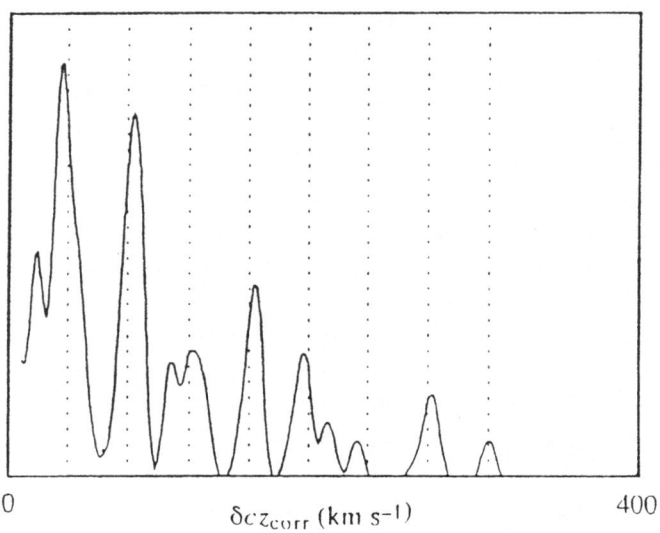

0   δcz_corr (km s⁻¹)   400

Figure 1. Distribution of weighted differential redshifts for 80 galaxies linked by group membership in the combined samples of 115 and 117 Local Supercluster galaxies (Paper II). The δcz have been corrected for the galactocentric solar vector $\mathbf{V}_\odot = (213$ km s^{-1}, $93°$, $2°$) obtained from Galactic modelling, but as the correction is differential the precise value of $\mathbf{V}_\odot$ is not critical. The vertical dotted lines represent, not the best fit, but the solution obtained from the earlier pilot study (period 37.5 km s^{-1}, phase 0 km s^{-1}). A Parzen smoothing has been applied.

of the observer, not the Galactic centre!

The periodicity can be seen in the data of several papers but, presumably because it is so unexpected, it is generally overlooked. As one recent example, Karachentsev & Makarov (1996) have determined a running apex for the Sun, Galaxy and Local Group with increasing volume out to 8 Mpc around the Sun; this study involves 103 galaxies with cz <500 km s^{-1}. They find the peculiar velocity dispersion of their sample to be remarkably constant, at ~72 km s^{-1}, independently of the volume sampled or the galaxy type (dwarf or giant). However such behaviour is clearly non–Gaussian, since in the Gaussian case the dispersion of the peculiar velocities should vary as $0.7\mu/\sqrt{n}$, yielding an expected 72±5 km s^{-1} at 8 Mpc (103 galaxies), 72±7 at 3.25 Mpc (50 galaxies) and 72±12 km s^{-1} at 0.7 Mpc (16 galaxies). An examination of the redshift residuals reveals that they are indeed non–Gaussian, clustering around values of ~ ±36, ~ ±75 and ~ ±114 km s^{-1}.

To sum up, over the range of redshifts tested, the redshift quantization has been consistently and reproducibly *observed* in all data sufficiently accurate to reveal it: our statistical analysis merely formalizes an empirical result.

WMN wishes to thank Bill Tifft and the Pima Community College for their invitation to attend the conference, and their hospitality and support during it.

References

Arp, H.: 1987, *J. Astroph. Astron.*, **8**, 241
Arp, H. & Sulentic, J. W.: 1985, *ApJ*, **291**, 88
Bottinelli, L., Gouguenheim, L., Fouqué, P. & Paturel, G.: 1990, *A&A Suppl.*, **82**, 391
Guthrie, B. N. G. & Napier, W. M.: 1990, *MNRAS*, **243**, 431
Guthrie, B. N. G. & Napier, W. M.: 1991, *MNRAS*, **253**, 533
Guthrie, B. N. G. & Napier, W. M.: 1996, *A&A*, **310**, 353
Karachentsev, I. D. & Makarov, D. A.: 1996, *AJ*, **111**, 794
Merrifield, M. R.: 1992, *AJ*, **103**, 1552
Tifft, W.G.: 1976, *ApJ*, **206**, 38
Tifft, W.G.: 1977, *ApJ*, **211**, 31
Tifft, W.G.: 1980, *ApJ*, **236**, 70
Tifft, W.G.: 1996, *ApJ*, **468**, 491
Tifft, W.G. & Cocke, W.J.: 1984, *ApJ*, **287**, 492 (TC)

THE DISTRIBUTION
OF GALAXY PAIR REDSHIFTS

T. E. NORDGREN, Y. TERZIAN AND E. E. SALPETER
Cornell University
Ithaca, New York 14853

Abstract. High signal to noise neutral hydrogen observations of a complete sample of 132 galaxy pairs show a velocity difference distribution which decreases monotonically from zero. There is no strong indication of redshift periodicity for the entire sample and no indication at all for a subset of 79 *isolated* galaxy pairs. In addition, when redshifts are corrected for the solar motion around the Galactic Center there is no indication of a redshift periodicity of 37.6 km s^{-1} in ΔV for galaxy pairs (as suggested by Guthrie and Napier 1996).

1. Introduction

Work on galaxy pairs at 21 cm first began at Cornell with Peterson (Peterson 1979) who used the 300 ft Green Bank radio telescope to study a sample of 279 galaxy pairs. The aim of this work was primarily to determine mass to luminosity ratios and to search for extended halos. This work was continued by Schneider (Schneider *et al.*1986) who extended the observations to small groups. Since much of the information derived from the study of galaxies in groups and pairs is statistical in nature, the validity of the results is only as good as the sample from which they are drawn.

Starting in 1990 a method was developed by Chengalur (Chengalur *et al.*1993) whereby galaxy pairs are chosen in a predetermined and systematic way from published optical redshift catalogs. Through the use of redshift information in the CfA catalog (Huchra *et al.*1983) and the Southern Sky Redshift Survey (da Costa 1988) galaxy pair candidates can be found solely on the basis of proximity in projected separation and velocity. Unlike in previous studies, no consideration is needed, or given, to choosing pairs on the basis of similar magnitude and angular separation being less than an

Astrophysics and Space Science **244:**65-71,1996.

arbitrary multiple of the diameter (Peterson 1979, Schneider *et al.*1986). In addition, by counting the number of galaxies in a truncated cone (with a base radius of 4.5 Mpc and depth of 335 km s^{-1}) about each galaxy in a catalog, galaxies can be separated or classified by the number density of galaxies around them (Chengalur *et al.*1993). Chengalur divided the galaxies in the catalogs into quartiles (i.e the number density below which one quarter of all the galaxies in the catalog can be found corresponds to the first quartile, the median number density corresponds to the upper limit of the second quartile, etc.). The information in redshift catalogs makes possible the evaluation of environmental effects on any derived galaxy pair sample.

Previous HI galaxy pair surveys using single dish telescopes have avoided pairs with separations small enough to cause confusion of the 21 cm lines (Chengalur 1994, Tifft & Cocke 1989). For the Green Bank 300 ft telescope (for example) all galaxy pairs with separations smaller than the beam FWHM of $\sim$10′will be excluded from the sample. In order to remove this selection bias from the current study, use is made of the high spatial resolution of aperture synthesis telescopes such as the Very Large Array. The pairs in this study are therefore made up of a close pair sample which has been imaged in HI using synthesis arrays, and a wide pair sample which has been observed using traditional single dish 21 cm techniques.

Nordgren *et al.*(1996) have completed the observation of all the pairs in the Chengalur wide and close pairs sample. In addition, observations were conducted at the VLA, Australia Telescope Compact Array and Westerbork Synthesis Radio Telescope which allow for comparison between close galaxy pairs in the lower two quartiles with pairs in higher quartiles.

2. The Close Pair Survey

The close galaxy pairs of the combined Chengalur-Nordgren database were determined by the following selection criteria:

1) To be a pair, each galaxy must have an independent listing in the redshift catalog. A galaxy pair in the very latest stage of merger which may only have one redshift listing will therefore not be included in this sample.

2) At least one member of the galaxy pair must have a number density of surrounding galaxies within the lower two quartiles.

3) Both galaxies in the pair must have redshifts between 1100 km s^{-1} and 4500 km s^{-1} for the CfA (or 5300 km s^{-1}for the SSRS). The lower velocity bound insures that no galaxies from the Virgo Cluster are included within our sample. For a magnitude or diameter limited survey, the incompleteness of the catalog is a function of redshift (Charlton & Salpeter 1991,

Chengalur *et al.*1993 and Chengalur 1994). The upper velocity limit insures that the catalog is complete to ten percent (Chengalur 1994).

4) In order to maximize the likelihood of detecting sufficient HI, both galaxies must be of type Sa or later.

5) The projected separation (r_p) must be less than 75 kpc (100 kpc for the SSRS).

6) The velocity difference (ΔV) must be less than 200 km s^{-1}.

7) No third galaxy within 1.0 Mpc or 400 km s^{-1}.

8) In order to insure detectable amounts of HI, the published integrated HI sum for the pair must be greater than 10 Jy km s^{-1}. (This requirement was relaxed for the SSRS due to the fewer number of galaxies for which there was published HI information).

In order to test the effect of environment on close galaxy pairs an additional sample has been observed where:

1) The restriction to pair members having densities in the lower two quartiles has been relaxed to allow for galaxies with densities in the third quartile.

2) $r_p$can be as great as 100 kpc.

3) ΔVcan be as great as 300 km s^{-1}.

4) The restriction to no neighbors within 1.0 Mpc and 400 km s^{-1}has been removed.

3. The Wide Pair Survey

The wide galaxy pairs of the combined Chengalur-Nordgren database were determined by the following selection criteria:

1) Each galaxy must have an independent listing in the redshift catalog.

2) At least one member of the galaxy pair must have a number density of surrounding galaxies in the lower two quartiles.

3) Both galaxies in the pair must have redshifts between 1100 km s^{-1} and 4500 km s^{-1} for the CfA (or 5300 km s^{-1} for the SSRS).

4) Both galaxies must be of type Sa or later.

5) The projected separation (r_p) must be less than 1.5 Mpc.

6) The velocity difference (ΔV) must be less than 250 km s^{-1}.

In order to account for the limited declination range of the Arecibo telescope, the declination of both galaxies in the CfA must be between 0°and 40°. Only limited observing time was available for observations at Parkes. Two further constraints were therefore placed on the SSRS wide pairs:

7) $r_p$must be less than 1.0 Mpc unless either galaxy member has a surrounding number density in the lowest density quartile, in which case $r_p$can be as large as 1.5 Mpc.

8) A declination range of -20°to -64.5°is imposed for those galaxies with surrounding number densities in the second quartile.

4. Observations

Observations were conducted using six telescopes from around the world (Table 1). 115 galaxies were observed using the Arecibo radio telescope while 81 galaxies were observed using the Parkes radio telescope. Due to the limited observing time available at Parkes, HI velocities from the literature were used whenever available. HI velocities for an additional 51 galaxies were obtained in this way. In total, neutral hydrogen was detected in 132 pairs (25 close pairs and 107 wide pairs). Table 1 lists the telescope, the minimum angular beam size (FWHM), the mean uncertainty in the derived 21 cm velocity and the number of pairs detected.

TABLE 1.   Telescope observations

Tel.	θ	$\sigma(\text{km s}^{-1})$	Num. Pairs
Arecibo	$3'$	4	48
ATCA	$30''$	5	9
Parkes	$14'$	7	58
VLA C	$30''$	5	6
VLA D	$45''$	5	8
WSRT	$30''$	5	2

Spectra for many of the wide pair galaxies have already been published (Chengalur et al.1993). Analyses of the close pair HI "moment-maps" can be found in Chengalur et al.1994 and Nordgren et al.1996.

5. Statistical Analysis

If galaxy pairs truly have some arbitrary range of real three-dimensional velocity differences, then randomly orienting them in the sky relative to the observer will produce a histogram of their redshift differences which is maximum at zero and decreases monotonically with increasing velocity difference. This can be easily seen by noting that a galaxy pair with a particular true velocity difference will, depending on its orientation to the observer, always be seen to have a velocity difference between zero and its true difference. Whatever the true velocity of a pair, oriented properly, it will always be able to be viewed as having a velocity difference of zero. In the past, histograms of ΔVfor samples galaxy pairs have not demonstrated this trend (Tifft & Cocke 1989, Schneider & Salpeter 1992, Figure 4.2 Chengalur 1994). Although the distribution is seen to be maximum at zero there is a second peak near 70 km s^{-1}. The motivation for including potential pairs with separations as large as 1.5 Mpc in the Chengalur-Nordgren sample is to address the hypothesis that the secondary peak in the ΔV distribution (and all non-zero velocity peaks) is the result of a selection bias in previous samples. If wide galaxy pairs tend to be on radial orbits (Schneider & Salpeter 1992) then by excluding the very widest pairs from a sample one will exclude a set of galaxies with potentially very small velocity differences. The suppression of this population of galaxy pairs could result in a non-zero peak in the distribution of ΔV.

Figure 1a shows the histogram of heliocentric ΔV for the entire sample of 132 galaxy pairs. Bin widths are chosen to be 20 km s^{-1} in width which is larger than the uncertainty in any individual HI measurement. Although the distribution of ΔV decreases rapidly from zero as velocity difference increases their are still non-zero peaks present. If the fluctuation in the number of pairs per bin is indicative of Poisson statistics then the uncertainty in the bins surrounding and including the non-zero peaks in Figure 1a is on the order of 4 pairs each. This uncertainty is comparable to the difference in counts between bins surrounding the non-zero peaks, and thus the observed distribution of ΔV is consistent with a monotonic function.

To investigate the effect of neighboring galaxies on the distribution of ΔV, we define a galaxy pair as being isolated if there is no other galaxy within 750 kpc and 250 km s^{-1} of at least one member of the pair. Figure 1b shows the observed velocity difference for this new set of 81 pairs. The shaded histogram is the sample of 81 isolated galaxy pairs. For comparison, the total sample of 132 galaxy pairs is shown in white. There is no longer any indication of a periodicity in the distribution of ΔV. With the exception of the second bin being higher than the first by only one pair the distribution of observed velocity differences monotonically decreases from zero velocity.

6. Galactocentric Velocities

Guthrie and Napier (1996) observe a 37.6 km s^{-1} periodicity in HI redshifts of spiral galaxies once heliocentric velocities have been corrected using a vector which is close to the Sun's galactocentric motion. In order to test the claim of periodicity we follow the prescription of Guthrie and Napier and use their solar vector of $l_\odot = 96°$, $b_\odot = -3°$ and $V_\odot = 210$ km s^{-1} to apply this galactocentric correction to the galaxies in the Chengalur-Nordgren sample. We will refer to velocities which have been corrected in this manner has being galactocentric. Due to the small size of the reported velocity period it is important to have velocity measurements of high precision (Guthrie &

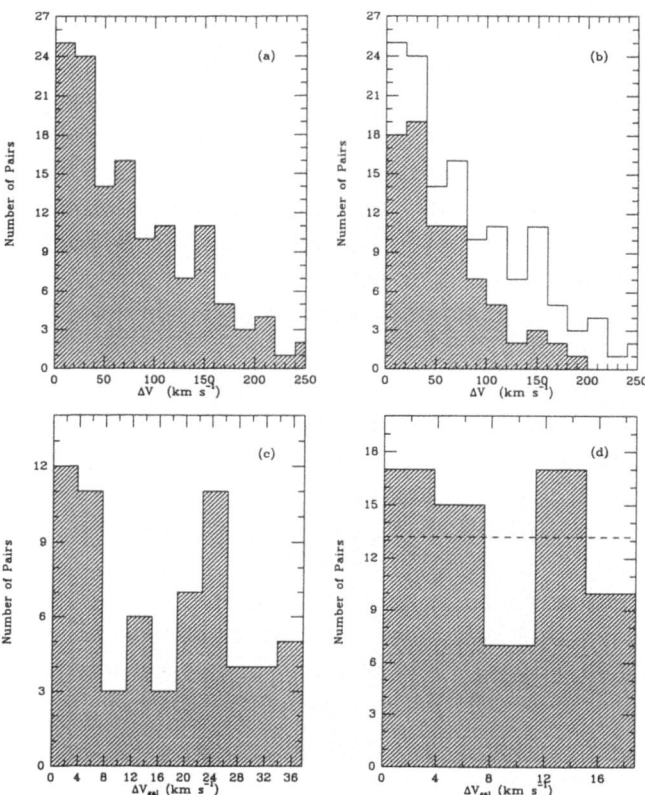

Figure 1. The distribution of velocity differences for the Chengalur-Nordgren galaxy pair sample. Figure (a) is for the full sample of 132 pairs. Figure (b) shows the full sample in white with the subset of isolated pairs shaded. Figure (c) is the wrapped distribution of 66 pairs where the velocities are now galactocentric. Figure (d) is the same sample wrapped over the range 0 to 18.8km s^{-1}. The dashed line is the mean of the five bins.

Napier 1996). To be certain of the quality of the HI velocities, we have used only those pairs where both galaxies were observed as part of the current study (i.e. no pairs were used where one galaxy measurement was quoted from the literature). This requirement resulted in a sample of 66 galaxy pairs being transformed into a galactocentric velocity reference system. Figure 1c shows the ΔVdistribution for the galactocentric pairs where the velocities have been wrapped over the range 0 to 37.6 km s^{-1}(e.g a galaxy pair with ΔV= 38.6 km s^{-1}will be binned as if it were 1.0 km s^{-1}, while a pair with ΔV= 74.2 km s^{-1}will be binned as if it were 36.6 km s^{-1}). Each bin is 3.76 km s^{-1}wide. A periodicity of 37.6 km s^{-1}will manifest itself as a peak at zero and a peak at 37.6 km s^{-1}. The peak at zero velocity observed in Figure 1c can be explained as the natural peak of the overall distribution at zero velocity. There is no corresponding increase at 37.6 km s^{-1}. Figure 1d is the same sample wrapped over the range zero to 1/2 the periodicity. Each bin is 3.76 km s^{-1}wide. A periodicity of 37.6 km s^{-1}will manifest itself as a peak at zero with a depression at 18.8 km s^{-1}. The dashed line is the mean number of pairs in the five bins: 13.2 pairs. There is no indication of a 37.6 km s^{-1}periodicity.

7. Summary

In order to minimize the effect of selection biases on the analysis of galaxy pair statistics we are engaged in a program of systematically compiling a complete sample of galaxy pairs. We have therefore created a database which includes the very closest galaxy pairs observed with aperture synthesis telescopes as well as the potentially very widest pairs out to separations of 1.5 Mpc. The observed velocity difference distribution for this sample decreases rapidly with increasing velocity and is monotonic well within the uncertainties in each bin. The distribution of $\Delta V$is therefore consistent with that produced by a randomly oriented sample of galaxy pairs. There is no indication in the full sample of pairs of statistically significant peaks at a periodicity around 70 km s^{-1}. Nor is there any indication at all for the isolated subset of any peaks in the distribution other than at low ΔV. Finally, no indication is seen of any 37.6 km s^{-1}periodicity for galaxy pairs transformed to a galactocentric velocity reference.

8. Acknowledgements

This work was supported in part by the National Astronomy and Ionosphere Center, which is operated by Cornell University under a cooperative agreement with the National Science Foundation.

References

Charlton, J. and Salpeter, E. E.: 1991, *ApJ*, **375**, 517
Chengalur, J. N.: 1994, *PhD thesis*, Cornell University
Chengalur, J. N., Salpeter, E. E., Terzian, Y.: 1993, *ApJ*, **419**, 30
Chengalur, J. N., Salpeter, E. E., Terzian, Y.: 1994, *AJ*, **107**, 1984
da Costa, L. N., Pellegrini, P. S., Sargent, W. L. W., Tonry, J., Davis, M., Meiksin, A., Latham, D. W., Menzies, J. W., Couson, I. A.: 1988, *ApJ*, **327**, 544
Guthrie, B. N. G. and Napier, W. M.:1996, *A&A*, in press
Huchra, J. P., Davis, M., Latham, D. W., Tonry, J.: 1983, *ApJ Supp*, **52**, 89
Nordgren, T. E., Chengalur J. N., Salpeter, E. E., Terzian, Y.: 1996, in preparation
Peterson, S.: 1979, *ApJ*, **232**, 20
Schneider, S. E., Helou, G., Salpeter, E. E., Terzian, Y.: 1986, *AJ*, **92**, 742
Schneider, S. E. and Salpeter, E. E.: 1992, *ApJ*, **385**, 32.
Tifft, W. G. and Cocke, W. J.: 1989, *ApJ*, **336**, 128

DENSITY FLUCTUATIONS ON SUPER-HUBBLE SCALES

LI-ZHI FANG
Department of Physics
University of Arizona, Tucson, AZ 85721, USA

AND

YI-PENG JING
Max-Planck-Institut für Astrophysik
Garching, Germany

Abstract.
   According to causality, the existence of density perturbations on scales larger than the present Hubble radius $y = 2c/H_0$ is crucial for discriminating between inflation and non-inflation models of the origin of inhomogeneity of the universe. Observations of the cosmic background radiation anisotropies favor a super-Hubble suppression on scales λ_{max} in the range $0.5 - 3.0y$. Many of non-inflation models are consistent with such a suppression. Inflation models are certainly not in conflict with this suppression; however one important parameter, the duration of the epoch of inflation, may need to be fine-tuned.

1. Introduction

The most popular theory of structure formation in the Universe is based on the assumption that the structures started from small amplitude density perturbations which grew by gravitational instability. The primeval density perturbations are assumed to arise as vacuum fluctuations of scalar fields during the inflation era [1]. It predicts that the fluctuation spectrum is scale invariant with index ~ 1. The subsequent evolution of these seeds of the density perturbations was brought about by gravitational interaction among baryonic and dark matter. The temperature fluctuations of the cosmic background radiation (CBR) and the formation of galaxies, clusters of galaxies, and even high-redshift objects were found to be consistent with

Astrophysics and Space Science **244**:73-80,1996.

this standard model if appropriate compositions of dark matter are taken into account. Nevertheless, there are still models for the origin of density fluctuations, such as cosmic strings and other defects of phase transitions, which cannot be discriminated by these observations[2].

A possible way to distinguish between the mechanisms of planting the perturbation seeds is to detect the density perturbations on scales larger than the Hubble radius H^{-1}, where H is the Hubble constant at the time being considered.

In the inflation era, the causal horizon grew exponentially, while the Hubble radius was almost unchanged. Since the onset of inflation, the causal horizon became much larger than the Hubble radius. The inflation scenario is then characterized by the existence of a special region of length scales larger than the Hubble radius, but less than the causal horizon. During inflation, sub-Hubble perturbations were stretched into super-Hubble scales by the expansion. Cosmologically interesting perturbations underwent an evolution from sub- to super-Hubble scales. Therefore, inflation can produce physically super-Hubble perturbations.

Perturbations on scales larger than the present Hubble radius have not reentered the Hubble radius yet. This kinematical feature is common for various versions of inflations regardless of their dynamical details. One can then typically describe the spectrum of inflation- generated perturbations by a power law

$$P(k) \propto k^n \tag{1}$$

without a cutoff related to the Hubble radius.

On the other hand, non-inflation evolution requires that the causal horizon is always about the same as the Hubble radius. No causally physical mechanism can seed perturbations on super-Hubble scales. All microphysics is impotent on scales larger than the Hubble radius. Even in the case that the seeds have scales larger than the horizon, causality will guarantee that there is no net density perturbation on superhorizon scales [3]. The spectrum of perturbations falls at least as fast as k^4 on superhorizon scales [4]. Instead of eq.(1), the spectrum on large scales should be suppressed as $P(k) \propto k^n f(k)$, with $f(k)$ being

$$f(k) \propto \frac{1}{1 + (k_{min}/k)^m}, \tag{2}$$

where the suppression index $m \geq 4 - n$, $k_{min} \sim \pi H/c$. The suppression factor $f(k)$ in the power spectrum has also been directly found in models like cosmic string plus cold or hot dark matters[5].

Therefore, the super-Hubble behavior of the perturbations is crucial to studying the origin of fluctuations. It has been shown that the existence of perturbations on scales larger than the Hubble radius before the recombination epoch can be tested by small-scale fluctuations of the CBR [6]. In this paper, we examine the perturbations on scales larger than the present Hubble radius. The existence of such super-Hubble scale perturbations can be tested by the CBR fluctuations on very large scales.

2. Fluctuations With Super-Hubble Suppression

Since COBE detected the CBR anisotropy, it was shown that the spectrum (1) can match the observations of the CBR anisotropy [7]. However, this success doesn't mean that the spectrum suppressed on scales larger than $2c/H_0$ is ruled out by the observations. In fact, right now, no statistical test has been done yet for these distinctive power spectra. It is worth studying whether the spectrum with long wavelength suppression is consistent with current observations.

Let's consider the large-scale fluctuations in the CBR temperature in a flat universe with a cosmological constant $\Lambda = 0$. The fluctuation in direction Ω is[8]:

$$\frac{\Delta T}{T}(\Omega) = -\frac{H_0^2}{2c^2} \sum_k \frac{\delta(k)}{k^2} e^{-ik \cdot y},$$ (3)

where $y = (2cH_0^{-1}, \Omega)$ is a vector of length $2cH_0^{-1}$ pointing to Ω on the sky. $\delta(k)$ is the Fourier amplitude of the density contrast $\delta(r)$. The power spectrum $P(k) = \langle |\delta(k)|^2 \rangle$ with super-Hubble suppression can be written as

$$P(k) = \frac{(2\pi)^3}{V_\mu} \Delta_0^2 \frac{1}{4\pi} \frac{k^n}{1 + (k_{min}/k)^m}$$ (4)

where V_μ is a large rectangular volume, and Δ_0^2 is a constant determined by the variance of the perturbed potential.

The observed temperature fluctuations of the CBR on the celestial sphere are usually expressed with spherical harmonics as $\Delta T/T(\Omega) = \sum_{lm} a_l^m Y_l^m(\Omega)$, where $Y_l^m(\Omega)$ are the spherical harmonic functions. Defining a rotationally invariant coefficient $C_l \equiv (1/4\pi) \sum_m \langle |a_l^m|^2 \rangle$, one finds from eqs.(3) and (4) that

$$C_l = \frac{H_0^4(2l + 1)}{4c^4} \sum_k \frac{P(k)}{k^4} j_l^2(ky),$$ (5)

where $j_l(x)$ is the spherical Bessel function. For spectrum (4), eq.(5) becomes

$$C_l = \frac{H_0^4(2l + 1)}{4c^4} \Delta_0^2 \int_0^\infty \frac{k^{n-2} j_l^2(ky) dk}{1 + (k_{min}/k)^m}.$$ (6)

Instead of using the amplitude Δ_0^2, one can also normalize the spectrum using the quadrupole amplitude defined by $Q = C_2^{1/2}T$, where $T = 2.726$ K is the mean temperature of CBR.

Strictly speaking, eqs.(3) and (6) were derived under the assumption that all the density perturbations are primordial, i.e. produced before recombination. They cannot be directly used to calculate the CBR fluctuations from density perturbations induced later. Many non-inflation models like those of phase transition defects do assume that partial, or even entire, perturbations were generated after recombination. The CBR fluctuations should be a mixture of the "primordial" anisotropies [eq.(6)] and those induced by perturbations generated later. Obviously, in this case the total CBR anisotropy will depend on the mechanism of the generation and evolution of perturbations after the last scattering.

However, for a very wide class of non-inflationary models, eq.(6) has been found to be still valid for describing the CBR anisotropy from late-time perturbations if the amplitude Δ_0^2 in eq.(6) is replaced by a factor describing the temporal dependence of the perturbations [9]. Therefore, if we treat Δ_0^2 and n in eq.(6) as phnomenological parameters, the results given by fitting the observational data to eq.(6) are valid for models of " primordial" and/or late-time induced anisotropies. Especially, the suppression factor $f(k)$ is model-independent, therefore, the fitting result of k_{min} should also be available to judge between models. Moreover, we will not consider the temperature fluctuations caused by tensor (gravitational wave) perturbations, because in most models CBR anisotropies are dominated by scalar perturbations, and the tensor perturbations also obey the suppression eq.(2). The power spectrum of tensor perturbations generally has a different index n from that of scalar perturbations eq.(1). Considering this point, we will investigate two extreme cases of the suppression on super-Hubble scales: 1) sharp cutoff, $m = \infty$; 2) softer cutoff, $m = 4 - n$.

3. Statistical Analysis

To test the models of eq.(6) with finite cutoff $\lambda_{max} = 2\pi/k_{min}$ we first use the COBE observations of the two-point angular correlation function $C(\theta)$ of the CBR temperature [7]. The two-point angular correlation function $C(\theta)$ is determined by C_l as

$$C(\theta) = \sum_l C_l W^2(l) P_l(\cos\theta), \qquad (7)$$

where $P_l(x)$ is the Legendre function, and $W(l) = \exp\{-1/2[l(l+l)/17.8^2]\}$ is a window function. The influence of cosmic variance [10] can be treated by the standard χ^2 technique [11]. For a given n, we estimate the goodness-of-fit of models with parameters of k_{min} and Q by minimizing χ^2 over the data:

$$\chi^2 = \sum_i \frac{[C_i - C(\theta_i)]^2}{\sigma_i^2 + \sigma_{cv}^2(\theta_i)}, \tag{8}$$

where C_i and σ_i are, respectively, the observed values and errors of the angular correlation at θ_i, and $\sigma_{cv}(\theta)$ is the 1σ cosmic variance of the $C(\theta)$.

In eq.(8) we assumed that the variances in different bins are mutually independent. Strictly speaking, this assumption does not consider the bin-bin correlation of cosmic variances, and we should use statistics applicable for a covariance matrix of C_i, such as a likelihood analyses. However, the approximation of eq.(8) is already suitable for our purpose: to estimate a statistical confidence of rejecting the supper-Hubble suppression model of eq.(6) by observations. Generally, considering more variances will lead to a lower probability of the rejection for a given model, and a higher confidence of the acceptableness of the model. Therefore, eq.(8) should safely provide an underestimated confidence limit of the consistence of the supper-Hubble suppression model with current observation. In fact, the likelihood analyses of a cubic toroidal (T^3) universe gave almost the same best-fit values for Q and k_{min} as eq.(8) does[11, 12]. This indicates that under the current precision of the data, the statistics of Q and k_{min} do not significantly depend on the nondiagonal part of the covariance matrix of C_i. Mathematically, testing the suppression scale λ_{max} of model (6) is completely the same as testing k_{min} of a T^3 universe. Therefore, it would be reasonable to apply the statistic eq.(8) to estimate the acceptableness of model with supper-Hubble suppression.

Since σ_{cv} is also proportional to Q, we adopt an iteration procedure to conduct the χ^2 minimization. First we assume a zero σ_{cv} and find out the best-fitting value of Q. Using this value we calculate the σ_{cv} based on 100 Monte Carlo realizations of C_l, and do the minimization again and find a new fitting value of Q. Using this new Q we repeat the minimization and find another more accurate value of $C_{2\ rms-PS}$. The iteration procedure is stopped until the differences of χ^2 and of Q between the two consecutive minimizations are less than 0.1%. The final Q and χ^2 are our desired values.

The goodness of the χ^2-fit of the sharp ($m = \infty$) and softer ($m = 4 - n$) cutoffs are shown in Figure 1, in which $P(> \chi_{min}^2)$ is the probability that the experimental data are drawn from a realization of the model. It can clearly be seen from the figure that there are remarkable peaks with respect to y/λ_{max}. It shows that the suppressed spectrum eq.(4) is acceptable, even improves the fitting to the observed CBR temperature fluctuations.

In Table 1, we list our fitting results of the λ_{max}/y ranges acceptable by the COBE observations at 95% confidence level. The results of λ_{max}/y do not sensitively depend on the suppression index m. For the cases of the

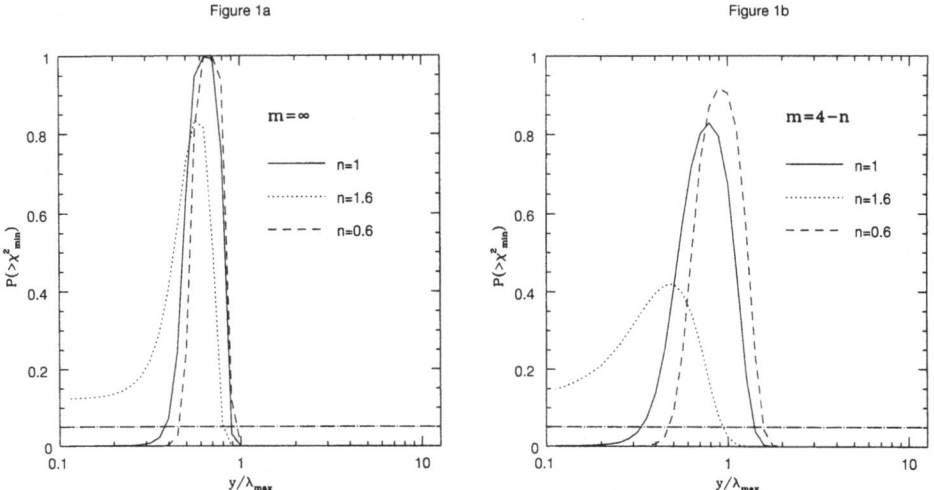

Figure 1. Probability $P(> \chi^2_{min})$, i.e. χ^2-fit goodness, as a function of y/λ_{max}. Here χ^2_{min} is calculated by fitting the spectrum with suppression (2) to the *COBE*-DMR two-point angular correlation function of the cosmic temperature fluctuations. The index n of the power spectrum is taken to be 0.6, 1.0 and 1.6. The dot-dash line denotes $P(> \chi^2_{min}) = 5\%$.

index $n \leq 1$ which are favored by current observations of galaxy distributions, the upper limits on λ_{max}/y are determined to be about 3. For $n=1.6$, although the most likely values of λ_{max}/y are about 2, no significant upper limit on λ_{max}/y can be deduced from the COBE observations. The upper or lower limits given above will change slightly if one includes the non-diagonal part of the error matrix in the fitting procedure. As mentioned above, the nondiagonal part will generally increase the confidence of the acceptableness.

TABLE 1. Super-Hubble suppression wavelength λ_{max}/y given by $C(\theta)$

n	m=0	m=4
0.60	1.05-2.20	0.64-2.13
1.00	1.11-2.70	0.77-3.03
1.60	> 1.25	> 1.11

Figure 2 plots the best-fitting quadrupole, i.e. $Q = C^{1/2}_{2\ rms-PS}$, as a function of y/λ_{max}. The thick line denotes the RMS quadrupole measurement and the dotted area is its 1σ region, i.e. $Q_{rms} = 6 \pm 3 \mu$ K [7]. For

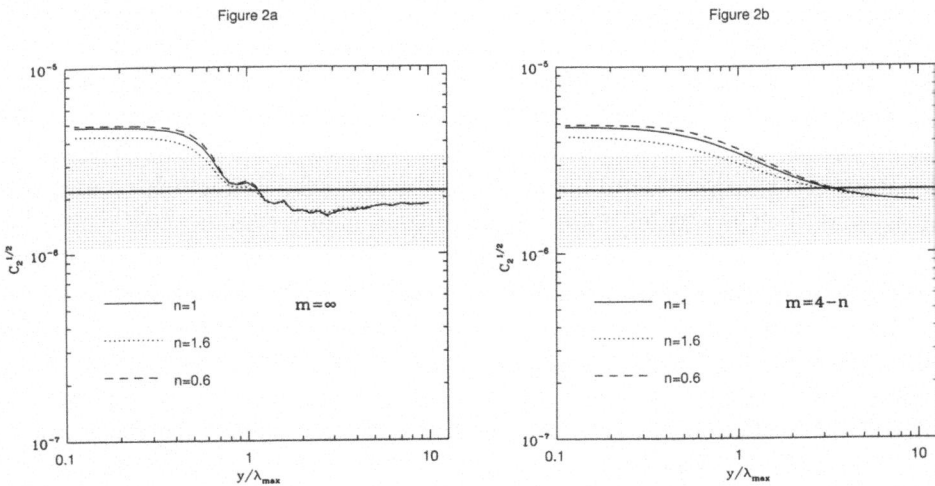

Figure 2. The best fitted quadrupole, i.e. $C_{2\ rms-PS}^{1/2}$, as a function of the wavelength of the suppression on super-Hubble radius, y/λ_{max}. The thick line denotes the RMS quadrupole measurement $C_{2\ rms}^{1/2}$, and the dotted area is its 1σ region.

the suppression scales allowed by the χ^2 tests, the best-fitting quadrupole agrees with the observed one. The result is weakly dependent on the power law index n and the suppression index m, though the $n = 1.6$ case without suppression gives a worse match of the best-fitting quadrupole with the observed one. Considering that the measures of $C(\theta_i)$ and Q_{rms} are independent, these results seem consistently to warn us of the existence of suppression on the super-Hubble scales.

4. Discussions and Conclusions

For a large class of non-inflation models like "late-time" cosmological phase transition [13], there is no mechanism for the growth of super-Hubble scale perturbations. It requires that the suppression scales λ_{max} should not be larger than about $1.5y$ [4]. For the models like topological defects, in which the perturbations are produced both before and after recombination, the super-Hubble suppression scale λ_{max} depends on the dynamics of the defects. For instance, λ_{max} is found to be in the range of $\sim 0.8 - 3.0y$ for models of cosmic string plus hot or cold dark matter [5]. From Table 1, the suppression scales for the $n \leq 1$ power spectra are consistent with these models. In the case of $n = 1.6$, the confidence level for a suppression scale to be larger than $1.5y$ is 80%, which is probably difficult to survive in some models of late-time phase transition. Overall, many of the non-inflation models are consistent with the two years of COBE data.

Our formal error analysis shows that the COBE data favor models with finite suppression λ_{max} on scales larger than but close to the Hubble radius. In the "standard" inflationary model, a super-Hubble suppression in the spectrum of primordial density perturbations can be given by the duration of the epoch of inflation. Because the longest wavelengths of the density perturbations should come from the perturbation which crosses the horizon at the time when inflation just began, a suppression scale would be determined by the number of e-folds of cosmic scale growth in the whole epoch of inflation. Therefore, in order to explain why the suppression scales in the spectrum of the primeval density perturbations are so close to the Hubble scale, the duration of the epoch of inflation may need to be fine-tuned. In the case of $n = 1.6$, the goodness of the fit of the $\lambda_{max} = \infty$ model with the COBE observations is comparable with models with finite λ_{max}, although a finite value of $\lambda_{max} \sim 2y$ is still favored. However, the current observations of galaxy distributions can hardly accommodate $n = 1.6$.

YPJ is supported by an Alexander-von-Humboldt research fellowship.

References

1. Kolb, E. W., and Turner, M. S.: 1990, *The Early Universe*, (Addison-Wesley, NY)
2. Bennett D. P., Stebbins, A. and Bouchet F. R.: 1992, *ApJ*, **399**, L5; Bennett, D. P. and Rhie, S. H.: 1993, *ApJ*, **406**, L7; Pen, U. L., Spergel, D. N. and Turok, N.: 1994, *Phys. Rev. D*, **49**, 692
3. Traschen, J.: 1985, *Phys. Rev. D*, **31**, 283; Veeraraghavan, S. and Stebbins, A.: 1990 *ApJ*, **365**, 37
4. Abbott L. F. and Traschen, J.: 1986 *ApJ*, **302**, 39; Robinson, J. and Wandelt, B. D.: astro-ph/9507043
5. Albrecht, A. and Stebbins, A.: 1992, *Phys. Rev. Lett.*, **68**, 2121 and **69**, 2615
6. Critenden, R. G. and Turok, N. G.: 1995, astro-ph/9505120; Albrecht, A., Coulson, D., Ferreira, P. and Magueijo, J.: 1995, astro-ph/9505030
7. Bennett, C. L. *et al.*: 1994 *ApJ*, **436**, 423; Bennett, C. L. *et al.*: 1996, astro-ph/9601067
8. Peebles, P. J. E.: 1982, *ApJ*, **263**, L1
9. Jaffe, A. J., Stebbins, A. and Frieman, J. A.: 1994, *ApJ*, **420**, 9
10. Abbott, L. F. and Wise, M. B.: 1984, *ApJ*, **282**, L47
11. Jing Y. P. and Fang, L. Z.: 1994, *Phys. Rev. Lett.*, **73**, 1882; Scaramella R. and Vittorio, N.: 1993, *MNRAS*, **263**, L17
12. Seljak, U. and Bertschinger, E.: 1993 *ApJ*, **417**, L9
13. Wasserman, I.: 1986 *Phys. Rev. Lett.*, **57**, 2234; Press, W. H., Ryden, B. and Spergel, D.: 1990 *Phys. Rev Lett.*, **64**, 1084; Fuller G. and Schramm, D. N.: 1992 *Phys. Rev. D*, **45**, 2595; Frieman, J. A., Hill C. T. and Watkins, R.: 1992 *Phys. Rev. D*, **46**, 1226

THE CHALLENGE OF LARGE-SCALE STRUCTURE

S. A. GREGORY
Dept. of Physics and Astronomy & Institute for Astrophysics
University of New Mexico
Albuquerque, New Mexico 87131

Abstract. The tasks that I have assumed for myself in this presentation include three separate parts. The first, appropriate to the particular setting of this meeting, is to review the basic work of the founding of this field; the appropriateness comes from the fact that W. G. Tifft made immense contributions that are not often realized by the astronomical community. The second task is to outline the general tone of the observational evidence for large scale structures. (Here, in particular, I cannot claim to be complete. I beg forgiveness from any workers who are left out by my oversight for lack of space and time.) The third task is to point out some of the major aspects of the field that may represent the clues by which some brilliant sleuth will ultimately figure out how galaxies formed.

1. Discovery

G. deVaucouleurs 1975 (and references therein) followed up suggestions made in 1937 by E. Holmberg and identified the Local Supercluster in the 1950s. The general significance of this work was quite controversial; many thought that this structure might just be a statistical local overdensity. Real progress could not be made until the 1970s when large, statistically complete redshift surveys could be conducted as a result of the introduction of image intensifying tubes. Among the first of these surveys was my dissertation (results published in Gregory 1975) in which all galaxies with $m_p < 15.7$ within a radial distance of $r < 3$ degrees from the Coma cluster center were surveyed spectroscopically. It was in an early discussion of this data in 1973 that Tifft pointed out the lack of a foreground. I suggest that this is the birth of the concept of cosmic voids. Tifft and I, along with Laird Thompson, further developed the observational status of large structures

Astrophysics and Space Science **244**:81-88,1996.
© 1996 *Kluwer Academic Publishers.*

(Tifft and Gregory 1976 [first wedge diagram] & 1978) culminating with the paper covering the Coma/A1367 region that is often cited as the first full presentation of voids and superclusters as the fundamental features of large-scale structure (Gregory and Thompson 1978).

A great deal of discovery-phase work was also done by Chincarini and Rood 1976, Joeveer and Einasto 1978, Tarenghi et. al 1979, and by Tifft, Hilsman, and Corrado 1975 who discovered the supercluster in Perseus (more on this later – the Rosetta Stone of galaxy formation?). I note that not much mention was made of the fact, but those of each of these groups that I talked to at the time all noted that the then-known superclusters had the interesting properties of having mean redshifts of approximately 3,000 km s^{-1} (Hydra - Centaurus), 5,000 km s^{-1} (Perseus), 7,000 km s^{-1} (Coma) and 9,000 km s^{-1} (Hercules). Chincarini and Rood went so far as to describe these structures as having a "fabric" nature. The regularity and our seeming location as the center of perhaps concentric shells were disturbing to the standard view.

Additional observations of note include Tully and Fisher 1987 (and references therein) who greatly elaborated on the structure of the Local Supercluster showing it to have sheetlike and linear features similar to those found in external superclusters and Kirshner, Oemler, and Schechter 1979 who found the Bootes void which was much larger than any previously known. Potentially important alignments were reported by Gregory, Thompson, and Tifft 1981 for galaxies in the Perseus supercluster and by Binggeli 1982 for the alignments of clusters with nearby clusters.

2. Bulk Motions

Rubin et al. 1976 studied ScI-II galaxies and found systematic motions, but Chincarini and Rood 1979 argued that these observations were also well interpreted as a mapping of large-scale structures. Aaronson et al. 1986 developed the IR Tully Fisher relation for spirals and Dressler et al. 1987 developed the $D_n - e$ relation for ellipticals. This body of work gave us two means of estimating distances that were independent of redshift. Hence, the difference between predicted Hubble flow distances and those found by the new methods enabled these two groups to investigate bulk motions. Various levels of refinements to these methods have yielded the concept of a Great Attractor located at a redshift distance of about 4500 km s^{-1} near to but not coincident with the Hydra and Centaurus clusters.

3. Intermediate Results and Implications

Batuski & Burns 1985 found unprecedentedly large structures in the distribution of Abell clusters. These include a supercluster (including the Perseus supercluster at the near end) of length approximately 1 billion light years and a void covering much of the northern galactic hemisphere that includes the Bootes void on one side. The CfA group (de Lapparent et al. 1986) showed a strip survey around Coma; the well defined boundaries of the voids suggested a bubble-like topology. In the Perseus supercluster Giovanelli, Haynes, and Chincarini 1986 found morphological segregation. They found that the filamentary nature of the supercluster was well defined by early morphological types but that later types showed increasingly diffuse structures.

A series of papers investigated the degree of emptiness of the Bootes void (Tifft, et al. 1986, Moody, et al. 1987, and Wiestrop and Downes 1988). These papers noted that one could find galaxies in the void by means of emission line surveys. However, the voids did not "fill up" with these galaxies; voids were still extremely underdense with respect to the mean.

A large amount of theoretical work has been generated by the observations reported above. I do not have the time or space to give justice to these important studies. I suggest that the reader look at a reasonable overview such as Deckel 1988 and more recent work. The concepts involved examine whether or not the Universe is dominated by hot dark matter (neutrinos) which accounts for the largest structures but has trouble with small groups and cold dark matter which seems to work in the opposite manner. Hybrid models seem promising, but perhaps Gaussian origins for the perturbations will be superseded by cosmic strings. Out of the bubble concept came explosive models for the amplification of density perturbations. These can account for small voids but have trouble at the larger scales.

4. Some Recent Work

A. In the early 1990's the COBE satellite (Wright et al., 1994 and references therein) found 1) the dipole nature of the background radiation and 2) the (probable) evidence for the seeds of large-scale structures.

B. Szalay et al., 1993 combined the results from 4 deep redshift surveys near both the NGP and SGP. They believe that they have found a significant periodic structure with a spacing of 12,700 km s^{-1}.

C. Praton and Schneider 1994 showed that wedge diagrams can have important artifacts induced by infall and transverse motion. Features of their diagrams appear very similar to bubbles in the slice survey, and they derive bulk motions that are "not consistent" with the COBE results. This work possibly explains the concentric appearance of features in redshift space.

D. The group including R. Kraan Kortevig and P. Henning 1996 conducted several HI and optical surveys/searches in the zone of avoidance.

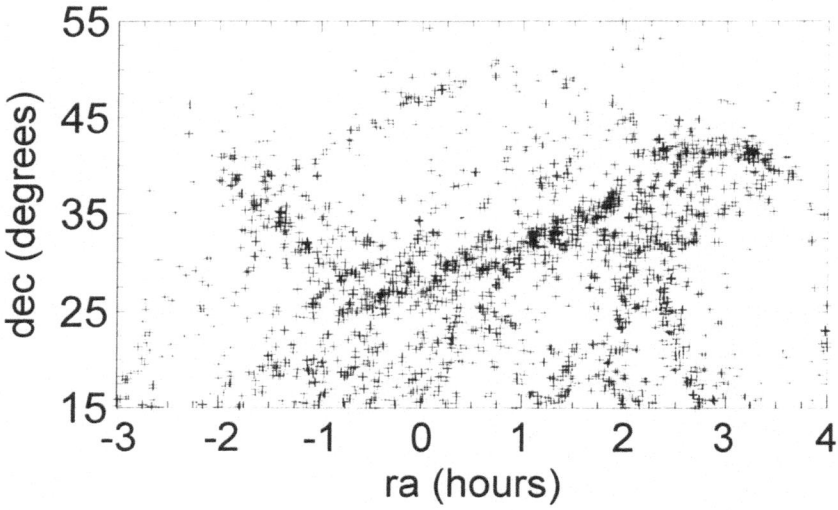

Figure 1. The 2-D distribution of Zwicky galaxies in the Perseus supercluster region. Note the obvious gently curving filamentary nature of the dominant structure.

There is very strong recent evidence for the discovery of a major cluster at the predicted position of the Great Attractor. This evidence comes from the discovery of a very large number of galaxian images on the sky survey that are strongly concentrated near the predicted position on the sky of the Great Attractor.

E. New results presented here: Results from the Arizona/New Mexico spectroscopic survey conducted by S. Gregory, W. Tifft, S. Hall, J. Moody, and M. Newberry

i. Details on filamentary structures in Perseus supercluster - We find that there are three intersecting filaments that are differentiated in 3-D. Figure 1 shows the general distribution of galaxies in the Perseus supercluster region. Figure 2 expands the western region that represents our new survey and shows that the two parallel filaments are separated in redshift space with the northern filament being on the near side and intersecting a third filament that extends to the northeast.

ii. Morphological Segregation - We confirm the results of Giovanelli, Haynes, and Chincarini with our new and differently defined data (we use the 4,000 A break to find morphologies). Figures 3, 4, and 5 show the same region as Figure 2 with different morphological mixes (Fig. 3 - all types, Fig. 4 - early types, and Fig. 5 intermediate and late types).

iii. Emission Line Incidence - If morphological segregation is thought of

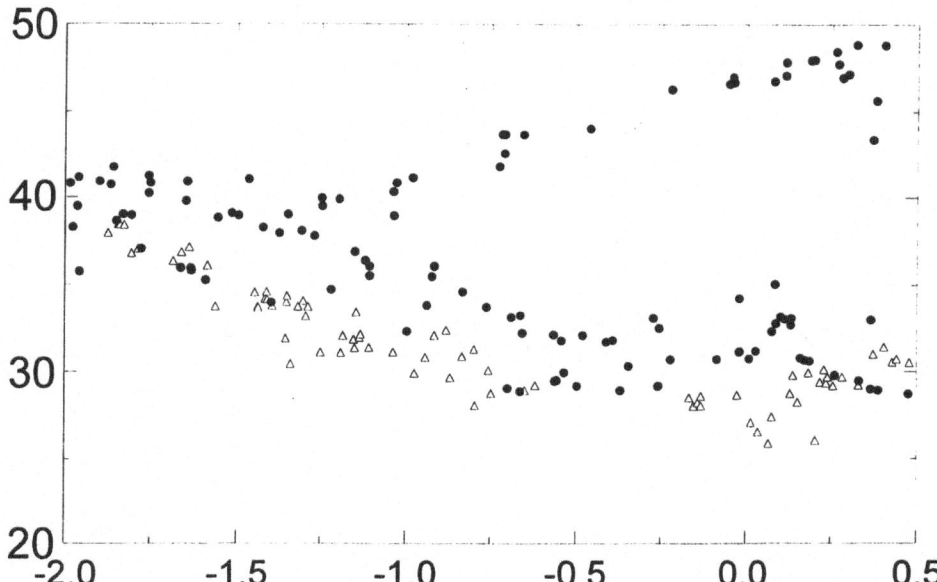

Figure 2. Here we have isolated the galaxies that lie in filaments that intersect the main supercluster. Circles indicate those galaxies with $cz < 6,000$ km s^{-1}, and triangle indicate those with $cz > 6,000$ km s^{-1}. The northern of the two parallel filaments intersects with the third filament that extends to the northeast. The two parallel filaments join with the main supercluster near the eastern border of this figure. The mean separation between the two parallel filaments is about $v = 1,600$ km s^{-1} in redshift space.

as a sequence of star formation epochs, then emission line activity can be thought of as an extension to late type absorption spectra (at least the HII type spectra). The distribution of emission line objects in the filaments is consistent with other indicators of morphological segregation. Also, one of the three filaments has no emission objects. Figure 6 shows the distribution of galaxies with emission line spectra.

5. Concluding Thoughts

Can we answer such questions as 1) what is a filament, 2) how empty are voids, 3) what are the topological properties of superclusters and voids – sheets, filaments, connectedness of these two, bubbles, sponges, and 4) what is the reason for morphological segregation (is it as simple as disk galaxies get turned into spheroidal systems by encounters?)

There appears to be an analogy: Spiral arm tracers in disk galaxies and filament tracers in superclusters. Can the analogy be extended? Is there an evolutionary sequence?

What is the nature of the observations of the regularity of the large-

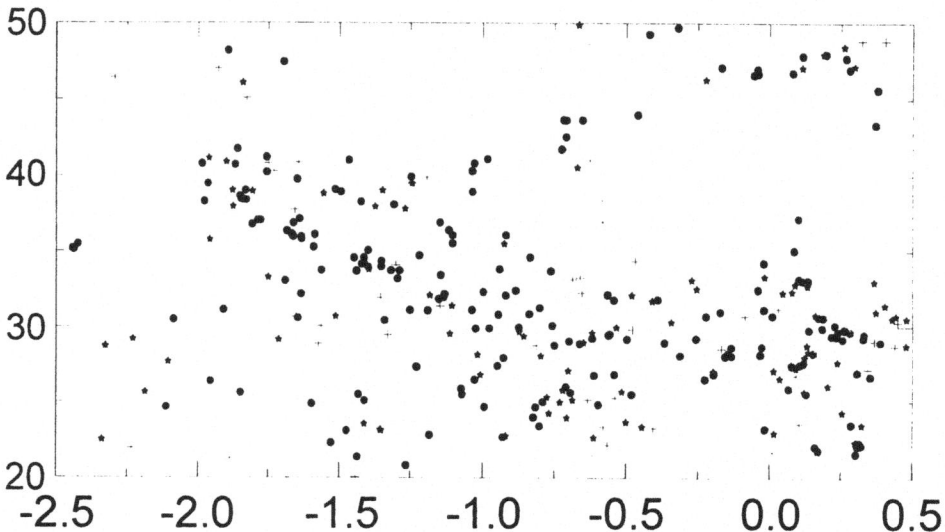

Figure 3. Here we show the distribution of all morphological types in our western Perseus supercluster survey region. Morphologies are estimated from the nuclear spectral type.

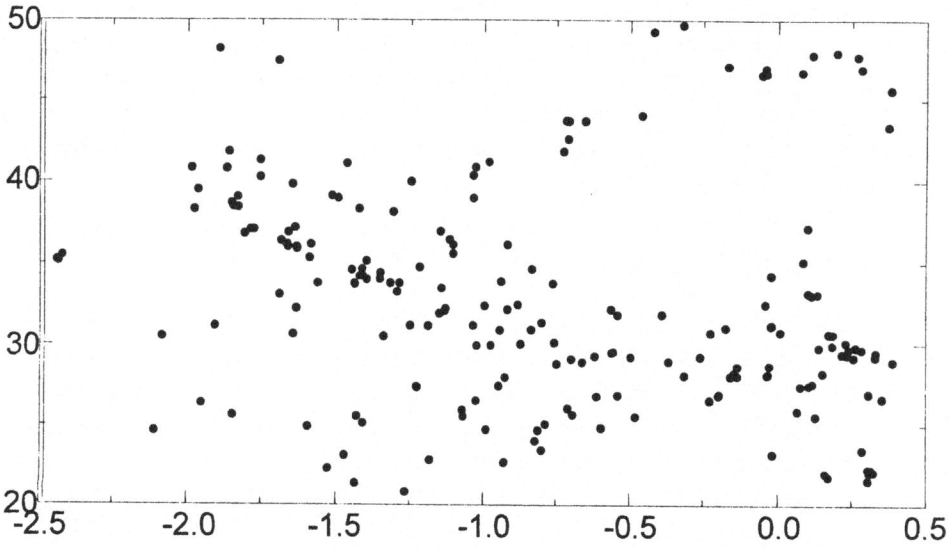

Figure 4. Same as Figure 3 except that early type galaxies are shown.

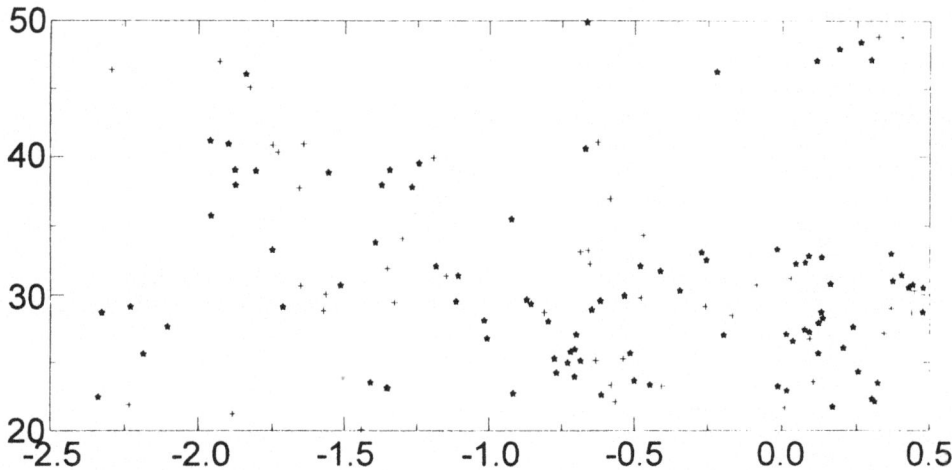

Figure 5. Same as Figure 3 except that intermediate (Sbc - Scd; star symbol) and late (Sd - Irr; cross symbol) type galaxies are shown.

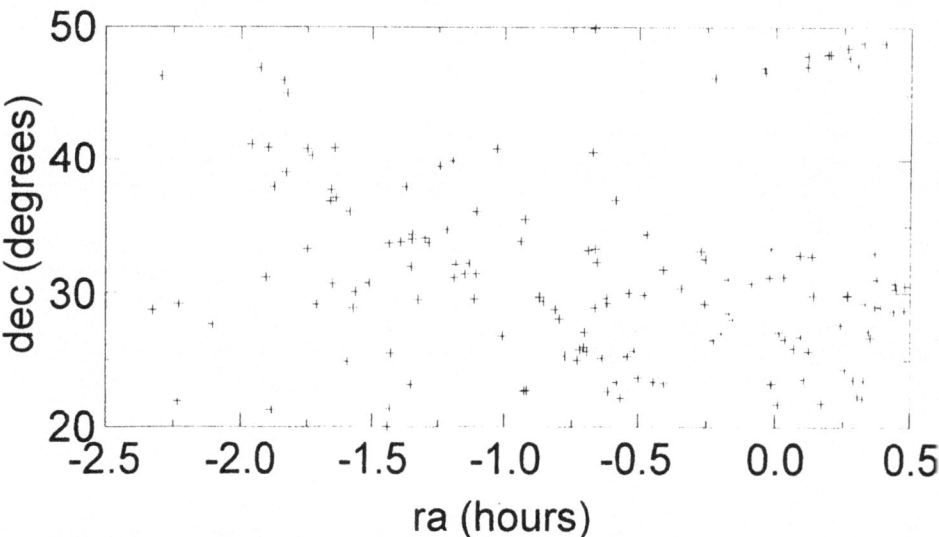

Figure 6. Same as Figure 3 except that those galaxies with emission lines in their spectra are shown. Note that the distribution is very diffuse, and that the filament extending to the northeast is largely missing from this diagram.

scale structures? Are the concentric shells real? Is there quantization at the level of 12,700 km s^{-1}?

What is the nature and significance of alignments?

References

Aaronson, M., Bothun, G. D., Mould, J., Huchra, J., Schommer, R., and Cornell, M.: 1986 *ApJ*, **302**, 536

Batuski, D. J. and Burns, J. O. 1985: *AJ*, **90**, 1413

Binggeli, B. 1982: *A&A.*, **107**, 338

Chincarini, G., and Rood, H. J. 1976: *ApJ*, **206**, 30

Chincarini, G., and Rood, H. J. 1979: *ApJ*, **230**, 648

Deckel, A.: 1988, in *Large-Scale Motions in the Universe*, ed. Rubin, V. C. and Coyne, G. V. (Princeton University Press)

de Lapparent, V., Geller, M. J., and Huchra, J. P.: 1986, *ApJ Lett.*, **302**, L1

Dressler, A., Faber, S. M., Burstein, D., Davies, R. L., Lynden-Bell, D., Terlevich, R. J., and Wegner, G. W.: 1987 *ApJ Lett.*, **313**, L37

Giovanelli, R., Haynes, M. P., and Chincarini, G. L.: 1986 *ApJ*, **300**, 77

Gregory, S. A.: 1975 *ApJ*, **199**, 1

Gregory, S. A. and Thompson, L. A.: 1978 *ApJ*, **222**, 784

Gregory, S. A., Thompson, L. A., and Tifft, W. G.: 1981, *ApJ*, **243**, 411

Henning, P. A.: 1996, *private communication*

Joeveer, M. and Einasto, J.: 1978, in *IAU Symp. 79* ed. J. Einasto and M. S. Longair (Dordrecht:Reidel), 241

Kirshner, R. P., Oemler, A., and Schechter, P. L.: 1978, *AJ*, **83**, 1549

Moody, J. W., Kirshner, R. P., MacAlpine, G. M. and Gregory, S. A.: 1987, *ApJ Lett.*, **314**, L33

Praton, E. A. and Schneider, S. E.: 1994, *ApJ*, **422**, 46

Rubin, V. C., Ford, W. K., Thonnard, N., Roberts, M. S., and Graham, J. A.: 1976 *AJ*, **81**, 687

Szalay, A. S., Broadhurst, T. J., Ellman, N., Koo, D. C., and Ellis, R. S.: 1993, *Proc. Nat. Acad. Sci. (USA)*, **90, No. 11**, 4853

Tarenghi, M., Tifft, W., Chincarini, G., Rood, H. J. and Thompson, L. A.: 1979, *ApJ*, **234**, 793

Tifft, W. G., Hilsman, K. A. and Corrado, L. C.: 1975, *ApJ*, **199**, 16

Tifft, W. G. and Gregory, S. A.: 1976, *ApJ*, **205**, 696

Tifft, W. G. and Gregory, S. A.: 1978, in *IAU Symp. 79* ed. J. Einasto and M. S. Longair (Dordrecht: Reidel), 267

Tifft, W. G., Kirshner, R. P., Gregory, S. A., and Moody, J. W.: 1986, *ApJ*, **310**, 75

Tully, R. B. and Fisher, J. R.: 1987, *Nearby Galaxies Atlas*, Cambridge University Press.

Vaucouleurs, G. de: 1975, in *Stars and Stellar Systems*, **Vol. 9**, pp. 557-600

Wiestrop, D. and Downes, R. A.: 1988, *ApJ*, **331**, 172

Wright, E. L., Smoot, G. F., Kogut, A., Hinshow, G., Tenorio, L., Lineweaver, C., Bennett, C. L. and Lubin, P. M.: 1994, *ApJ*, **420**, 1

ELECTRIC SPACE:
EVOLUTION OF THE PLASMA UNIVERSE

ANTHONY L. PERATT
Los Alamos National Laboratory
Los Alamos, New Mexico
Scientific Advisor,
Office of Research and Development,
United States Department of Energy, Washington D.C.

Abstract. Contrary to popular and scientific opinion of just a few decades ago, space is not an 'empty' void. It is actually filled with high energy particles, magnetic fields, and highly conducting plasma. The ability of plasmas to produce electric fields, either by instabilities brought about by plasma motion or the movement of magnetic fields, has popularized the term 'Electric Space' in recognition of the electric fields systematically discovered and measured in the solar system. Today it is recognized that 99.999% of all observable matter in the universe is in the plasma state and the importance of electromagnetic forces on cosmic plasma cannot be overstated; even in neutral hydrogen regions ($\sim 10^{-4}$ parts ionized), the electromagnetic force to gravitational force ratio is 10^7.

An early prediction about the morphology of the universe is that it be filamentary (Alfvén, 1950). Plasmas in electric space are energetic (because of electric fields) and they are generally inhomogeneous with constituent parts in motion. Plasmas in relative motion are coupled by the currents they drive in each other and nonequilibrium plasma often consists of current-conducting filaments. This paper explores the dynamical and radiative consequences of the evolution of galactic-dimensioned filaments in electric space.

1. Introduction

Contrary to popular and scientific opinion of just a few decades ago, space is not an 'empty' void. It is actually filled with high energy particles, mag-

Astrophysics and Space Science **244**:89-103,1996.
© 1996 *Kluwer Academic Publishers.*

netic fields, and highly conducting plasma. The ability of plasmas to produce electric fields, either by instabilities brought about by plasma motion or the movement of magnetic fields, has popularized the term 'Electric Space' in recognition of the electric fields systematically discovered and measured in the solar system. Today it is recognized that 99.999% of all observable matter in the universe is in the plasma state and the importance of electromagnetic forces on cosmic plasma cannot be overstated; even in neutral hydrogen regions ($\sim 10^{-4}$ parts ionized), the electromagnetic force to gravitational force ratio is 10^7.

Among the earliest predictions about the morphology of the universe is that it be filamentary (Alfvén, 1950, 1981, 1990). Plasmas in electric space are energetic (because of electric fields) and they are generally inhomogeneous with constituent parts in motion. Plasmas in relative motion are coupled by the currents they drive in each other and nonequilibrium plasma often consists of current-conducting filaments. This paper explores the dynamical and radiative consequences of the evolution of galactic-dimensioned filaments in electric space.

In the laboratory and in the Solar System, filamentary and cellular morphology is a well-known property of plasma. As the properties of the plasma state of matter is believed not to change beyond the range of our space probes, plasma at astrophysical dimensions must also be filamentary.

Additionally, transition regions have been observed that delineate the 'cells' of differing plasma types (Eastman, 1990). On an astrophysical scale, these transition regions should be observable at radio wavelengths via transition radiation signatures.

The suggestion that the universe be filamentary and cellular was generally disregarded until the 1980s, when a series of unexpected observations showed filamentary structure on the Galactic, intergalactic, and supergalactic scale. By this time, the analytical intractibility of complex filamentary geometries, intense self-fields, nonlinearities, and explicit time dependence had fostered the development of fully three-dimensional, fully electromagnetic, particle-in-cell simulations of plasmas having the dimensions of galaxies or systems of galaxies. It had been realized that the importance of applying electromagnetism and plasma physics to the problem of radiogalaxy and galaxy formation derived from the fact that the universe is largely a *plasma universe*.

Any imbalance in the constitutive properties of a plasma can set it in motion [if, in fact, it has not already derived from an evolving, motional state (Bohm, 1979)]. The moving plasma, *i.e.*, charged particle flows, are currents that produce self magnetic fields, however weak. The motion of any other plasma across weak magnetic fields produces and amplifies electromotive forces, the energy of which can be transported over large distances via currents that tend to flow along magnetic lines of force. These 'field-aligned currents,' called *Birkeland currents* (Cummings and Dessler 1967) in planetary magnetospheres, should also exist in cosmic plasma. The dissipation of the source energy from evolving or moving plasma in localized regions can then lead to pinches and condense states. Where double layers form in the pinches, strong electric fields can accelerate the charged particles to high energies, including gamma ray energies (Alfvén, 1981). These should then display the characteristics of relativistic charged particle beams in laboratory surroundings, for example, the production of microwaves, synchrotron radiation, and non-linear behavior such as periodicities and 'flickering.'

2. Filamentation by Birkeland Currents

An electromotive force $\int \mathbf{v} \times \mathbf{B} \cdot \mathbf{dl}$ giving rise to electrical currents in conducting media is produced wherever a relative perpendicular motion of plasma and magnetic fields exist (Akasofu, 1984; Alfvén, 1986). An example of this is the (nightside) sunward-directed magnetospheric plasma that cuts the earth's dipole field lines near the equatorial plane, thereby producing a potential supply that drives currents within the auroral circuit. The discovery of these Birkeland currents in the earth's magnetosphere in 1974 (Dessler, 1984) has resulted in a drastic change in our understanding of aurora dynamics, now attributed to the filamentation of Birkeland charged-particle sheets following the earth's dipole magnetic-field lines into vortex current bundles.

3. Galactic Dimensioned Birkeland Currents

Extrapolating the size and strength of magnetospheric currents to interstellar space leads to the suggestion that confined current flows in interstellar clouds assists in their formation (Alfvén, 1981).

As a natural extension of the size hierarchy in cosmic plasmas, the existence of galactic dimensioned Birkeland currents or filaments was hypothesized (Alfvén & Fälthammar, 1963; Peratt, 1986a).

A galactic magnetic field of the order $B_G = 10^{-9} - 10^{-10}$T associated with a galactic dimension of $10^{20} - 10^{21}$m suggests the galactic current be of the order $I_G = 10^{17} - 10^{19}$A.

In the galactic dimensioned Birkeland current model, the width of a typical filament may be taken to be 35 kpc ($\approx 10^{21}$m), separated from neighboring filaments by a similar distance. Since current filaments in laboratory plasmas generally have a width/length ratio in the range $10^{-3} - 10^{-5}$, a typical 35 kpc wide filament may have an overall length between 35 Mpc and 3.5 Gpc with an average length of 350 Mpc. The circuit, of course, is closed over this distance (Peratt, 1990).

4. The Large Scale Structure of the Plasma Universe

Surface currents, delineating plasma regions of different magnetization, temperature, density, and chemical composition give space a cellular structure (Alfvén & Fälthammar, 1963). As current-carrying sheet beams collect into filaments, the morphology of the surface currents is filamentary.

For the case of tenuous cosmic plasmas, the thermokinetic pressure is often negligible and hence the magnetic field is force-free. Under the influence of the electromagnetic fields the charged particles drift with the velocity

$$v = (\mathbf{E} \times \mathbf{B}) / E^2 \tag{1}$$

The overall plasma flow is inwards and matter is accumulated in the filaments which, because of their qualitative field line pattern, are called "magnetic ropes". Magnetic ropes should therefore tend to coincide with material filaments that have a higher density than the surroundings. The cosmic magnetic ropes or current filaments are not observable themselves, but the associated filaments of condensed matter can be observed by the radiation they emit and absorb.

It is because of the convection and neutralization of plasma into radiatively cooled current filaments (due to synchrotron losses) that matter in the plasma universe should often display a filamentary morphology.

5. Synchrotron Emission from Pinched Particle Beams

One of the most important processes that limit the energies attainable in particle accelerators is the radiative loss by electrons accelerated by the magnetic field of a betatron or synchrotron. This mechanism was first brought to the attention of astronomers by Alfvén and Herlofson (1950); a remarkable suggestion at a time when plasma, magnetic fields, and laboratory physics were thought to have little, if anything, to do with a cosmos filled with isolated "island" universes (galaxies). Synchrotron radiation is characterized by a generation of frequencies appreciably higher than the cyclotron frequency of the electrons; a continuous spectra (for a population of electrons) whose intensity decreases with frequency beyond a critical frequency (near intensity maxima); increasing beam directivity with increasing relativistic factor γ $\left(\gamma = (1 - \beta)^{-1/2}\right)$; and polarized electromagnetic wave vectors.

Z-Pinches are among the most prolific radiators of synchrotron radiation known. In this regard, the Bennett-pinch (Bennett 1934), or Z-pinch, as a synchrotron source has been treated by Meierovich (1984) and Newberger et. al. (1984).

TABLE 1. Simulation derived parameters based on
the radiation properties of the double radio galaxy
Cygnus A.

Parameter	Simulation Value
Galactic current, I_G	2.4×10^{19} A
Galactic magnetic field, B_θ	2.5×10^{-4} G
Galactic magnetic field, B_z	2.0×10^{-4} G
Plasma temperature, T	$2.0 - 32.0$ keV
Plasma density, n_e	1.79×10^{-3} cm^{-3}
Electric field strength, E_z	62 mV/m
Synchrotron power, P_{syn}	1.16×10^{37} W
Radiation burst duration	1.28×10^{14} s
Total energy	6.3×10^{62} J

The radiation produced from the two nearest plasma filaments in interaction replicates both the isophotal and power spectra from double radio
galaxies (Figure 1). Table I delineates the basic parameters used in the
interacting galactic filament simulation.

Because the highly relativistic electrons depicted in Figure 1 flow in
direction outwards from the plane of the figure, the synchrotron radiation
is also beamed in this direction (Johner, 1988).

The monochromatic power of quasars and double radio galaxies span
a range of about $10^{33}W - 10^{39}$W (Peratt, 1986b). For example, the "pro-
totype" double radio galaxy Cygnus A has an estimated radio luminosity of 1.6-4.4 $\times 10^{37}$W. Together with the power calculated, the simulation
isophotes are very close to those observed from this object (Peratt, 1986a).
The upper row of Figure 1 suggests that previously apparently unrelated
double radio galaxies all belong to the same species but are simply seen at
different times in their evolution.

6. Confining and Interacting Forces Between Cosmic Currents

If the cosmic current is cylindrical and in a rotationless, steady-state condition, it is described by the *Carlqvist Relation*:

$$\frac{\mu_0}{8\pi}I^2(a) + \frac{1}{2}G\bar{m}^2N^2(a) = \Delta W_{Bz} + \Delta W_k \tag{2}$$

for a current of radius $r = a$ where μ_0 is the permeability of free space, G is
the gravitational constant, $\bar{m}$ is the mean particle mass, N is the number
of particles per unit length, and ΔW_{Bz} and ΔW_k are the differential beam

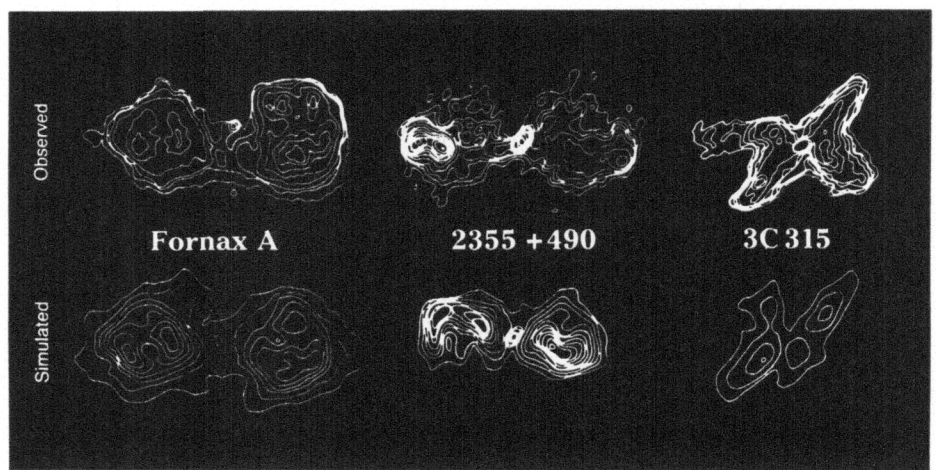

Figure 1. (top) Synchrotron isophotes (various frequencies) of double radio galaxies, (bottom) Simulation analogs at time 10.4 Myr to 58.7 Myr. Time increases from left to right.

magnetic and kinetic energies, respectively (Peratt, 1992a)[1]. Thus, whether or not a current or beam is gravitationally balanced, electromagnetically balanced, or force-free, depends on the magnitude of the individual terms in Eq.(2). Applications of the Carlqvist Relation are presented in this journal (Verschuur, 1995).

In contrast to the gravitational and electromagnetic forces that determine the characteristic of an individual beam, interactions between beams are always dominated by electromagnetic *Biot-Savart* forces,

$$\mathbf{F_{21}} = \int \mathbf{j_2} \times \mathbf{B_{21}} d^3 r \qquad (3)$$

for all space, where $\mathbf{j_2} \times \mathbf{B_{21}}$ is the Lorentz force between the field B_{21} induced by a current I_1 on the current density j_2 at current I_2.[2]

Parallel axial currents within the filaments are long-range attractive, while circular (helical) currents within the filaments (as the electrons gyrate along the axial magnetic field) are short-range repulsive. If the axial currents are able to bring the filaments close enough together so that the

---

[1]When current rotation and transient phenomena are important, the *Generalized Bennett Condition* may be used in place of Eq.(2) (Peratt, 1992a, Chap. 2)

[2]The Biot-Savart force varies as r^{-1} and thus dominates gravitational attraction which varies as r^{-2}. 'Great Attractors', often attributed to gravitational forces between 'missing masses' display Biot-Savart, not mass attraction, characteristics.

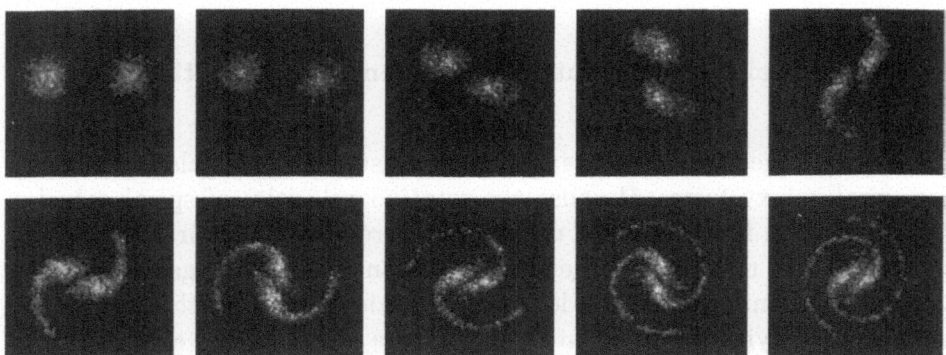

Figure 2. Single frame stills of plasma in the simulation of two adjacent Birkeland filaments: ω_c/ω_p=3.0, $T_{e0} = T_{i0}$=32 keV, E_{z0}=62 mV/m. Total time elapsed: $\approx 10^9$ yr. The initial dimensions in frame 1 (top, lefthand corner) are: radius of filaments $r_{filament}$=17.5 kpc, distance between filaments $d_{filaments}$=80 kpc. The length over which E_{z0} exists in the filaments is taken to be $\approx$10 kpc.

repulsive component of the Lorentz force becomes important, the circular currents repulse and brake, and release energy in the form of synchrotron radiation.

While a complete description of the evolution of interacting galactic currents is given elsewhere (Peratt, 1992a,b), it is useful to reproduce the evolutional sequence in this paper. Figure 2 illustrates the cross-sections of the filaments over a 10^9 yr period.

7. Rotation Velocities

Rotational velocities of spiral galaxies are found by measuring the doppler shift of the Hα line emitted by neutral hydrogen in the spiral arms. If the galaxy is at a cant towards earth, the emission-line in the arm moving away from earth is red-shifted while the line in the arm moving towards earth is blue-shifted. Measurements of the outer rotation curves using radio techniques indicated that these were flat, rather than following the Keplerian law. If galaxies were gravitationally bound systems, their outer mass should follow Kepler's laws of motion and be slower than the inner mass. The flat rotation curves of galaxies has been cited as the strongest physical evidence for the existence of dark matter. In this scenario a massive halo of dark matter has been evoked to produce the flat rotation curves. However, the rotation curves are not really flat; they show appreciable structure representative of an instability mechanism within the arms. This instability questions the existence on any external halo of matter around galaxies that, while making the rotation curves flat, would also dampen any instability growth.

8. The Association of Neutral Hydrogen with Galactic Magnetic Fields

Neutral hydrogen distributions are characteristic of spiral galaxies but not pre-spiral galaxy forms. Because the rotation velocities of spiral galaxies are determined by the motion of neutral hydrogen, it is desirable to know the process for neutral hydrogen accumulation in late–time galaxies.

In the plasma universe model, spiral galaxies form from the interaction of current-carrying filaments at regions where electric fields exist. The individual filaments are defined by the *Carlqvist Condition* that specifies the relationship between gravitational and electromagnetic constraining forces (Verschuur, 1995). In this model, whether or not neutral hydrogen and other neutral gases form from hydrogenic plasma depends of the efficiency of convection of plasma into the filament.

When an electric field is present in a plasma and has a component perpendicular to a magnetic field, inward convection of the charged particles occurs. Both electrons and ions drift with velocity

$$\mathbf{v} = (\mathbf{E} \times \mathbf{B})/\mathrm{B}^2$$

so that the plasma as a whole moves radially inwards. The material thus forms as magnetic ropes around magnetic flux tubes. Magnetic ropes thus contain material filaments that have a higher density than the surrounding plasma.

When a plasma is only partly ionized, the electromagnetic forces act on the non–ionized components only indirectly through the viscosity between the ionized and non–ionized constituents. For a filament, the inward radial velocity drift is

$$v_r = E_z/B_\phi$$

for the case of an axial electric field and azimuthal magnetic field (induced by the axial current I_z). Hence, at a large radial distance r, the rate of accumulation of matter into a filament is

$$\frac{dM}{dt} = 2\pi r v_r \rho_m = (2\pi r)^2 \, \rho_m \frac{E_z}{\mu_0 I_z} \tag{4}$$

Marklund (1979) found a stationary state when the inward convections of ions and electrons toward the axis of a filament was matched by recombination and outward diffusion of the neutralized plasma (Figure 3). The

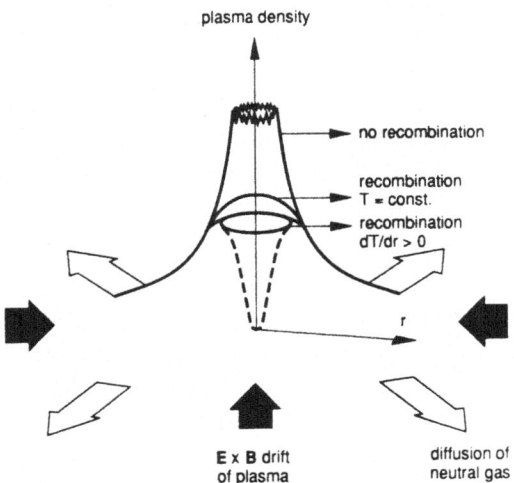

Figure 3.   Plasma density profile as a function of radius shown qualitatively for three cases: No recombination, recombination with temperature T = constant, and recombination with a lower central temperature.

equilibrium density of the ionized component normally has a maximum at the axis. However, because of the radiated loss of energy, the filament cools and a temperature gradient is associated with the plasma. Because of this hollow cylinders, or modifications of hollow cylinders of matter, will form aout the flux tubes.

Because the radial transport depends on the ionization potential of the element, elements with the lowest ionization potentials are brought closest to axis. The most abundant elements of cosmical plasma can be divided into groups of roughly equal ionization potentials: He(24 eV); H, O, N(13 eV); C, S(11 eV); and Fe, Si, Mg(8 eV). These elements can be expected to form hollow cylinders whose radii increase with ionization potential. Helium will make up the most widely distributed outer layer; hydrogen, oxygen, and nitrogen should form the middle layers, while iron, silicon, and magnesium will make up the inner layers. Interlap between the layers can be expected and, for the case of galaxies, the metal–to–hydrogen ratio should be maxium near the center and decrease outwardly. Both the convection process and luminosity increase with the field E_z.

For the case of a fully ionized hydrogenic plasma, the ions drift inwards until they reach a radius where the temperature is well below the ionization potential and the rate of recombination of the hydrogen plasma is considerable. Becouse of this "ion pump" action, hydrogenic plasma will be evacuated from the surroundings and neutral hydrogen will be most heavily deposited in regions of strong magnetic field.

When this process was discovered in the laboratory, there existed some debate as to whether galaxies possessed magnetic fields at all. Today, it is known that large-scale magnetic fields do exist in spiral galaxies. For example, the Effelsberg radio telescope has collected polarization data from about a dozen spiral galaxies at 6 to 49 cm wavelengths (Beck, 1986, 1990). Rotation measures show two different large-scale structures of the interstellar fields: Axisymmetric-spiral and bisymmetric-spiral patterns (Krause et al, 1989).

The orientation of the field lines is mostly along the optical spiral arms. However, the uniform field is often strongest outside the optical spiral arms. For example, in the case of galaxy IC 342, two filamentary structures were found in the map of polarized intensity (Peratt, 1992a, Chap. 3). Their degree of polarization of $\approx$ 30 percent indicates a high degree of uniformity of the magnetic field on the scale of the resolution ($\approx$ 700 pc). These filaments extend over a length of $\approx$ 30 kpc and hence are the most prominent magnetic-field features detected in normal spiral galaxies so far.

A detailed analysis of the rotation measure distribution in a spiral arm southwest of the center of the Andromeda galaxy M31 (Beck, 1990) shows that the magnetic field and a huge HI cloud complex are anchored together. The magnetic field then inflates out of the plane outside the cloud. The tendency for the magnetic field to follow the HI distribution has been noted in several recent observations. Circumstantial evidence has accumulated which suggests that there is a close connection between rings of CO and Hα seen rotating in some galaxies and the magnetic fields in the nuclear regions. This is particularly apparent in observations of spiral galaxies viewed edge-on. This scenario has also been invoked for our Galaxy (Wielebinski, 1989).

Neutral hydrogen is detected from galaxies via the van de Hulst radio-emission line at 21.11 cm (1.420 GHz). High-resolution observation of neutral hydrogen in irregular and spiral galaxies usually reveal extended HI distributions. Contour maps of the HI typically show a relative lack of HI in the cores of spiral galaxies but high HI content in the surrounding region, usually in the shape of a "horseshoe". This region is not uniform but may have two or more peaks in neutral hydrogen content. Figure 4 (right–side) shows an example of the HI distribution in a spiral galaxy.

9. Simulation Results

Figure 4 (left–side) shows the plasma spiral formed in this simulation overlayed on its magnetic field line (squared) isobars. The diameter of the spiral is about 50 kpc with a mass of 10^{41} kg, i.e., a size and mass of that ob-

SIMULATION   **NGC4151**

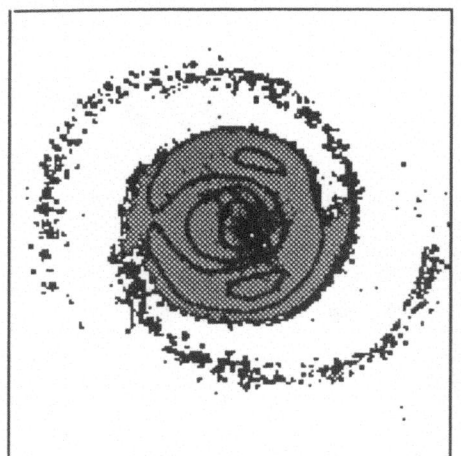

Figure 4. (right) HI distribution superimposed on an optical photograph of NGC 4151. (left) Simulation magnetic energy density superimposed on simulation galaxy. In both cases a 'horse–shoe' shaped cusp, opening towards a spiral arm, surrounds a magnetic field/HI minima core. Within the cusp, two magnetic field/HI peaks are observed.

served from spiral galaxies. A direct comparison to observations is made by superimposing the HI distribution in NGC 4151 on its optical photograph. The observation shows two peaks in neutral hydrogen surrounding a void. The void is orientated towards one of the arms. The simulation allows the two peaks to be traced back to their origin. Both are found to be the remnants of the originally extended components, i.e., cross-sections of the original Birkeland filaments. The hydrogen deficient center is the remnant of an elliptical galaxy formed midway between the filaments, in the magnetic null.

Since E_z is out of the plane of the page, the column electrons spiral downward in counter-clockwise rotation while the column ions spiral upward in clockwise rotation. A polarization induced charge separation also occurs in each arm, which, as it thins out, produces a radial electric field across the arm. Because of this field, the arm is susceptable to the diocotron instability (Peratt, 1992a). This instability appears as a wave motion in each arm and is barely discernable in the single frame photographs in Figure 2 at late times. However, the instability is readily apparent in the simulation spiral rotational velocity curve (Figure 5).

The velocity consists essentially of a linearly increasing component due to a central body undergoing rigid rotation, with two 'flat' components on

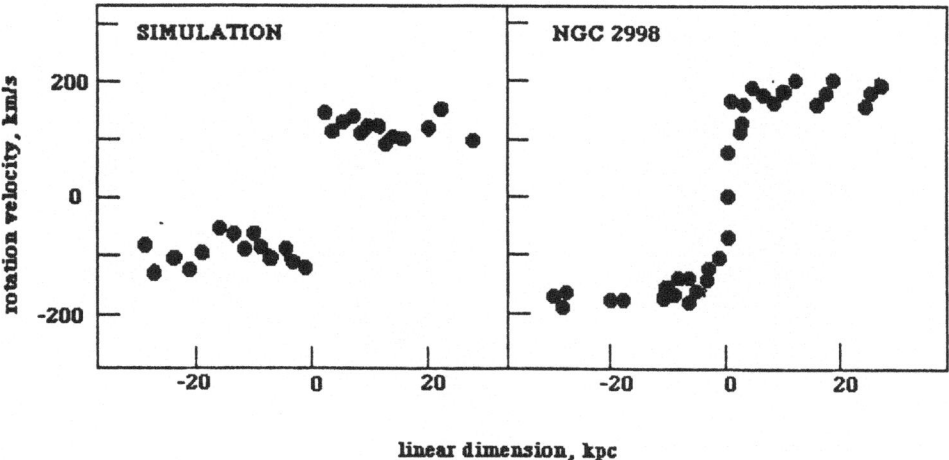

Figure 5. Spiral galaxy rotational velocity curves. Note the well–defined structure on the 'flat' portions of the curves.

either side of $r = 0$ due to the trailing arms. The diocotron instability modulates the 'flat' components at the strong-magnetic-field, low-density instability wavelength $\lambda \approx 2.5\Delta r$, where Δr is the width of an arm.

Figure 6 shows the magnetic field orientations of the vertical (or 'parallel' to the line–of–sight) and the circumferential magnetic fields in an Sc type galaxy. Because of the acute reversals of the circumferential magnetic field, whether or not an observer sees axisymmetric or bisymmetric patterns in the synchrotron radiation depends on the location of the parallel electric fields in the radiating plasma with respect to the circumferential magnetic field.

10. Conclusions

The importance of applying electromagnetism and plasma physics to the problem of radio galaxy, galaxy and star formation derives from the fact that the universe is largely matter in its plasma state, i.e., a universe of plasma. The motion of this plasma in local regions can lead to pinches and ultimately condense states of matter. Where double layers form, strong electric fields can accelerate particles to high energies. The intensity and patterns of synchrotron radiation observed in the model simulations are in excellent agreement with those observed from double radio galaxies.

This paper has summarized previous research relating to the morphol-

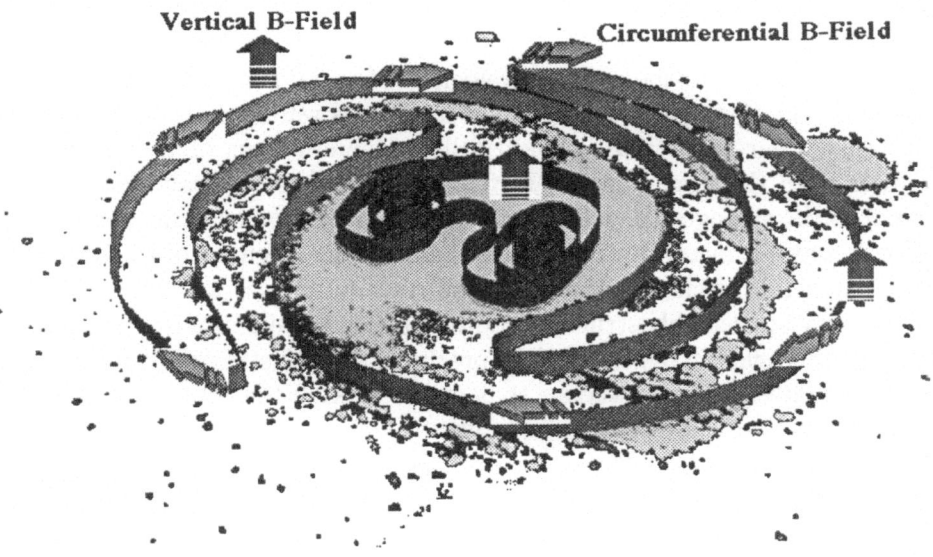

Figure 6. Simulated magnetic fields in an Sc type galaxy.

ogy and large–scale dynamics of a plasma universe. It has also addressed the special case of the radiation seen by an observer when the observer happens to be located in the directed pattern of a synchrotron source. Many sources with this orientation can be expected in various regions of the sky from the "spaghetti" of radiating filaments surrounding the viewer. The background spectrum caused by an extremely large number of synchrotron radiating filaments, when the observer is not in the directed beam, has been treated by Peter & Peratt (1988).

In the late 1970s, plasma simulations of the interacting plasma filaments suggested a scenario that double radio galaxies and quasars were embryonic precursors of galaxies. The simulations also suggested that highly-ordered magnetic fields should exist in galaxies, that would stretch for tens of thousands of light years. The strengths of the magnetic fields appearing in the simulations also suggested that appreciable amounts of nearly neutral hydrogen, known as HI regions, should collect around the field lines.

Simulation plots of the magnetic fields compared nicely with maps of observed HI regions, both showing a "horse-shoe" shaped distribution of gas with either an axisymmetric or a bisymmetric pattern, the pattern type dependent on the direction of the circumferential magnetic field in the dominant synchrotron radiating plasma.

At about the same time, radio astronomers at the Max-Planck-Institute for Radio Astronomy in Bonn, started to measure ordered magnetic fields in galaxies. This work had by 1988 shown unequivocally that large scale magnetic fields do exist in galaxies and do trace the distribution of neutral hydrogen.

With respect to the rotation velocities of spiral galaxies measured from the rotation of the neutral hydrogen: If galaxies were gravitationally bound systems, their outer mass should follow Kepler's laws of motion and be slower than the inner mass. The flat rotation curves of galaxies has been cited as the strongest physical evidence for the existance of dark matter. In this scenario a massivive halo of dark matter has been envoked to produce the flat rotation curves. However, the rotation curves are not really flat; they show appreciable structure representative of an instability mechanism within the arms. This instability precludes the existance of any external halo of matter around galaxies that, while making the rotation curves flat, would also dampen any instability growth.

The best agreement between the particle–in–cell maps, both magnetic field and neutral hydrogen, to the radio telescope data, and the replication of the optical features of spiral galaxies by the simulation, occurs when the observable galaxy mass is used. No dark matter is needed to explain the detailed features of a galaxy if electromagnetic forces are present.

An important question unaddressed in this paper is the existence of periodicities in the redshifts of cosmic objects. The solution of this problem is of paramount importance in all cosmologies, including the electric-space, plasma universe model (Tifft, 1995).

11. Acknowledgments

This project was supported by the U.S. Department of Energy.

References

Akasofu, S.-I.: 1984, in *Magnetospheric Currents*, T. A. Potemra, Ed. (Geophysical Monograph No.28), Washington, DC: Amer. Geophys. Union, pp.29–48
Alfvén, H.: 1950, *Cosmical Electrodynamics* Oxford, London
Alfvén, H.: 1981, *Cosmic Plasma* D. Reidel, Dordrecht
Alfvén, H.: 1986, *IEEE Trans. on Plasma Sci.*, bf PS-14, 779
Alfvén, H.: 1990, *IEEE Trans. Plasma Sci.*, **18**, 5
Alfvén, H. and Herlofson, N.: 1950, *Phys. Rev.*, **78**, 616
Alfvén, H., Fälthammar, C.-G.: 1963, *Cosmical Electrodynamics*, Oxford University Press
Beck, R.: 1986, *IEEE Trans. Plasma Sci.*, **14**, 470
Beck, R.: 1990, *IEEE Trans. Plasma Sci.*, **18**, 33
Bennett, W. H.: 1934, *Phys. Rev.*, **45**, 890
Bohm, D. J.: 1979, in *A Question of Physics: Conversations in Physics and Biology*, Buckley, P., Peat, F. D. (eds.), University of Toronto Press, Toronto
Cummings, W., Dessler, A. J.: 1967, *J. Geophys. Res.*, **72**, 1007

Dessler, A. J.: 1984, in *Magnetospheric Currents*, T. A. Potemra, Ed. (Geophysical Monograph No.28), Washington, DC: Amer. Geophys. Union, pp.26–32

Eastman, T.: 1990, *IEEE Trans. Plasma Sci.*, **18**, 18

Johner, J.: 1988, *Phys. Rev. A*, **36**, 1498

Krause, M., Beck, R., Hummel, E.: 1989, *Astron. Astrophys.*, **217**, 4

Marklund, G. T.: 1979, *Nature*, **277**, 370

Meierovich, B. E.: 1984, *Phys. Reports*, **104**, 259

Newberger, B. S., M. I. Buchwald, R. R. Karl, D. C. Moir, and Starke, T. P.: 1984, *Bull. Am. Phys. Soc.*, **29**, 1435

Peratt, A. L.: 1986a, *IEEE Trans. on Plasma Sci.*, **PS-14**, 639

Peratt, A. L.: 1986b, *IEEE Trans. on Plasma Sci.*, **PS-14**, 763

Peratt, A. L.: 1990, *IEEE Trans. on Plasma Sci.*, **18**, 26

Peratt, A. L.: 1992a, *Physics of the Plasma Universe*, Springer-Verlag, New York

Peratt, A. L.: 1992b, *Sky and Tel.*, **February**, 136

Peter, W., Peratt, A. L.: 1988, *Laser and Particle Beams*, **6**, Part 3, 493

Tifft, W. G.: 1995, *Astrophys. Space Sci.*, **227**, 25

Verschuur, G. L.: 1995, *Astrophys. Space Sci.*, **227**, 187

Wielebinski, R.: 1989, "Magnetic Fields in the Galaxy", Max-Planck-Institut für Radioastronomie preprint series no. 347

GAMMA-RAY BURSTS: SHOULD COSMOLOGISTS CARE?

J. G. LAROS
Lunar and Planetary Laborotory
Univ. of Arizona
Tucson, Arizona 85721

Abstract. Gamma-Ray Burst (GRB) locations are distributed isotropically on the sky, but the intensity distribution of the bursts seems clearly incompatible with spatial homogeneity. Of the scenarios that attempt to provide an explanation, there are two that enjoy current popularity: (1) GRBs are produced by high-velocity neutron stars that have formed an extended ($\sim$100 kpc) spherical halo or "corona" around our galaxy. (2) The bursters are at cosmological distances, with redshifts near unity for the weaker events. The major evidence used to argue for or against each of these scenarios remains inconclusive. Assuming, not unreasonably, that the cosmological scenario is correct, one can discuss the advantages and disadvantages of studying GRBs as opposed to other objects at moderate redshift. We find that the advantages of GRBs–high intensity, penetrating radiation, rapid variability, and no expected source evolution–are offset by observational difficulties pertaining to the extraction of cosmological information from GRB data. If the cosmological scenario proves to be correct and if the observational difficulties are overcome, then cosmologists certainly should care.

1. Introduction

Cosmic Gamma-ray bursts (GRBs), discovered in 1973, remain one of the great "unsolved mysteries" of astrophysics. Amazingly, with approximately two thousand bursts observed and new bursts being detected at the rate of nearly one per day, astrophysicists remain divided over absolutely the most basic issue of whether the bursters are galactic or extragalactic (or some of each). A recent AIP Conference Proceedings volume devoted exclusively to GRBs (Fishman et al. 1994) contained 67 pages on cosmological models

Astrophysics and Space Science **244**:105-109,1996.

vs. 63 pages on galactic models! The crux of the problem is that, lacking identified burster counterparts, we are forced to interpret the results of distribution analyses. The two main results are: (1) The sky distribution (top half of Figure 1) is highly isotropic. With over 1000 bursts localized to several-degree accuracy, any deviations from isotropy of ~10% would have been detected. (2) The intensity distribution (bottom half of Figure 1) is inconsistent with spatial homogeneity. Observational effects notwithstanding, there is a clear deficiency of weak events relative to the number expected from a homogeneous distribution. However, the distribution is consistent with homogeneity over more than two decades in intensity.

These results do not lead to a definite conclusion about the GRB distance scale. However, it is generally agreed that if the bursters are associated with our galaxy they are in a giant halo at ~100 kpc distances; if extragalactic they are at cosmological distances corresponding to redshifts up to z~1 (Fishman and Meegan 1995, and references therein). The galactic scale is set by the combination of the observed isotropy, which does not allow distances of the same order as the dimensions of the visible galaxy or distances that approach M31, and the rollover in the intensity distribution, which does not allow smaller distances. The extragalactic scale is determined primarily by using only the Hubble flow to explain the shape of the intensity distribution, although the isotropy and non-identification with bright galaxies also rules out "nearby" extragalactic models. There are very few proponents of other distance scales, although the far reaches of the Solar System and the region between the Local Cluster and the distance of the Virgo Cluster have been mentioned. In any case, it appears that there is a reasonable probability that GRBs might constitute an important new observational tool for cosmologists. In this paper we will provide a two-part assessment of that probability. First, we will examine the primary evidence for and against the cosmological and galactic scenarios. Next, assuming a cosmological origin for GRBs, we will discuss their strengths and weaknesses as cosmological probes.

2. Cosmological vs. Galactic Scenarios

The basic premise behind the most popular Galactic scenario is that neutron stars born with high velocities (~1000 km/s) have populated an extended (~100 kpc) spherical Galactic halo, or "corona". There is no direct physical evidence for such a corona, but a recent finding (Lyne and Lorimer 1994) that radio pulsar velocities seem to average 450 km/s supports such a notion. Furthermore, although the evidence is less than overwhelming, GRB time scales, energetics, and certain spectral features have long suggested a neutron star origin. The intense 1979 March 5 event has recently

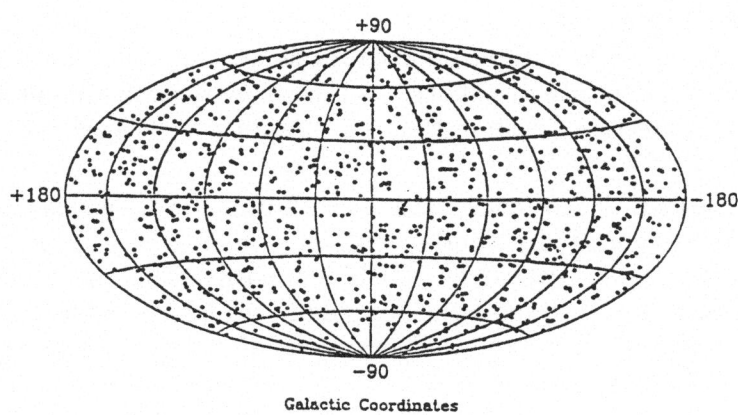

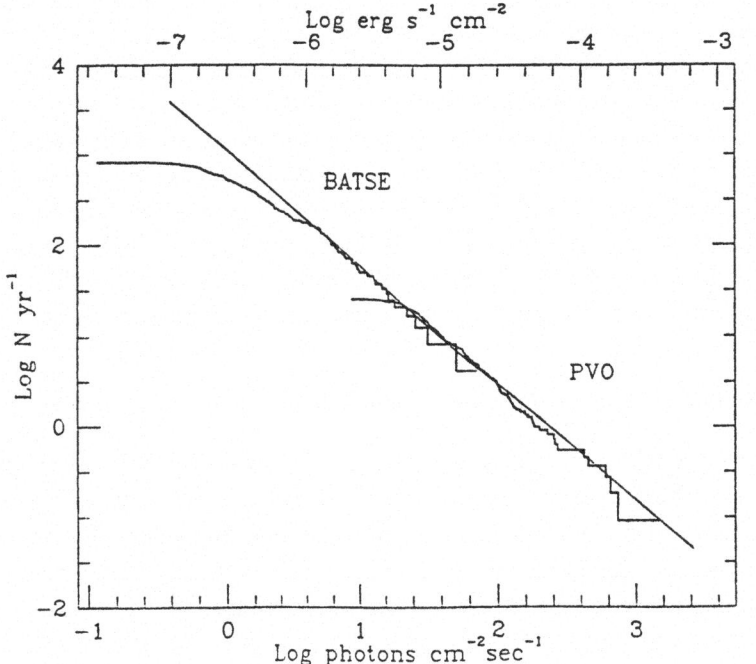

Figure 1. GRB Skymap and Intensity Distribution. Data are from the BATSE experiment on CGRO (Fishman and Meegan 1995).

been shown to be more like typical GRBs than had been thought (Fenimore, Klebesadel, and Laros 1996). The source of this burst was almost certainly within an LMC supernova remnant at a distance of 55 kpc, thereby supporting the 100 kpc distance scale. On the negative side, Galactic scenarios seem to require too much fine tuning and too many ad hoc assumptions. The observed degree of isotropy is unnatural, requiring both the correct luminosity and a lack of bursts from the slower moving neutron stars. The shape of the intensity distribution is even more difficult to reproduce with Galactic sources. A "custom tailored" beaming or delayed turn-on of the sources seems to be required.

In contrast, the cosmological scenario automatically yields isotropy and the correct intensity distribution with only one free parameter, the intrinsic (assumed standard-candle) GRB luminosity. The gamma-ray energy requirement is $\sim 10^{51}$ ergs, which is readily available in collisions involving compact objects. The predicted frequency of collisions involving neutron stars ($\sim$1/galaxy/Myr) is in accord with the observed numbers of GRBs. Under standard cosmologies, time dilation and reddening of the weaker GRBs is also predicted. Such effects apparently have been detected, but it is not clear that the measurements are quantitatively in accord with expectations based on the intensity distribution. Also, it has been pointed out that similar correlations could be intrinsic to the sources. The main difficulty with the cosmological scenario is the tremendous energy output that appears almost exclusively as gamma-rays with a clearly nonthermal spectral distribution. The problem is not the amount of energy, but explaining how it can be emitted as gamma-rays with the correct time scale and without the accompanyment of intense thermal radiation. It has been speculated that some exotic mechanism–involving, e.g., cosmic strings, white holes, or strange stars–might provide a way around this difficulty.

3. Potential as Cosmological Probes

If GRBs are at the cosmological distances implied by their isotropy and intensity distribution, they have certain advantages over other yardsticks that have been used (Trimble 1994). Gamma-rays are an entirely new wavelength regime for this purpose, and they are sufficiently penetrating that absorption corrections are not likely to be important. The bursts themselves are intense enough (supernova-like output in seconds) to probe rather large distances. Like supernovae, they would not be expected to show source evolution. And, their rapid variability allows unique time dilation studies. On the negative side, the apparent lack of lines in GRB spectra means that individual redshifts perhaps cannot be accurately determined. The situation is somewhat akin to measuring redshifts by using only optical colors. Also, GRBs can be quantitatively much different from one another, and they have resisted efforts at classification. This large intrinsic scatter in burst properties indicates that they may not be good standard candles, and it creates major difficulties in quantifying correlations such as time dilation vs. intensity. Finally, the GRB intensity distribution is as likely to have been modified by density evolution as would be the case for any other cosmological probe.

4. Summary and Conclusions

Based on our present state of knowledge, GRBs are probably about as likely to be at cosmological distances as they are to be anywhere else. Counterparts will probably be needed to settle the issue conclusively, but higher sensitivity gamma-ray measurements or almost any GRB detection at radio or optical wavelengths might tip the balance one way or the other. GRBs have interesting potential as cosmological probes, but observational difficulties presently stand in the way. We need to find spectral lines–perhaps through better gamma-ray measurements or detections at optical wavelengths–in order to determine individual redshifts, and we need to develop a classification scheme or some other means of combating the large intrinsic scatter in burst properties. A favorable outcome of the cosmological vs. Galactic controversy combined with observational breakthroughs will be needed for an affirmative answer to the question: Should cosmologists care?

References

Fenimore, E.E., Klebesadel, R.W., and Laros, J.G.: 1996, *ApJ*, **460**

Fishman, G.J., and Meegan, C.A.: 1995, *Ann. Rev. Astron. Astrophys.*, **33**, 415

Fishman, Brainerd, and Hurley, Eds. 1994,: *Gamma-Ray Bursts*, (Proceedings of Gamma-Ray Burst Workshop - 1993, Huntsville, AL), AIP Conf. Proc. 307 (AIP Press- New York)

Lyne, A. and Lorimer, D.: 1994, *Nature*, **369**, 127

Trimble, V.: 1994, *Gamma-Ray Bursts* (Proceedings of Gamma-Ray Burst Workshop - 1993, Huntsville, AL), AIP Conf. Proc. 307 (AIP Press- New York), 717

TESTING FOR QUANTIZED REDSHIFTS. II.
THE LOCAL SUPERCLUSTER

W. M. NAPIER
Armagh Observatory
College Hill, Armagh BT61 9DG, N. Ireland

AND

B. N. G. GUTHRIE
5 Arden Street
Edinburgh EH9 1BR, Scotland

Abstract. Samples of 97 and 117 high–precision 21 cm redshifts of spiral galaxies within the Local Supercluster were obtained in order to test claims that extragalactic redshifts are periodic ($P \sim 36$ km s^{-1}) when referred to the centre of the Galaxy. The power spectral density of the redshifts, when so referred, exhibits an extremely strong peak at 37.5 km s^{-1}. The signal is seen independently with seven major radio telescopes. Its significance was assessed by comparison with the spectral power distributions of synthetic datasets constructed so as to closely mimic the overall properties of the real datasets employed; it was found to be real rather than due to chance at an extremely high confidence level. The signal was subjected to various tests for robustness such as partitioning of data, increase of strength with precision and size of sample, and stability of the correcting vector. In every respect tested, it behaved like a physically real phenomenon. The periodicity is particularly strong within small groups and associations of galaxies, showing no sign of an intrinsic spread $\gtrsim 3$ km s^{-1}.

1. Introduction

As explained in our introductory paper at this conference (Paper I), we initiated a project at the Royal Observatory, Edinburgh to test the 20–year old claims (Tifft 1976, 1977, 1980) that extragalactic redshifts are 'quantized'. For this purpose we are using recent, high-precision 21 cm

Astrophysics and Space Science **244**:111-126,1996.
© 1996 *Kluwer Academic Publishers.*

data in the literature and rigorous statistical procedures. The importance of this problem lies in the revolutionary consequences which would follow if the conventional interpretation of redshifts turned out to be wrong. A corollary is that, for 'quantized redshifts' to be accepted, most astronomers would probably require the evidence to be at a level beyond that normally regarded as sufficient in cosmology.

For our project we first require a properly formulated hypothesis based on the quantization claims. This hypothesis should be fixed at the outset and make specific predictions with regard to redshift periodicity; second, it should be tested against the null hypothesis of no periodicity using independent data (*i.e.* data not employed in the original formulation of the hypothesis). If the new data are culled from a catalogue, the culling should be unbiased with regard to the hypothesis and done prior to the analysis. Before accepting a periodicity as real, we have to consider other possibilities, *e.g.* the effects of a non-uniform overall distribution of redshifts, and clustering. Finally, when the original hypothesis has been tested, we may use the data to formulate a modified or 'improved' version (since all hypotheses, including statistical ones, eventually fail!); the improvement then has to be tested against a new dataset using the above procedure.

The early claims for redshift quantization amount to a statement that there is a local periodicity of $\sim$72 km s^{-1} within binaries, groups and clusters (Tifft 1976, 1977, 1980; Arp & Sulentic 1985), and a related global periodicity of $\sim$24 or $\sim$36 km s^{-1} for field galaxies when the heliocentric redshifts are corrected for the solar motion with respect to the Galactic centre (Tifft & Cocke 1984, hereinafter TC). The global periodicity is claimed to be 24.2 km s^{-1} for galaxies with narrow HI profiles and 36.2 for broadline galaxies. For testing, we formulated the 'local periodicity' claim as that of a single redshift periodicity somewhere in the range 70–75 km s^{-1} (*cf.* Tifft 1976), and we searched for this in new samples of spiral and irregular galaxies in the Virgo cluster. A significant periodicity of 71.0 km s^{-1} was found for the corrected redshifts of 48 spirals in low-density regions of the cluster, but there was no sign of any periodicity in the prescribed range for the irregulars (Guthrie & Napier 1990); the distinction between high– and low–density regions has not yet been tested for spirals in another cluster. A significant periodicity ($P \simeq 37.5$ km s^{-1}) of galactocentric redshifts was also found for a sample of nearby field spirals, although once again not for the irregulars (Guthrie & Napier 1991). These pilot studies, reported in Paper I, encouraged us to embark on a major analysis (Guthrie & Napier 1996) to test whether the global redshift periodicity proposed by TC holds for spirals throughout the Local Supercluster (LSC).

Since the solar vector $(V_\odot, l_\odot, b_\odot) = (233.6$ km s^{-1}, 98.6°, 0.2°) for which TC claimed the periodicities emerge is close to estimates of the galactocentric solar vector, we formulated the hypothesis to be tested as:

> Extragalactic redshifts, when corrected for a velocity vector equal or close to the Sun's galactocentric motion $\mathbf{V}_\odot$, tend to occur in multiples of $\sim$24.2 or $\sim$36.3 km s^{-1}.

We took the galactocentric solar motion $\mathbf{V}_\odot$ to be $(213\pm10$ km s^{-1} $93\pm3°$, $2\pm5°)$ following recent modelling of the Galactic HI distribution by Merrifield (1992). The hypothesis was first tested for a sample of 97 LSC spirals with accurate redshifts from an HI database, and the reproducibility of the results was then checked by examining a further sample of 117 LSC spirals with accurate HI redshifts obtained with the 300-foot Green Bank telescope.

2. The Technique

The technique most commonly used to test data for periodicity is power spectrum analysis (PSA), in which the given set $\{V_i\}$ of N numbers is circularly transformed with respect to a trial period P, and a statistic $I = 2R^2/N$ is calculated. R represents the magnitude of the vector sum of the unit vectors l_i $(i = 1, 2, \cdots N)$ whose directions make angles $\phi_i = 2\pi(V_i \bmod P)$ with the x–axis: essentially, it represents the distance walked by the classical drunk man taking unit steps in random directions (in the absence of a signal), while I represents this distance normalized. Thus $R = \sqrt{C^2 + S^2}$, where $C = \Sigma \cos \phi_i$ and $S = \Sigma \sin \phi_i$; the mean phase $\bar{\phi}$ is $\tan^{-1}(S/C)$ if $C > 0$, or $\tan^{-1}(S/C) + \pi$ if $C < 0$. A power spectrum is a plot of $I(\nu)$ against frequency ν, where $\nu = 1/P$. Against a white noise background, the probability p of obtaining a value I greater than some value I_0 by chance for a single trial period is $\sim \exp(-I_0/2)$ (the exact solution is tabulated by Webster 1995).

PSA, although an established technique of great utility in the analysis of time series, suffers from some more or less well–known limitations (*e.g.* Newman & Terzian, these proceedings). Edge effects may feed spurious power into sub–multiples of the series length. Bias may arise from secular trends, leading also to departures from the exponential distribution. I is an inconsistent estimator of power, its variance equalling its mean. This latter is illustrated in Fig. 1, which shows the normalized I–distributions resulting from PSA of synthetic datasets simulating the real dataset of 97 LSC spirals in overall redshift distribution, but with the inclusion of a periodicity of 37.5 km s^{-1} at various dispersions σ. It can be seen that signals of identical strength as measured by σ/P (the underlying physical process) may be detected with widely different confidence levels as measured by I (the statistical estimator). Because of these and other difficulties, the unwindowed periodogram has often been regarded as a spectral estimator to be avoided

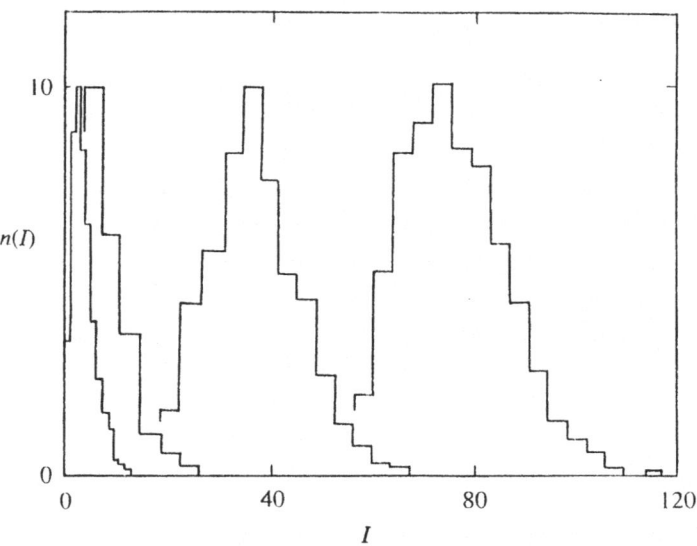

Figure 1. An inconsistent statistic: normalized power distributions of power I for synthetic datasets, each comprising 97 redshifts with periodicity 37.5 km s^{-1} and dispersions (left to right) σ=32, 16, 8 and 6 km s^{-1} respectively.

(Thomson 1990), and various smoothing windows have been proposed to ameliorate its problems. There does not seem to be a fully–developed analytical theory to handle all these difficulties, however. A further difficulty in the present application is that a true periodicity for a prescribed solar vector generally yields false or 'ghost' peaks in I at other vectors and periods, the latter often close to harmonics of the true one.

As an alternative to window carpentry, we tested the redshift quantization hypothesis by purely numerical procedures. The 'ghost peak' phenomenon was explored through whole–sky searches for peaks, taking wide ranges of solar speeds $V_\odot$ and periods P. For each assumed vector $V_\odot$ subtracted from the heliocentric redshift set, an unwindowed periodogram was constructed and the highest peak I in the search range of P recorded. The power estimator employed was not an individual I but the overall incidence of high I–values in the volume of $(V_\odot, P)$–space prescribed by the periodicity hypothesis; thus conclusions were based on the overall behaviour of the power in the region rather than the height and location of a single high peak. The distribution of high I–values for the real dataset was then compared with those obtained for a large number of synthetic datasets constructed so as to simulate the real dataset in every respect except the periodicity under test (to construct each synthetic dataset, a small random displacement, just large enough to smear out the periodicity under test, was added to each redshift in the real dataset, which was otherwise unchanged). Since all variables other than the periodicity under test were thereby frozen, any significant difference between the I–distribution of the real dataset and those of the synthetic ones could only be due to the presence of the periodicity. This procedure appears to be robust, since the use of extreme–value statistics is avoided, the issues of bias, consistency *etc.* become irrelevant (Newman & Terzian, these proceedings), and the results are relatively insensitive to subsequent revisions of the adopted galactocentric solar vector.

3. The Sample

Using HI data for 6439 galaxies from Bottinelli *et al.* (1990; hereinafter BGFP), we took galaxies with galactocentric redshifts cz_{GC} <2600 km s^{-1} (sometimes considered to be the limit of the LSC) and quoted redshift errors σ_{cz} <4 km s^{-1}. Eliminating possible members of the Virgo cluster (galaxies within 12° of M87), non–spirals, and spirals previously used by TC, the list was reduced to 247 spirals, of which 97 have 'more accurate' redshifts (82 with $\sigma_{cz} \leq 3$ km s^{-1} in the BGFP catalogue, along with BGFP redshifts for 15 other galaxies adopted as redshift calibrators by Baiesi–Pillastrini & Palumbo (1986) on the basis of at least five HI line measurements). The sample of 97 galaxies is biased towards nearby and high–luminosity spirals, and it includes the 40 nearby spirals with 'more accurate' redshifts used in our pilot study (Guthrie & Napier 1991).

To see whether the $(\mathbf{V}_\odot, P)$ found by TC might simply reflect their limited search region, we varied $\mathbf{V}_\odot$ in direction over the whole sky, and in speed from 140 to 300 km s^{-1}; for each vector we carried out a PSA over the period range 20–200 km s^{-1} and recorded the maximum I–value. The ten highest peaks provide remarkable support for the TC solution. Five of them are for periods 24±3 km s^{-1}; the other five have essentially the same period (37.6±0.3 km s^{-1}), and three of these have vectors close to the galactocentric solar motion (see figure 2 in Guthrie & Napier 1996). Thus the TC hypothesis is a reasonable one to test in that only solutions of the sort claimed by TC appeared in this initial reconnaissance.

The test was carried out as follows:

First, the power structure in a neighbourhood encompassing both the galactocentric solar vector and the prescribed period was examined. Making generous allowance for the uncertainty in the prescribed $(\mathbf{V}_\odot, P)$, we carried out a coarse–grid exploration over a 30° × 30° region centred at $l_\odot = 93°, b_\odot = 2°$, varying $V_\odot$ from 203 to 263 km s^{-1} and P from 34 to 39 km s^{-1}. Thus 10 800 I–values were obtained (not all of which were independent). Of these, 25 were >20 and 76 were >15, i.e. $n_{20}=25$ and $n_{15}=76$. Since the values of n_{20} and n_{15} are somewhat dependent on the exact po-

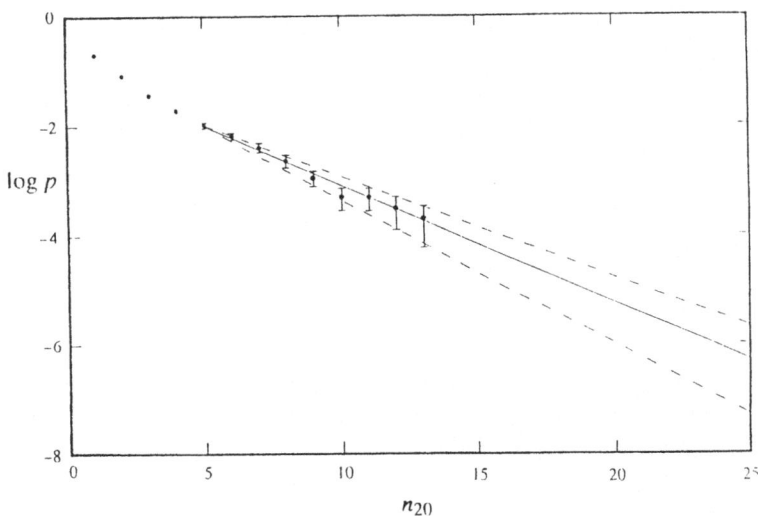

Figure 2. Probability p that a set of 97 randomized redshifts, constructed as described, would yield more than n_{20} values of $I \geq 20$.

sitioning of the coarse grid, fine–grid values for the same region were also obtained, and scaled to give the appropriate values, $n_{20}=17.2$ and $n_{15}=52.4$, for comparison with coarse–grid values for synthetic datasets.

Second, the power distribution expected for suitable random datasets was obtained. In constructing these synthetic datasets, the sky positions of the 97 galaxies were preserved and each real redshift was randomized by adding to it the difference of two random numbers in the range 0 to 50 km s^{-1} (σ=20 km s^{-1}), large enough to smear out the periodicities under test (*cf.* Fig. 1) but small enough to preserve the overall redshift distribution, including any clustering. The synthetic datasets were therefore identical to the real dataset in all respects except for the distribution of fine redshift structure. Coarse–grid searches for 10 000 synthetic datasets yielded no values of $n_{20} >17.2$ or $n_{15} >52.4$.

Finally, the cumulative distributions of n_{20} and n_{15} were used to give the single–trial probability p of exceeding any prescribed value of n_{20} or n_{15}. Thus, in Fig. 2, $\log p$ is plotted against n_{20}; an extrapolation yields the 1σ range for the probability of obtaining $n_{20} >17$ in a single trial as $6 \times 10^{-6} \lesssim p(n_{20}) \lesssim 6 \times 10^{-5}$. Similarly, for $n_{15} \geq 52$, the 1σ range is $1 \times 10^{-4} \lesssim p(n_{15}) \lesssim 4 \times 10^{-4}$. Thus, even with a generous allowance for uncertainty in the prescribed ($\mathbf{V}_{\odot}, P$), the hypothesis of redshift periodicity is supported at a very high confidence level ($C \sim 1 - 10^{-4}$).

A power spectrum of the 97 redshifts corrected for the solar vector yielding the highest peak (217 km s^{-1} 95°, –12°) is shown in Fig. 3. The

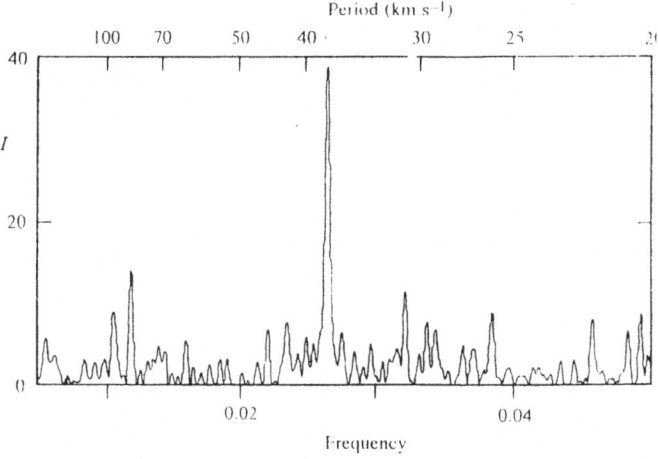

Figure 3. Power spectrum associated with the solar vector $\mathbf{V}_\odot=(217 \text{ km s}^{-1}, 95°, -12°)$ for the 97 spirals.

sharpness of the peak at $P=37.5$ km s^{-1} shows that the redshift structure is indeed a periodicity, rather than say clustering on scales $\gtrsim 50$ km s^{-1} which would produce broad 'humps' in the power spectrum.

The robustness of the result was examined in several ways. We varied the upper limit of cz_{GC} in steps from 1 000 to 4200 km s^{-1}, and compiled a sample of galaxies as before for each upper limit. A fine–grid search was carried out for each sample, covering the same $(\mathbf{V}_\odot, P)$ region as that used in the test. The values of I, and the degrees of quantization R/N, for the optimum solar vectors are listed in Table 1 for the two strongest peaks. The periods P and vectors $\mathbf{V}_\odot$ for these peaks are stable. Also, $(R/N)_{max}$ does not vary much as the upper limit of cz_{GC} is increased from 1 400 to 4200 km s^{-1}, and so the adopted cut–off at 2600 km s^{-1} is not critical. There is nevertheless a downward trend in $(R/N)_{max}$, consistent with a signal which weakens with increasing separation between galaxies; (or it could be that the redshifts of more distant galaxies are being less accurately measured). Other modifications to the sample (*e.g.* exclusion of the 15 redshift calibrators or inclusion of 19 Virgo cluster galaxies) had little effect on the result. We also examined the individual HI redshift measurements for each of the 52 spirals $> 12°$ from M87 with $cz_{GC} <1000$ km s^{-1} and $\sigma_{cz} <3$ km s^{-1} in the BGFP database (including galaxies previously used by TC), and found that seven major radio telescopes (Arecibo, Effelsberg, Green Bank 140 and 300 ft., Jodrell Bank, Owens Valley and Westerbork) independently revealed the periodicity (see figure 9 in Guthrie & Napier 1996). Thus the result survived all the tests for robustness.

Although the existence of a periodicity is strongly supported by the

W. NAPIER, B. GUTHRIE

TABLE 1. The two main peaks, each with periodicity 37.6 ± 0.2 km s^{-1}, as a function of sample size. Notable features are (i) the stability of the solutions, each vector varying by only ±2 km s^{-1} and $\pm1°$ as the sample size is more than doubled; and (ii) a slight downturn in intrinsic signal strength $(R/N)_{max}$ as the redshift limit (and the mean separation between galaxies) increases.

cz_{max}	N	I_{max}	$(R/N)_{max}$	$V_\odot$	$l_\odot$	$b_\odot$
1 000	51	30	0.55	207	95	−7
1 400	72	30	0.46	207	95	−7
1 800	86	33	0.44	209	94	−7
2 600	97	37	0.44	209	94	−7
3 400	105	38	0.42	203	94	−8
4 200	111	35	0.40	205	94	−7
1 000	51	30	0.54	215	93	−13
1 400	72	31	0.46	213	94	−13
1 800	86	36	0.46	215	94	−13
2 600	97	38	0.44	217	95	−12
3 400	105	32	0.39	217	95	−13
4 200	111	31	0.37	215	95	−13

above analysis, the presence of several high peaks makes it difficult to know which if any of them represents the 'real' vector. We tried to determine the solar vector for the periodicity more accurately by examining the patterns of peaks yielded by artificial datasets generated from various trial vectors, and selecting the vector which gave the best match to the pattern of peaks yielded by the real data. Such pattern–matching was carried out for the 37.5 km s^{-1} periodicity for the 97 LSC spirals, and also for the 71.0 km s^{-1} periodicity which we had earlier detected in the 48 spirals in low–density regions of the Virgo cluster. The results are listed in Table 2, together with the current best estimate of the galactocentric solar vector obtained from Galactic modelling. The agreement between the three vectors is remarkably good, and is most unlikely to have arisen by chance.

Thus, assuming that we have correctly identified the solar vector, the project to this point has identified (a) a strong periodic signal from redshifts within the search region for the 97 LSC spirals, (b) a periodic signal from the Virgo cluster, and (c) error–box coincidence between the solar galactocentric vector and the correcting vectors of both these samples. Each of these factors separately has chance probability $p \sim 10^{-4}$ or less, and so the hypothesis under test (redshift periodicity with respect to the galactocen-

TABLE 2.   Optimizing vectors derived by pattern-matching of peaks as compared with the galactocentric solar vector from Galactic modelling

Method	$V_\odot$ km s^{-1}	$l_\odot(°)$	$b_\odot(°)$
Peaks for LSC 97	210±7	96±3	–3±4
Peaks for Virgo 48	194±14	100±3	–
Galactic modelling	213±10	93±3	2±5

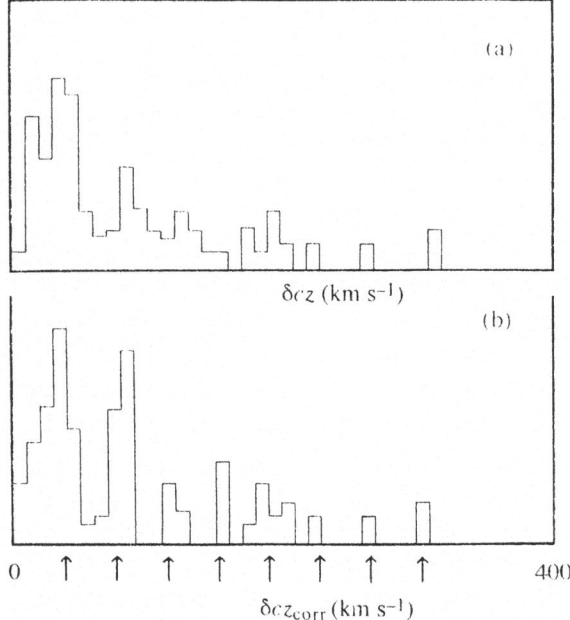

Figure 4.   Weighted differential redshifts for the 53 galaxies linked by group membership (histograms with binwidth 10 km s^{-1}). (a) Uncorrected. (b) After subtraction of the velocity component due to the galactocentric solar motion $\mathbf{V}_\odot$=(213 km s^{-1}, 93°, 2°). The vertical arrows mark a periodicity of 38 km s^{-1} with zero phase.

tric solar vector) is supported at an extremely high confidence level.

The systematic decline in $(R/N)_{max}$ nevertheless suggests that the periodic signal might be weakening with increasing distance from the Sun, or increasing separation between the galaxies. Conversely, it is reasonable to ask whether the quantization is stronger between adjacent galaxies. More than half of the galaxies in our sample of 97 LSC spirals belong to loose groups containing a few bright galaxies. Using the catalogue of groups (excluding binaries) compiled by Fouqué *et al.* (1992), we first examined the periodicity as a function of redshift accuracy and group membership for a

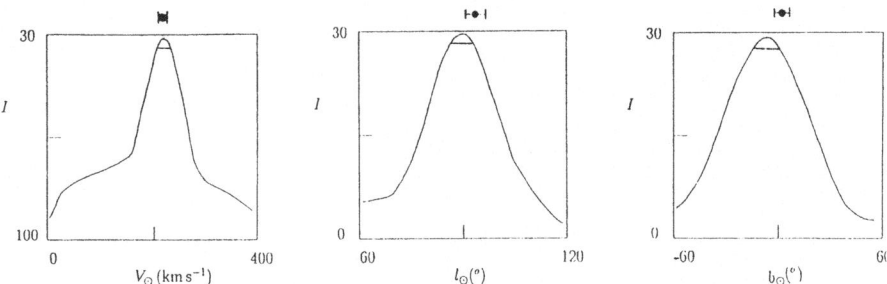

Figure 5. Distribution of power I around the optimum solar vector for the unweighted differential redshifts. Horizontal lines represent the width to half–power (in probability terms). The dots with error bars represent the $\mathbf{V}_\odot$ solution from Galactic HI data.

sample of 261 spirals. A complication is that there is a strong tendency for the more accurately measured galaxies to belong to groups (chance probability $p \sim 0.002$). Allowing for this, we found a weak tendency for the signal to be stronger in the more accurate data ($p \sim 0.043$), as would be expected for a real signal. We then constructed an exploratory sample of 115 spirals with 'more accurate' redshifts (the 97 above and 18 used by TC): there are 18 groups in this set, comprising 53 galaxies linked by group membership and yielding 59 differential redshifts in all within the groups. The weighted, uncorrected differential redshifts δcz reveal no periodicity (Fig. 4a). However, if the galactocentric solar vector (213 km s^{-1} 93°, 2°) is subtracted from the individual redshifts, a phase–zero periodicity of $\sim$38 km s^{-1} is clearly seen in the differential redshifts δcz_{corr} (Fig. 4b). Differential redshifts may, in principle, be used to determine $(\mathbf{V}_\odot, P)$ with less ambiguity from ghosts and harmonics but with some loss of accuracy. Applying PSA to the unweighted differential redshifts, and varying $(\mathbf{V}_\odot, P)$ over a wide range, a single, broad but clearly defined peak was found for the vector (208 km s^{-1} 90°, −4°) – see Fig. 5 – which again coincides with the galactocentric solar vector; the weighted differential redshifts yield a similar result. 10 000 sets of properly conditioned synthetic data failed to yield a signal of this strength and proximity to the TC periodicity, and the probability of a chance result is $\sim 4 \times 10^{-5}$. For the 53 individual (not differential) redshifts of the group–linked spirals, a grid search yielded $I_{max} \sim 41$ for P=37.8 km s^{-1} and $\mathbf{V}_\odot$= (216 km s^{-1} 93°, −13°), a remarkably strong signal for the size of sample. This suggests either that the redshifts of group–linked galaxies have been more accurately determined, or that the periodicity itself resides largely in group–linked galaxies.

To sum up, the above considerations appear to establish the existence of the redshift periodicity. However they also lead to the conjecture that the periodicity, even if global, may concentrate in galaxies belonging to groups. The latter represents a significant modification of the hypothesis under test which, in keeping with standard methodology, has to be tested against a further sample of galaxies.

4. A Further Sample of 117 Spirals

A further sample of LSC spirals with accurate redshift measurements was drawn from tables in Tifft & Cocke (1988) and Tifft (1990, 1992). All the data were obtained with the 300–foot Green Bank telescope. From a list of LSC galaxies for which HI profiles with signal–to–noise ratios $S/N > 10$ had been obtained we eliminated non–spirals and very late–type spirals ($T=8$ or 9), possible members of the Virgo cluster, members of our previous sample of 97 LSC spirals, and galaxies previously used by TC. Where there was overlap with the sample of 97, the agreement between redshifts was generally very good although there were a few spectacular discordances (the ability to determine systemic redshifts to within a few km s^{-1} in some spirals seems to arise from the steep fall in the HI profiles near the disc edge). The resulting independent sample of 117 LSC spirals matches the previous one of 97 with regard to distribution of HI linewidths and morphological type, but the mean galactocentric redshift (1511 km s^{-1}) is significantly higher than that for the previous sample (997 km s^{-1}). A further distinction is that whereas 40 of the 97 galaxies are group–linked, only 12 of the 117 are group–linked. Thus the galaxies in the new sample are much more widely separated than those in the previous sample.

We first tested whether the new sample of 117 spirals has a redshift periodicity consistent with that found for the previous sample of 97 spirals, i.e. having the same $(\mathbf{V}_\odot, P)$, but not necessarily the same strength, as the periodicity for the 97 spirals. For a given degree of quantization R/N, a linear increase of I–values with sample size N is expected. However, a fine–grid search for the combined value of 214 spirals (varying $V_\odot$ from 200 to 230 km s^{-1}, $l_\odot$ from 80° to 110°, $b_\odot$ from –20° to +10°, and P from 36 to 39 km s^{-1}) yielded a maximum I–value of only 27 (cf. $I_{max}=38$ for the 97 spirals alone), which is unremarkable considering the number of independent trials involved. Nevertheless for the sample of 117 spirals alone, a grid search for signals in the period range 20 to 200 km s^{-1} yielded a significant preponderance of peaks in the range 36.5 to 37.5 km s^{-1} and around 24 km s^{-1}; the I_{max} distribution was indistinguishable from those of datasets with $P=37.5$ km s^{-1} and $\sigma \simeq 9$ km s^{-1} (cf. $\sigma \simeq 7$ km s^{-1} for the 97 spirals). Thus, while a galactocentric $P \sim 37.5$ km s^{-1} may still be present in this more widely dispersed set, it is significantly weaker than that in the set of 97, and more work would be required to determine its precise confidence level, and whether it is consistent with the downward trend in $(R/N)_{max}$ discovered in the sample of 97. It cannot be excluded, for example, that the signal resides entirely in the groups within the sample.

In the same way we tested the modified hypothesis of redshift periodicity for group–linked spirals. Excluding the triplet in the Ursa Major cluster, there are 50 group–linked galaxies in the exploratory sample of 115 spirals. Combining this sample with the new one of 117 spirals, we obtained 30 additional group–linked spirals. (Ten of the 80 group–linked spirals had been used by TC, but not in the context of a discussion of periodicity in groups.) Using the same grid as before, we found I_{max}=48 for the 80 group–linked spirals, as against I_{max}=42 for the 50 group–linked spirals alone. From 200 grid searches in which the heliocentric redshifts of the 30 additional group–linked spirals were randomized, we found that the already strong periodic signal in the 50 galaxies would be enhanced with the 30 additional redshifts, by chance, with a probability of only $\sim 5 \times 10^{-4}$. Thus the group–linked galaxies in the new sample continue to enhance the signal already obtained from the group–linked galaxies in the earlier one. This is in contrast to the situation for the field galaxies as a whole, wherein the signal appears to decline with increasing separation.

To test whether this local periodicity is coherent in phase from one group to the next, each group was artificially shifted by adding the difference of two random numbers in the range 0 to 50 km s^{-1} to the redshifts of all the galaxies in the group. This is equivalent to shifting the groups by up to 200–400 kpc radially with respect to the Sun, thereby destroying any global periodicity within the LSC but preserving the internal periodicities. For each of 160 synthetic LSCs so constructed, a grid search was carried out as above. The distribution of the resulting 160 values of n_{20} is shown in Fig. 6, and it is clear that all the values of n_{20} are substantially less than that of 10 984 for the real LSC. According to the cumulative distributions of n_{15} and n_{20} for the 160 synthetic LSCs, the probability that the real groups, by chance, would be so placed as to give the illusion of a strong global phenomenon is of order 10^{-4}. These simulations therefore suggest that the periodicity is coherent over regions much larger than a Fouqué et al. group.

The grid search using the real data for the 80 group–linked spirals yields two remarkably strong peaks, whose periods and vectors are essentially identical to those which we found in our pilot sample of 89 field galaxies (Guthrie & Napier 1991), although only 33 galaxies are common to both samples. It is clear that in this case, unlike that where the entire sample of 97 was examined (Table 1), the solutions found in the earlier pilot study

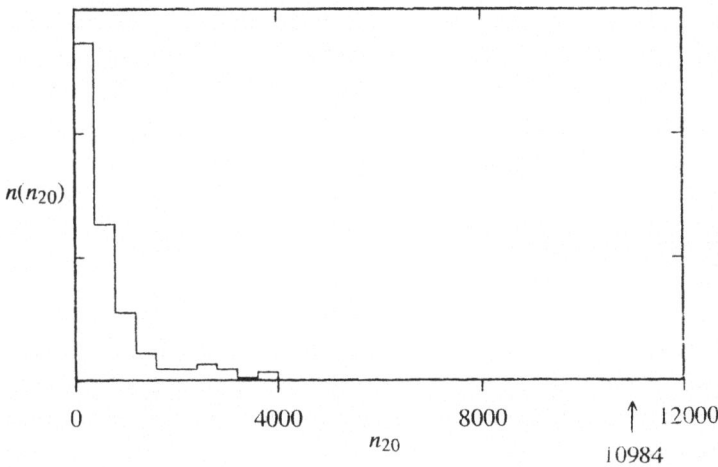

Figure 6. Relative n_{20} distribution from 160 grid searches on synthetic 'Local Super-clusters' constructed by randomly displacing the groups with the 80 linked galaxies while preserving their internal relative redshifts. $n_{20}=10\,984$ for the real LSC is shown. The periodicity is seen to be global (group to group) rather than a purely differential effect within groups.

TABLE 3. Continuity of solution: the two strongest peaks ($I=28$ and 29) found in the pilot study continue to increase in strength ($I=48$ and 42) as the region explored extends from $cz_{max}=1000$ km s^{-1} to 2600 km s^{-1}, while the corresponding vectors are maintained with remarkable stability.

I	P km s^{-1}	$V_{\odot}$ km s^{-1}	$l_{\odot}$ °	$b_{\odot}$ °
48	37.9	212	92	−13
28	37.5	212	94	−13
42	37.2	224	97	−4
29	37.2	228	99	−3

($cz_{max} \leq 1000$ km s^{-1}) continue to hold and strengthen. Our modified hypothesis, that the quantization exists but resides most strongly in the group–linked galaxies, is thus strongly supported by the further sample of 117 Green Bank redshifts (Table 3).

The 80 group–linked galaxies provide 90 differential redshifts. Applying the phase–zero constraint and using unweighted differential redshifts, we carried out a whole–sky grid search for periodicity peaks for solar speeds 150 to 300 km s^{-1} and periods 20 to 100 km s^{-1}. The highest peak found occurs at $P \sim 38$ km s^{-1}, for a vector $\mathbf{V}_\odot = (215 \pm 5$ km s^{-1}, $90° \pm 5$, $0° \pm 5)$ consistent with the galactocentric one. There is also a triplet of high peaks at $(P \sim 38.4$ km s$^{-1})$, but far from the galactocentric vector.

5. Concluding Remarks

A study of this type is prone to errors due to faulty statistical procedures, observational selection effects, and instrumental and other artefacts. We have tried to avoid these problems by using standard procedures and methodology, by adopting simple criteria in the selection of samples, and by checking that the periodicity is revealed independently by the major radio telescopes. The effect has survived several tests for robustness and behaves in every respect tested like a real phenomenon. There are two arguments to indicate that the periodicity is physically real rather than some obscure artefact: first, it is galactocentric rather than observer–centric; second, the original claims for quantization were based on optical data, and it is hardly likely that an artefact would span different telescopes, data selection and reduction procedures to yield an identical, spurious (galactocentric!) periodicity.

Although PSA is a standard and widely used technique, Newman & Terzian (these proceedings) have reiterated that the 'power' I of the periodogram is a biased, inconsistent and slowly convergent statistic, and that the high tail of the I-distribution departs from an exponential distribution (cf Webster 1995). However their comments are largely irrelevant in the present context. Firstly, we employed Monte Carlo simulations *explicitly* to circumvent these well-known limitations of PSA (Guthrie & Napier 1996). Thus (for example) no assumptions had to be made about the asymptotic form of the I-distribution. Further, to the extent that the statistic I is (say) prone to bias in the real dataset, it is in equal measure prone to bias in the synthetic datasets. Since the latter were constructed so as to be identical to the real set in all respects except for the periodicity under test, they therefore provide a proper basis for a *comparative* assessment of the observed power. Further, these comparisons were made using not I associated with isolated high peaks but aggregates of I-values over velocity space, so providing a robust measure and avoiding extreme value statistics. It remains true that the quantitative assessment of confidence limits from these trials involves a large extrapolation (e.g. Fig. 2) since the extraordinarily high peaks ($I \sim 40$–50) found in the real datasets examined have never been attained in the tens of thousands of simulated ones generated in the course of this study. However in that situation one is already extrapolating from high improbability. In general, although the theoretical concerns expressed by Newman & Terzian are very proper, they have nowadays largely been superseded by the widespread use of purely numerical techniques. Secondly, Newman & Terzian also fail to address the strongly galactocentric nature of the signals in the datasets examined so far: the coincidence has probability of order 10^{-4} and can hardly be ascribed to limitations in PSA! Finally, we note that the quantization is ultimately a matter of observation rather than statistics: in every high-precision dataset examined so far, it is easily seen by eye.

It is not yet clear that the phenomenon is a simple global periodicity applying to all galaxies; instead, it may hold only on a regional basis and possibly only for luminous galaxies. Our samples (the Virgo cluster and nearby field galaxies) covered limited regions, and the 97 LSC spirals contain a high proportion of group–linked galaxies. Also, we treated the bright spirals and dwarf irregulars separately in our pilot studies, and the sample of LSC spirals is strongly biased towards nearby and high–luminosity spirals. Our trials on synthetic datasets also revealed that spurious periodicities, often close to harmonics of the one 'fed in', arise in directions far from the original vector. This complicates the issue of determining a unique solution and may be relevant to the current claims of sub–periodicities, COBE–based corrections and so on: in our view these latter issues are still open.

To sum up, our study constitutes by far the most severe test yet applied to the quantized redshift claims, and has involved the analysis of over 250 new, accurately measured redshifts. At a very high confidence level, we confirm the existence of a strong, consistent, easily seen galactocentric redshift periodicity of ~ 37.5 km s^{-1}. The quantized redshift claims do not sit comfortably with standard, and successful, cosmological theories, and have been largely ignored for twenty years. However high–precision redshift data have been accumulating rapidly over the last decade, and we may now be reaching the stage where the issue is forced by the weight of these new observations.

References

Arp, H. & Sulentic, J. W.: 1985, *ApJ*, **291**, 88
Baiesi-Pillastrini, G. C. & Palumbo, G. G. C.: 1986, *A&A*, **163**, 1
Bottinelli, L., Gouguenheim, L., Fouqué, P. & Paturel, G.: 1990, *A&A Suppl.*, **82**, 391 (BGFP)
Fouqué, P., Gourgoulhon, E., Charmaraux, P. & Paturel, G., 1992. *A&A Suppl.*, **93**, 211
Guthrie, B. N. G. & Napier, W. M.: 1990, *MNRAS*, **243**, 431
Guthrie, B. N. G. & Napier, W. M.: 1991, *MNRAS*, **253**, 533
Guthrie, B. N. G. & Napier, W. M.: 1996, *A&A*, **310**, 353

Merrifield, M. R.: 1992, *AJ*, **103**, 1552
Thomson, D. J.: 1990, *Phil. Trans. R. Soc. London A*, **330**, 601
Tifft, W. G.: 1976, *ApJ*, **206**, 38
Tifft, W. G.: 1977, *ApJ*, **211**, 31
Tifft, W. G.: 1980, *ApJ*, **236**, 70
Tifft, W. G.: 1990, *ApJ Suppl.*, **73**, 603
Tifft, W. G.: 1992, *ApJ Suppl.*, **79**, 183
Tifft, W. G. & Cocke, W. J.: 1984, *ApJ*, **287**, 492 (TC)
Tifft, W. G. & Cocke, W. J.: 1988, *ApJ Suppl.*, **67**, 1
Webster, A. S.: 1995, *A&A Suppl.*, **114**, 191

POWER SPECTRUM ANALYSIS AND REDSHIFT DATA

The Statistics of Small Numbers

W. I. NEWMAN

Departments of Earth and Space Sciences,
Physics and Astronomy, and Mathematics
University of California, Los Angeles, CA 90095-1567

AND

Y. TERZIAN

Department of Astronomy, and
National Astronomy and Ionosphere Center
Cornell University, Ithaca, NY 14853-6801

1. Introduction

We provide a formal mathematical analysis of the "Power Spectrum Analysis" (PSA) method by Yu and Peebles (1969), including illustrative controlled numerical experiments, to better understand their properties. The PSA method generates a sequence of random numbers from observational data which, it was claimed, is exponentially distributed with unit mean and unit variance. Although the derived variable may be reasonably described by an exponential distribution over much of its range, the tail of the distribution is far removed from that of an exponential, thereby rendering statistical inference and confidence testing based on the tail of the distribution completely unreliable. We show that a recently constructed method due to Guthrie and Napier (1996) is formally equivalent, and offers no new insights.

Since astronomers often employ descriptive approaches in their analysis of data, we also review some salient issues relating to the preparation of frequency histograms, particularly with respect to the "bin width" (properly, the frequency class interval). We show that common usage in the astronomy community violates the general principles established by statisticians and, accordingly, can lead to incorrect inferences as to the possible existence of pattern in the underlying data.

Astrophysics and Space Science **244:**127-141,1996.
© 1996 *Kluwer Academic Publishers.*

We begin this paper by providing a review of each of the assumptions made in the Power Spectrum Analysis method and critically examine each of these, showing that each of these is formally incorrect. Although widely believed to be "analytic" in its derivation, the PSA combines a set of approximations, some of which are simply not correct, with the remaining ones being very weak. We provide illustrations of the underlying mathematics to make these points clear to the reader. Indeed, this method was originally developed in the statistics literature by Bartlett (1963), albeit this seems to have been unknown to astronomers, and was later discarded due to its many deficiencies (Bartlett, 1978). The essential issue here is that this method cannot be used for statistical inference in its present form. Similarly, the issue of descriptive statistics and its application to redshift data, among other problems, require significant refinement if such visual representations are to have any meaning. We will also describe some recent developments in statistics that could be exploited by those seeking to develop valid tests for periodicity in redshift data.

2. Power Spectral Analysis Method

The power spectrum analysis (PSA) method of Yu and Peebles (1969) is widely employed in the analysis of redshift data, and is widely thought to be analytic and *exact*. The PSA method makes several assumptions which are wrong plus several approximations which are known to be poor among mathematicians (especially the weak convergence of the Central Limit Theorem and its inapplicability in the statistical regime under consideration). These issues were described by Newman, Haynes, and Terzian (1989, 1994), but requires significant elaboration as the PSA method continues to be used in the astronomical community—see Cocke and Tifft (1996), Guthrie and Napier (1996), and Tifft (1996)—owing to a lack of appreciation for some of the mathematical underpinnings of the method and its limitations.

Although our focus here will be largely mathematical, there are two other kinds of consideration that must be given in the analysis data. First, redshift data from clusters is subject to contamination by sources that overlap the cluster that is being examined. Second, redshift data is inherently noisy and an appropriate model for the noise—and a rigorous quantitative descriptor for it (e.g. its second moment or "error bars")—must be specified. These two physically-motivated factors play an important role in the redshift problem, yet are often neglected in the analysis of data. The focus of this paper remains the mathematical problems associated with the PSA method. Although the efforts made by Yu and Peebles were admirable, their method is nevertheless ill-suited for this purpose and the conclusions derived from it are generally invalid.

Following Yu and Peebles (1969), consider N points x_j distributed in the interval 0 to 2π, and let (where n is an integer)

$$``z_n = N^{-1/2} \sum_{j=1}^{N} e^{inx_j} . \; [1]"$$

(We use square brackets to identify Yu and Peebles' equation numbers.) They suggested that we may regard the z_n as an ancillary series that has the appearance of a Fourier transform. Further, they suggested that if the measured data x_j were clustered, particularly around uniformly spaced points separated by $\Delta x < 2\pi$, then z_n would be large when $n \approx (\Delta x)^{-1}$. (Note, as mentioned above, that this description completely ignores the role of noise. Issues of contamination or "censorship" of the data are also ignored.) In their analysis, the variable n has the role of a "frequency" and, in that sense, the power spectrum for the distribution of points is $|z_n|^2$, $n = 1, 2, \ldots$. Now, we proceed to look at each of the various incorrect and weak assumptions in the PSA method.

2.1. CLAIM #1

Yu and Peebles went on to say

"... if the points x_j are distributed at random in the interval, the ensemble average of z_n (when $n \neq 0$) is

$$\langle z_n \rangle = N^{-1/2} \sum \left\langle e^{inx_j} \right\rangle = N^{-1/2} \sum \int_0^{2\pi} \frac{dx_j}{2\pi} e^{inx_j} = 0 . \; [2]"$$

Correct result: Suppose the x_j are identically distributed and independent deviates (i.i.d.) with some distribution $\mathcal{P}(x)$. As a note for non-specialists, we use the (cumulative) distribution in contrast with its derivative, namely the probability density distribution, since the latter is not always well-posed (particularly if Dirac δ-functions are involved). We require that $\mathcal{P}(x)$ be non-decreasing, with $\mathcal{P}(-\infty) = 0$ and $\mathcal{P}(\infty) = 1$. Then we define the *characteristic function* or generator $\mathcal{F}(n)$ according to

$$\mathcal{F}(n) = \left\langle e^{inx} \right\rangle = \int_{-\infty}^{\infty} e^{inx} d\mathcal{P}(x) . \tag{1}$$

Then, it follows that

$$\langle z_n \rangle = N^{1/2} \mathcal{F}(n) . \tag{2}$$

The assumption that $\langle z_n \rangle$ vanishes is a common error made by physicists, sometimes called the "random phase approximation" in statistical mechanics. The essential point here is that the word "random" necessarily refers to some statistical distribution function. Implicit to this claim by Yu and Peebles is the assumption that the distribution is *uniform* over the interval from 0 to 2π; it is easy to prove that no other distribution function will produce these results. There generally is a sensitivity to the underlying distribution function $\mathcal{P}(x)$, e.g. if x is normally distributed (i.e. Gaussian) with a mean μ and a variance σ^2, then Eq. (1) shows that

$$\langle z_n \rangle = N^{1/2} \exp\left(-\frac{n^2\sigma^2}{2}\right) . \tag{3}$$

This example is particularly significant since many astronomers believe that redshifts within a cluster have relaxed to a Maxwellian or Gaussian distribution.

2.2. CLAIM #2

In the same spirit, Yu and Peebles went on to say

"Similarly, the ensemble average value of the square of the absolute value of z_n is for a random distribution,

$$\left\langle |z_n|^2 \right\rangle = \frac{1}{N}\sum_j \left\langle \left|e^{inx_j}\right|^2 \right\rangle + \frac{1}{N}\sum_{k\neq j} \left\langle e^{in(x_k - x_j)} \right\rangle = 1 . \text{ [3]}"$$

Correct result: Employing the notation described earlier, we obtain that

$$\left\langle |z_n|^2 \right\rangle = 1 + (N-1)\,\mathcal{F}(n)\,\mathcal{F}^*(n) \geq 1 . \tag{4}$$

In the case of normally distributed x, we find that

$$\left\langle |z_n|^2 \right\rangle = 1 + (N-1)\,e^{-n^2\sigma^2} . \tag{5}$$

Again, we are witnessing the implicit assumption built into Yu and Peebles methodology of a *uniform distribution* in x. As it happens, there are applications of statistics where uniform probability distributions for random variables are to be expected, e.g. in isotropic environments in two dimensions where all angles are expected. This was the basis for Mardia (1972) developing a test similar to that of Bartlett and to Yu and Peebles, where the test was explicitly designed to test for uniformity.

Interestingly, the quantity $|z_n|^2$ is intimately connected to the two-point redshift correlation function that can be constructed for a cluster. In particular, we note that $\mathcal{F}(n)\,\mathcal{F}^*(n)$ is the spectrum since $\mathcal{F}(n)$ is by Eq. (1) just the Fourier transform of the redshift distribution. Accordingly, the use of the redshift correlation function, such as that employed by Guthrie and Napier (1996), does not provide any information that is not already contained in $|z_n|^2$.

2.3. CLAIM #3

Yu and Peebles also argue

"When N is large, and the points are distributed at random, the real and imaginary parts of z_n will have approximately normal distributions. Furthermore, the real and imaginary parts of z_n will be statistically independent."

There are two problems implicit to this statement. Statistical independence is not assured unless the distribution $\mathcal{P}(x)$ is uniform, as we show below. Let us define a quantity ζ_n according to

$$\zeta_n = \langle \Re(z_n) \Im(z_n) \rangle = \frac{1}{N} \sum_{i,j=1}^{N} \langle \cos(nx_i) \sin(nx_j) \rangle \ . \qquad (6)$$

For the second part of Yu and Peebles statement to be valid, we would require that ζ_n be identically zero. However, it immediately follows that

$$\zeta_n = \frac{1}{2N} \sum_{i=1}^{N} \langle \sin(2nx_i) \rangle + \frac{1}{N} \sum_{j \neq k} \langle \cos(nx_j) \rangle \langle \sin(nx_k) \rangle \qquad (7)$$

$$= \frac{1}{2} \Im[\mathcal{F}(2n)] + \frac{(N-1)}{2} \Im\left[\mathcal{F}^2(n)\right] \ . \qquad (8)$$

In general, this quantity will not vanish unless the distribution is symmetric about the origin; any skew in the distribution will guaranty that the imaginary part of the characteristic function does not vanish.

The issue emerging from the first part of their statement is very subtle, and emerges directly from the Central Limit Theorem. Suppose that $x_j, j = 1, ..., N$ are i.i.d. random variables described by some distribution function $\mathcal{P}(x)$, and assume that the mean μ and variance σ^2 defined by

$$\mu = \int_{-\infty}^{\infty} x \, d\mathcal{P}(x) \ \text{and} \ \sigma^2 = \int_{-\infty}^{\infty} (x-\mu)^2 \, d\mathcal{P}(x) \qquad (9)$$

both exist (i.e. are finite). (Interestingly, this condition is not met by all probability distribution functions encountered in astrophysics, e.g. the Lorentz distribution has an infinite variance.)

2.3.1. *Central Limit Theorem*

Consider now the random variable y defined by

$$y = \frac{1}{\sqrt{N}} \sum_{j=1}^{N} \frac{x_j - \mu}{\sigma} \ ; \qquad (10)$$

it is straightforward to show that y has a mean of zero and a variance of unity. Because of its importance to our discussion, we now briefly sketch a proof of the Central Limit Theorem by defining the characteristic function $\mathcal{F}_y(n)$ for the distribution of y, analogous to $\mathcal{F}(n)$ in Eq. (1), but where n is not restricted to being an integer, namely

$$\mathcal{F}_y(n) = \left\langle e^{iny} \right\rangle = \left\langle e^{in\frac{1}{\sqrt{N}}\sum_{j=1}^{N}\frac{x_j-\mu}{\sigma}} \right\rangle = \left\langle \Pi_{j=1}^{N}\exp\left(in\frac{x_j-\mu}{\sqrt{N}\sigma}\right) \right\rangle . \quad (11)$$

[More rigorous proofs of the Central Limit Theorem, and its generalization the Feller-Lindeberg Theorem, can be found in Volume II of Feller (1968).] Assuming that the x_j are i.i.d., the latter can be expressed as a product of individual terms, namely

$$\mathcal{F}_y(n) = \left\langle \exp\left(in\frac{x-\mu}{\sqrt{N}\sigma}\right) \right\rangle^N = \left\{ \int_{-\infty}^{\infty}\exp\left(in\frac{x-\mu}{\sqrt{N}\sigma}\right) dP(x) \right\}^N . \quad (12)$$

Assuming for fixed $x-\mu$ that N is taken to be arbitrarily large, we can expand the exponential as a Taylor series, and we obtain

$$\mathcal{F}_y(n) = \left\{ \int_{-\infty}^{\infty}\left[1 + in\left(\frac{x-\mu}{\sqrt{N}\sigma}\right) - \frac{n^2(x-\mu)^2}{2!N\sigma^2} + \mathcal{O}\left(n^3N^{-3/2}\right)\right] dP(x) \right\}^N .$$
$$(13)$$

Here, we employ $\mathcal{O}(...)$ to denote the *order* of the remainder term, i.e. the error in the approximation. We now introduce Eq. (9) defining the mean and variance to obtain

$$\mathcal{F}_y(n) = \left[1 - \frac{n^2}{2!N} + \mathcal{O}\left(n^3N^{-3/2}\right)\right]^N = \left[1 - \frac{n^2}{2!N}\right]^N + \mathcal{O}\left(n^3N^{-1/2}\right) ,$$
$$(14)$$

where we have exploited the Binomial Theorem to obtain the remainder term $\propto n^3N^{-1/2}$. [Statisticians employ a more sophisticated mode for describing the convergence, in the sense of distributions, but this sketch will suffice for our purpose. Guthrie and Napier (1996) claim without proof or citing references that the convergence rate varies as $\ln(N)/N$; their claim, as we have just shown, is incorrect.] What is particularly significant here is that the error term diminishes very slowly with respect to N (the number of datum) and can increase rapidly with respect to n (the "wavenumber"). To conclude our sketch, recall that

$$\lim_{N\to\infty}\left[1 - \frac{\alpha}{N}\right]^N = \exp(-\alpha) , \quad (15)$$

so that

$$\mathcal{F}_y (n) = \exp \left(-\frac{n^2}{2} \right) + \mathcal{O} \left(N^{-1/2} \right) . \tag{16}$$

Since the Fourier transform of a Gaussian is a Gaussian, this shows, in the limit of fixed y and N approaching infinity, that the newly constructed variable is normally distributed.

This is not the situation at hand with redshift data—we need a methodology predicated on fixed and finite N (typically $100 - 300$) where y is very large. Cramér (1938) extended the Central Limit Theorem for y "large" but with $N \to \infty$. This topic has received significant attention in the mathematics community—see, for example, Borovkov (1985)—but the issue relevant to the present discussion with N fixed and finite and y becoming very large is only now being scrutinized (Frisch and Sornette, 1996) in applications to statistical mechanics. The weak convergence of the Central Limit Theorem is further complicated by the first two incorrect claims; the second of these can be remedied in part by the use of the Feller-Lindeberg Theorem (Feller, 1968) instead, but the problem of weak convergence remains.

2.3.2. *Weak Convergence to Normality*

In order to make these abstract ideas more concrete, it is useful to consider a methodology once employed by IBM in generating Gaussian random numbers on its mainframe computers, the so-called "sum of uniform deviates method." Suppose we have a set i.i.d. random variables U_i, $i = 1, ..., N$ that are uniformly distributed on the interval $[0, 1]$. Then, it is easy to show that the ancillary variable X defined by

$$X = \sqrt{\frac{12}{N}} \left\{ \sum_{i=1}^{N} U_i - \frac{N}{2} \right\} \tag{17}$$

is asymptotically normal with zero mean and unit variance; the case of $N = 12$ is particularly trivial to implement. The maximum errors in the normal deviate (Abramowitz and Stegun, 1965) are 0.009 for $|X| < 2$, but rise to 0.9 for $2 < |X| < 3$, and become dramatically worse as we proceed into the tails of the distribution.

In Fig. 1, we illustrate this by showing a "probability plot" using IBM's old algorithm for normal distribution generation; we employ 128 random numbers U generated in this way, a quantity typical of astronomical data sets. In Fig. 2, we show the same type of plot where we employ 1024 random numbers U. This helps illustrate that, for fixed U, the distribution function becomes progressively closer to a Gaussian as N is increased. However, we see that, for fixed N, the distribution function becomes progressively further from a Gaussian as U is increased.

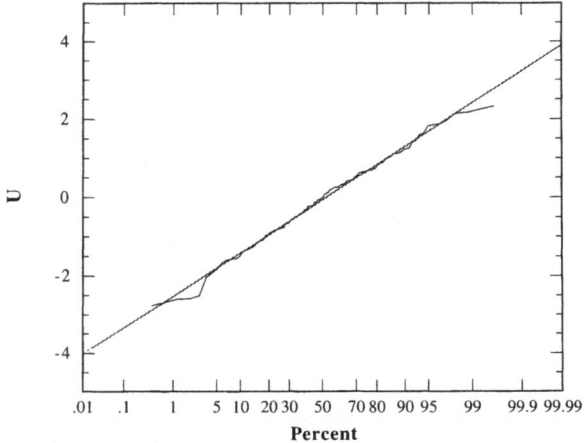

Figure 1. Probability plot employing 128 samples.

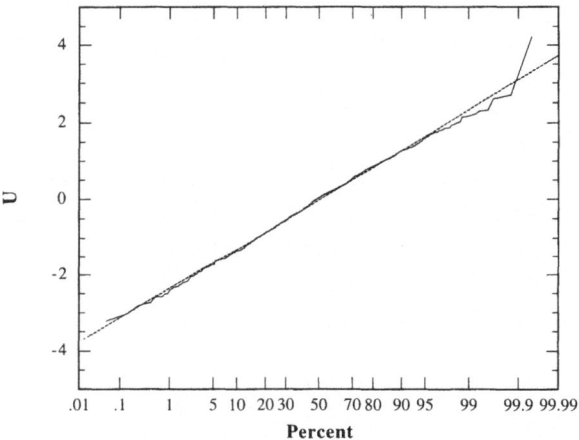

Figure 2. Probability plot employing 1024 samples.

2.4. CLAIM # 4

Yu and Peebles elaborate by saying

"Since $|z_n|^2$ is the sum of the squares of two independent variables, each normally distributed, $|z_n|^2$ must have an exponential distribution with the width fixed by equation [3]. Thus we conclude that, when N is large, and the points x_j are distributed at random in the interval $0 \leq x \leq 2\pi$, the a priori probability for finding a value of $|z_n|^2$ greater than x is

$$P\left(|z_n|^2 > z\right) = e^{-x} \text{ (random). [5]"}$$

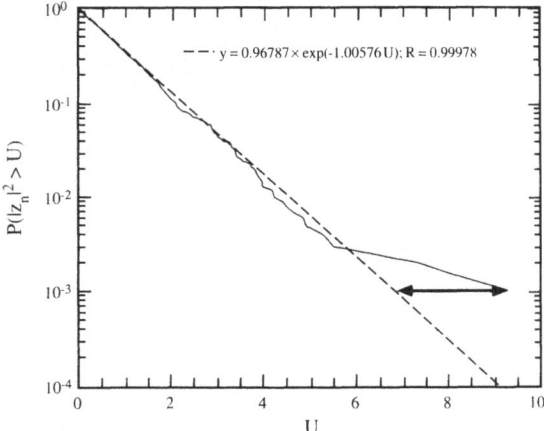

Figure 3. Demonstration of global quality of exponential fit, which degenerates in the tail of the distribution.

Correct claim: $|z_n|^2$ has an asymptotically exponential distribution provided that the bias in the mean (claim # 1) and the variance (claim # 2) are properly accommodated. Nevertheless, the basic problem remains that the tail of the distribution for $N < \infty$.

In Fig. 3, we provide an illustration of this problem where we employ 128 approximately Gaussian-distributed samples to derive the $|z_n|^2$ variables whose distribution we plot. We observe that the plot does very well so long as $U \leq 1$ and begins to deteriorate showing a dramatic departure from our exponential expectation in its tail. (The inset equation is provided to show how close the agreement is with Yu and Peebles' Eq. [5] for small values of $|z_n|^2$ which dominates the exponential curve fit.) To reiterate, the point here is that the Central Limit Theorem approximation is not valid in the tail of the distribution for finite N; a methodology based on the theorems of Cramér (1938) or of Frisch and Sornette (1996) is required. *Any* utilization of the tail of the $|z_n|^2$ distribution is particularly vulnerable and is in no way robust.

2.5. CLAIM #5

Yu and Peebles then comment that:

> "It is readily seen ... that the coefficients z_n are statistically independent in the sense that the ensemble average $\langle z_n z_m \rangle$ vanishes when $n \neq m$."

Correct claim: A quick calculation shows that

$$\langle z_n z_m \rangle = \mathcal{F}(n-m) + (N-1)\, \mathcal{F}(n)\, \mathcal{F}^*(m) \neq 0 . \tag{18}$$

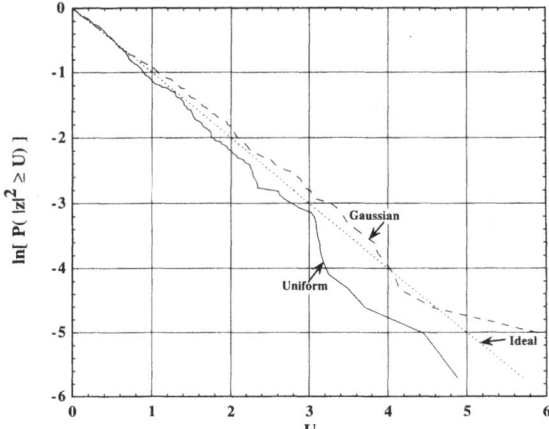

Figure 4. Further comparisons using Yu and Peebles' methodology with underlying uniform and normal distributions for artificial "redshift" data.

This result again has profound implications on the quality of the *approximation* Yu and Peebles had derives for the exponential distribution, particularly in its tail. The outcome, once again, is that we must remain suspicious of any quantities computed from the estimated probabilities in the tail of the distribution.

In Fig. 4, we elaborate on the theme of the departure of the tail of the distribution of the $|z_n|^2$ from that of an ideal exponential distribution. We again create artificial redshift data, employing both Gaussian *and* uniform deviates. As before, we observe that the closeness of the fit to the exponential distribution is very good for small deviates. An important lesson that we must derive from the formal analysis and illustrated by these figures, which are representative of what occurs with "ideal" artificial data, is that *the failure of the various assumptions implicit to the Power Spectrum Analysis method renders inconclusive any conclusions generated from this method from the tail of the distribution.*

2.6. CLAIM #6

Users of the PSA method generally assume that

$$P_N(x) = 1 - [1 - \mathcal{P}(x)]^N$$

describes the probability that the *extreme* value of z_n so obtained is consistent with random data.

Correct claim: If Yu and Peebles' Eq. [5] were strictly correct, it would be highly improbable ($< 0.1\%$) that excursions of 5 or more in the power spectral amplitude could occur. In Fig. 5, we show the power spectra associated

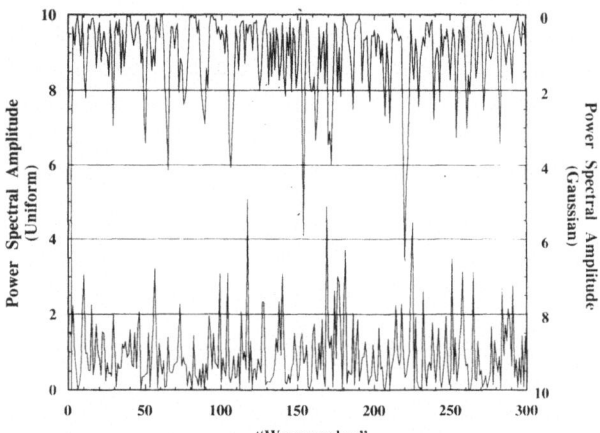

Figure 5. Power spectrum analysis plots for uniformly distributed data over the interval $[0, 2\pi]$ and for normally distributed data with unit variance and a mean of π. The first point on this plot is off scale with an amplitude of 106.41, an artifact of incorrect claim #1.

with the data described in Fig. 4. (Note that the bottom and right axes are used for uniformly distributed data, while the upper and left axes are used for normally distributed data.) In viewing these figures, we are looking at the outcome of a *set* of experiments as we are identifying the extremum of a distribution of individual spectral estimates. We must introduce the "statistical theory of extremes" (Galambos, 1978) to quantitatively assess the significance of these figures.

The maximum value x that can emerge from a set of experiments is described by a probability distribution that is very different from the distribution $\mathcal{P}(x)$ given in Yu and Peebles' Eq. [5]. In particular, if $\mathcal{P}(x)$ is the probability that in a given experiment the observed value $\geq x$, then $1 - \mathcal{P}(x)$ is the probability that in a given experiment the observed value $\leq x$. If we have N *independent* experiments, the probability that none of the observed results exceeds x is $[1 - \mathcal{P}(x)]^N$. Finally, the probability $P_N(x)$ that at least one of the observed results exceeds N-independent results exceeds x is given by

$$P_N(x) = 1 - [1 - \mathcal{P}(x)]^N \ . \tag{19}$$

While this result is rigorous and general for independent deviates, two fundamental problems emerge in this application. As already shown, the $|z_n|^2$ deviates are *not* statistically independent (since their correlation function does not vanish), making the number of independent experiments very difficult to assess—it could be very different (i.e. smaller) than N. Even more important, the strong nonlinearity in this expression can amplify the error in $\mathcal{P}(x)$ (which we have already observed in Fig. 3 and Fig. 4) by orders of magnitude. See Scott (1991) for a discussion of this equation and its history in an astronomical context; Newman, Haynes, and Terzian (1994) give other analytic examples showing how extreme value statistics can give erroneous estimates of probabilities. The magnification in the error in $\mathcal{P}(x)$ is evident from the Binomial Theorem which shows that the latter expression gives

$$P_N(x) \approx N\mathcal{P}(x) \ ; \tag{20}$$

moreover, this sensitivity is also shown in the strict bound

$$P_N(x) = 1 - [1 - \mathcal{P}(x)]^N > 1 - \exp[-N\mathcal{P}(x)] \ . \tag{21}$$

Here, it is profoundly evident that any uncertainty or error in N or in $\mathcal{P}(x)$ will grossly affect the estimate.

To conclude this section on the Power Spectrum Analysis method, we recall its origins as an important attempt to identify and assess the statistical significant of clustering. However, we have shown that the PSA is analytically flawed in its assumptions and usage and does not provide for reliable hypothesis testing. We have pointed to some recent developments in statistics and probability theory (notably due to Cramér and to Frisch and Sornette) that might help in eliminating this dilemma. For those astronomers who believe that clustering is present in redshift data, it is incumbent upon them to use more appropriate assumptions in order to develop a mathematically rigorous scheme for hypothesis testing.

3. Frequency Histogram Analysis

In many instances, astronomers continue to employ descriptive statistics to present their data as claimed evidence for clustering. The principal tool used is the frequency histogram, whose appearance can vary dramatically according to user's selection of class intervals (or, more commonly in astronomy, bin size). Two sets of "empirical" rules have emerged in the statistics community for producing histograms. Sturges' (1926) rule together with later refinements—see Newman, Haynes, and Terzian (1989, 1991)—recommends that "the optimal number of classes is $1 + \log_2(N)$" so that a binomial sequence (which is essentially normal) is not subdivided into intervals containing fewer than one sample. Freedman and Diaconis (1981a,b) devised a criterion $\propto N^{1/3}$, although it was not meant to be used as an "algorithm" for class interval selection. Typical astronomical data sets contain $N = \mathcal{O}(10^2 - 10^3)$ and should contain no more than ten frequency intervals. Otherwise, anomalous pattern (with many pedagogical examples being given in the statistics literature) emerges. In Fig. 6, we show the frequency histogram for an artificial normally distributed data set with 300

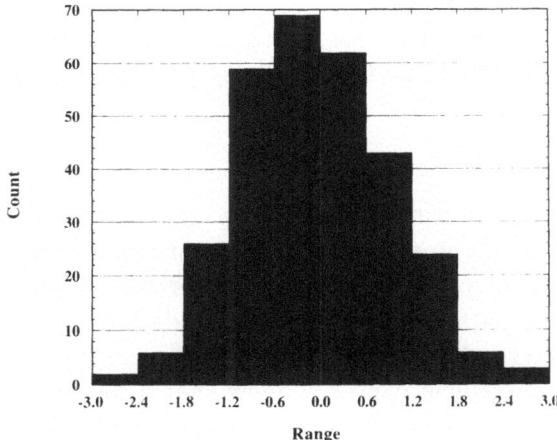

Figure 6. Frequency histogram for 300 Gaussian random deviates with 10 class intervals (following Sturges' empirical rule).

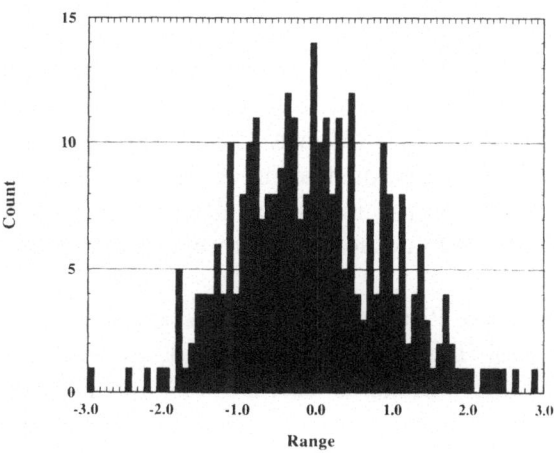

Figure 7. Same as Fig. 6, but with 72 class intervals (following prevailing astronomical usage).

points employing 10 class intervals. This figure appears rather boring; there is no visible asymmetry (allowing for the statistics of small numbers that must follow for each of the class intervals). In Fig. 7, we plot the same data employing 72 class intervals, so that the bin size is comparable to that employed in the astronomical literature. Moreover, we have selected the center and width of the class intervals to provide a seeming pattern of alternating boxes in the diagram, a feature evident in many claimed empirical "proofs of clustering." Any apparent pattern consistent with clustering in this latter diagram is completely illusory and an artifice of selecting the "binning" to maximize the picket-fence appearance—the data employed are from a random number generator which has been guaranteed by its creators (`gasdev` from Press et al., 1992) to be free from any periodicities or correlations.

4. Epilogue

The potential for clustering in cosmology remains a possibly important problem. Since the outcome of clustering, particularly what has been called "quantization" might possibly require an overhaul of the laws of physics, it is essential that particularly robust methods for testing models be developed by observers. A fundamental dilemma emerges from the principles of hypothesis testing—you can only show that a model is *not* satisfied at some confidence level. However, this requires that the *mathematical* model and method be correct in all details, and not require that any asymptotic conditions be met.

We have shown above, adding to work we have presented elsewhere (Newman, Haynes, and Terzian, 1989 and 1994), that the widely used Power Spectrum Analysis method is analytically flawed and inappropriate to the type of hypothesis testing where it is commonly applied. While the effort made by Yu and Peebles in constructing this method was admirable (also that made by Bartlett in the statistical literature), the methodology nevertheless is unreliable. Our demonstration of the problems faced by the PSA is *analytic*; we have employed numerical illustrations solely as a device to demonstrate how the various anomalies present in the technique can present itself.

Cocke, DeVito, and Pitucco (1996, these proceedings) also discuss the statistical validity of Tifft's periodicities and, in addition, they perform some Monte-Carlo simulations. Cocke, DeVito, and Pitucco conclude that, even though Tifft's procedures may be criticized, the claimed periodicities appear to have statistical significance. However, we note that the data sample employed is too small and restrictive to arrive at any conclusions. Indeed Cocke, DeVito, and Pitucco (1996) also state "situation is exacerbated by the complex nature of the RSP's (redshift periodicities)."

A key ingredient for rectifying this problem is the use of methods predicated on a different set of assumptions, namely that the number of data N is fixed and finite and that the excursions in some variable are large. This is exactly the opposite, as we have shown above, of the assumptions implicit to the Central Limit Theorem which played a pivotal role in the Yu and Peebles' analysis. Concomitant with this constraint is that the sample size, in many seemingly anomalous situations, may be too small to permit reliable hypothesis testing.

While cosmologists continue to advance the frontiers of science, it is essential that they work closely with statisticians in developing robust tests that genuinely capture the essence of the phenomena they believe reside in the data.

5. Acknowledgments

We wish to thank E.E. Salpeter for some useful discussions. One of us (W.I.N.) was partly supported by NASA grant NAGW 3132 to the University of California, Los Angeles. Part of this work (Y.T.) was supported by the National Astronomy and Ionosphere Center, which is operated by Cornell University under a cooperative agreement with the National Science Foundation.

References

Bartlett, M. S.: 1963, *J. R. Statist. Soc. B*, **25**, 264
Bartlett, M. S.: 1978, *Introduction to Stochastic Processes*, 3rd ed, Cambridge: Cambridge Univ Press
Borovkov, A. A. ed.: 1985, *Limit Theorems for Sums of Random Variables*, New York: Springer-Verlag
Cocke, W. J., DeVito, C. L., Pitucco, A.: 1996, *Astrophys. Space Sci.*, to appear
Cocke, W. J. and Tifft, W. G.: 1996, *Astrophys. Space Sci.*, **239**, 35
Cramér, H.: 1938, *Actualités Scientifiques et Industrielles*, **736**, pp. 5–23.
Feller, W.: 1968, *An Introduction to Probability Theory and Its Applications*, **Vol. 1 and 2**. New York: Wiley
Freedman, D. and Diaconis, P.: 1981a, *Z. Wahrscheinlichkeitstheorie verw. Gebiete*, **57**, pp. 453–476
Freedman, D. and Diaconis, P.: 1981b, *Z. Wahrscheinlichkeitstheorie verw. Gebiete*, **58**, pp. 139–167
Frisch, U. and Sornette, D.: 1996, 'Stretched Exponentials from Multiplicative Cascades', in preparation
Guthrie, B. N. G. and Napier, W. M.: 1996, *Astron. Astrophys.*, **310**, 353
Mardia, K.V.: 1972, *Statistics of Directional Data*, New York: Academic Press
Newman, W. I., Haynes, M. P., and Terzian, Y.: 1989, *Ap. J.*, **344**, 111
Newman, W. I., Haynes, M. P., and Terzian, Y.: 1994, *Ap. J.*, **431**, 147
Sturges, H. A.: 1926, *J.A.S.A.*, 65
Tifft, W. G.: 1996, *Ap. J.*, **468**, 491
Yu, J. T., and Peebles, P. J. E.: 1969, *Ap. J.*, **158**, 103

STATISTICAL ANALYSIS OF THE OCCURRENCE
OF PERIODICITIES IN GALAXY REDSHIFT DATA

W. J. COCKE

Steward Observatory
University of Arizona
Tucson, AZ 85721, USA

C. L. DEVITO

Department of Mathematics
University of Arizona
Tucson, AZ 85721, USA

AND

A. PITUCCO

Department of Physics
Pima Community College
Tucson, AZ 85721, USA

Abstract.

We investigate some of Tifft's recent statistical analyses of periodicities in extragalactic redshift samples. The values of the periodicities are refinements of those predicted by Lehto. The redshifts have been corrected for the apparent motion of the solar system relative to the cosmic background radiation and have been filtered by applying criteria such as 21 cm profile width and redshift. In all cases except one, our Monte-Carlo simulations show general agreement with Tifft's results. However, we find that one of his analyses is weakened by applying an inappropriate Bernoulli-trials statistic. We apply a new, more straightforward statistic that shows high statistical significance for some of the periodicities. We conclude that although some of Tifft's procedures seem to be open to some criticism, the periodicities are present at a level that is statistically significant.

Astrophysics and Space Science **244**:143-157,1996.

1. Introduction

The concept of extra-galactic redshift periodicities (RSP), first introduced by Tifft (1976), has been discussed for some time now. The idea violates standard cosmology, which interprets the redshift as a continuous velocity. Objections to the work of Tifft and others are usually based on the contention that the statistical methods used are incorrect. In a recent paper on the structure of RSP, Tifft discusses a proposed periodic velocity rule (Lehto 1990, Tifft 1996) and uses it to predict a specific set of redshift periodicities. It is the purpose of this paper to investigate the statistical significance of certain correlations reported in Tifft (1996) between observed periods and those predicted by this formula.

The paper is structured as follows: (a) using the results readily available in the astronomical literature, we check Tifft's data samples; (b) we apply the statistical tests used by Tifft in order to check his numerical results; (c) a Monte Carlo technique is employed to investigate these statistical tests; the tests are examined critically and, where necessary, replaced by more suitable tests; (d) a new, simple test is devised which avoids using one of the free parameters in one of his tests.

2. Preliminary treatment of the data samples

All redshifts used in Tifft's (1996) paper and in this paper are from 21 cm observations. Uncertainties in such data are small. At signal-to-noise levels of 10 or greater, measurements can be repeated to within a fraction of a km s^{-1}. Since most of the discussion is concerned with periods in excess of 4 km s^{-1}, uncertainty in the data is of little consequence.

Tifft (1996) applied two corrections to the catalogued redshifts, which are all given in the literature as heliocentric. He first transformed the redshifts to a galactocentric frame of reference using a Lorentz transformation. The motion of the solar system about the galactic center was assumed to be represented by the velocity (θ, π, z), where θ is the tangential component, positive in the direction of galactic rotation toward $\ell = 90°$. The radial component, π, is taken to be positive inward toward $\ell = 0°$, and z is the component toward the north galactic pole. The numerical values used are $(232.2, -36.5, 0.2)$ km s^{-1}.

Tifft (1996) has stated that spectral shifts relative to the CBR may not reflect actual kinematic motions. Accordingly, he treated the transformation from the galactocentric frame to the CBR as a Galilean transformation, not a Lorentz transformation. For the data samples discussed here, however, the differences are very slight.

For this transformation Tifft used the velocity $(-243, -31, 275)$ km s^{-1}. The COBE value (Smoot et al, 1992) is $(-245, -23, 275)$ km s^{-1}. For peri-

ods in excess of $9 - 10$ km s^{-1} this difference between the value used and
the COBE value is of no consequence.

Redshifts were further corrected for cosmological effects. If V_0 is the
velocity observed in a given rest frame, and if we define $z \equiv V_0/c$, then the
corrected velocity V_Q is (Tifft 1991)

$$V_Q = 4c[(1+z)^{1/4} - 1] - 2c\left(q_0 - \frac{1}{2}\right)\left[(1+z)^{1/4} + \frac{1}{3}(1+z)^{-3/4} - \frac{4}{3}\right] + \cdots,$$
$$(1)$$

where q_0 is the deceleration parameter.

When $q_0 = \frac{1}{2}$ one obtains the closed-form relationship between V_Q and
z (Cocke & Tifft 1996, Tifft 1996).

These transformations seem *ad hoc*, especially the Galilean nature of the
CBR transformation. However, Tifft has used them all in previous papers,
with nearly the same parameters. See Cocke & Tifft (1996).

Tifft's major, and controversial, contention is that the values of V_Q, after
the application of a Fourier transform, exhibit predictable peaks or "periods". Moreover, the periods observed are given by the following empirical
equation (Tifft 1996):

$$P = P(N) = P(D, T) = c2^{-\left(\frac{9D+T}{9}\right)} \qquad (2)$$

where c is the velocity of light, D and T are non-negative integers, and
$0 \leq T < 9$. Also, $N \equiv 9D + T$. Observe that these periods are known with
an accuracy equal to that with which we know c; i.e., they are accurate to
six significant figures.

Let us stress a number of points about Tifft's contention and his attempts to justify it: (1) The peaks one sees depend strongly upon the type
and character of the galaxies involved. The statistical results appear most
strongly in homogeneous samples. Tifft's methods of achieving this homogeneity are discussed below in connection with the individual data samples;
(2) one cannot expect that V_Q will be equal to an integral multiple of $P(N)$
for a given galaxy. Mathematically this means that $V_Q/P(N)$ is not, in general, an integer. We can, of course, write $V_Q = (k + \gamma)P(N)$, $\quad 0 \leq \gamma < 1$,
where k is an integer. The quantity γ is a measure of the position of the redshift within a periodic cycle. We refer to this number as the "phase". Tifft
has found that, in most cases, γ is a simple fraction; e.g., $0, \frac{1}{2}, \frac{1}{4}$, etc. (3)
Quantization is evaluated by determining the degree of concentration of a
galaxy sample within a narrow redshift phase interval. The basic technique
used is power spectrum analysis (Lake & Roeder 1972, Yu & Peebles 1969)
with power distributed exponentially, for which the probability of finding a
power greater than a given w at a specified frequency is e^{-w}. The formula

used is

$$w(P) = \frac{1}{N} \left| \sum_{j=1}^{N} \exp\left[\frac{2\pi i V_{Q_j}}{P}\right] \right|^2. \tag{3}$$

3. Definition of the Samples and Sub-Samples

Tifft (1996) discusses several sets of galaxy redshift data. In the present paper, we focus on his treatment of three of these sets, referred to as "Virgo", "Perseus," and "TCF."

Sample homogeneity is usually achieved by restricting the range of the 21 cm profile width W. But in large survey samples, which cover a wide range of redshifts, it is often necessary to restrict the redshift range as well in order to isolate spatial structures. For example, in the Cancer sample (Tifft (1996)), two clearly defined galaxy clouds are contained in the subsample with redshifts between 3000 km s^{-1} to 7000 km s^{-1}. Two additional clouds are contained in the overlapping subsample whose redshifts are between 5000 km s^{-1} and 10,000 km s^{-1}. These two clouds have centers near 6500 km s^{-1} and 8500 km s^{-1}.

Data quality was controlled by using signal-to-noise information. No S/N filtering was used for the Virgo or TCF data, but for the Perseus data the filtering was done by using the ratio $(F/W)/rms$, where F is the total 21 cm flux and rms is the noise. For the Perseus data, Tifft also used the ratio F/W itself as a filter, since a low value for this ratio tends to select galaxies with high total luminosity.

We now discuss the individual samples in more detail.

a) Virgo

A survey of the Virgo cluster dwarf galaxies was compiled by Hoffmann et al. (1987). One hundred and fifty three galaxies were observed at Arecibo between 1983 and 1987. Tifft used the 137 galaxies from this sample with redshifts less than 3500 km s^{-1}. The purpose of this restriction was to eliminate background galaxies. In this subsample the profile widths W range from 40 km s^{-1} to 250 km s^{-1} with most less than 175 km s^{-1}. Tifft subdivided the subsample by profile width and considered the effect of certain low redshift galaxies, but otherwise no objects are excluded.

It should be mentioned that Guthrie and Napier (1991) reported no obvious periodicities in a similar subsample containing 77 galaxies. Their data were transformed to galactocentric coordinates, and their subsample was chosen to minimize redshift uncertainty. For the longer periods that Tifft considers, these restrictions serve only to reduce the sample size with no detectable effect on its character. By using the full sample he is able to demonstrate that the periodicities depend on profile width.

b) Perseus

Tifft's source was two major studies by Giovanelli and Haynes (1985, 1989). Galaxies here cover a very wide redshift range approaching 20,000 km s^{-1}. A set of 154 galaxies with profile widths greater than 400 km s^{-1} was studied. This sample contains all wide-profile galaxies in the two references that are not flagged as contaminated.

As stated above, he used the S/N ratio $(F/W)/rms$ to filter the data. Specifically, he used S/N ≥ 4. Also, he used $W \geq 400$ km s^{-1} and $F/W \leq 0.01$. These criteria define a sample containing 53 galaxies and have been used before in other data sets to define subsamples (see, for example, Cocke & Tifft (1996)). They do not seem to have been introduced *ad hoc* to improve the significance of the periodicities artifically.

c) TCF

The source of this sample is high-precision surveys made with the Green Bank 300-ft telescope between 1984 and 1986 by Tifft & Cocke (1988). These surveys are mostly of galaxies in the Local Group, and there was much planned overlap with the Fisher-Tully catalogue (1981). Signal-to-noise information is contained in the catalogue itself. The restriction $170 \leq W \leq 250$ km s^{-1} gives a subsample containing 92 galaxies. The velocity range of this subsample is about 0 - 2500 km s^{-1}.

In analyzing some of the data samples, Tifft's approach was to use a Bernoulli-trials statistic in order to compare observed RSP's with those predicted by equation (1). To do this, he sorted the observed data into bins of different profile width by using specific criteria established in earlier published work. Much criticism has been directed against the use of these criteria. In order to address these criticisms we first checked the Virgo data and applied Tifft's methods to these data to confirm his numerical results. Further discussion of our re-examination of the statistics applied by Tifft is given in the next paragraph. Here we want to remind the reader of the specific statistical tests that he performed. In the Virgo sample the power spectra were computed and analyzed as follows: Peaks with power w greater than four were considered significant. These often corresponded to velocities close to those predicted by eq. (2).

Note that the periods given by eq. (2) are separated by a factor of 1.08006, the ninth root of two. Most period agreements are within a factor of 1.005 and, as Tifft states, do not appear to be a random set of the predicted $P(N)$. Here we focus on Tifft's application of the Bernoulli statistic to the Virgo data set. He used several width intervals W; namely 60 – 100, 70 – 110, 60 – 120, and 40 – 150, all more or less appropriate for dwarf galaxies (Tifft & Cocke (1984)). These were combined with the velocity filters $-500 \leq V \leq 3500$ and $500 \leq V \leq 3500$ km s^{-1}. Over a period range

$20 \leq P \leq 100$ km s^{-1}, he found a total of 8 peaks with $w \geq 4$ that for some combination of these W and V intervals satisfy

$$\frac{\text{Peak location with } w \geq 4}{P(N)} = 1.000 \pm 0.005 \equiv 1 \pm \alpha, \qquad (4)$$

where Tifft defines α as the deviation of the Peak/Period ratio from unity. Since $2^{1/9} \approx 1.08$, $\alpha = 0.005$ includes roughly 1/8 of the total range on the P axis. The total number of peaks with $w \geq 4$ in this combination of filters was 14.

The Bernoulli trials probability that, out of a total number of trials N_{Tot}, the number of heads be greater than or equal to N_h is given by

$$\text{Prob}(\text{heads} \geq N_h | N_{\text{Tot}}, p_h) = \sum_{k=N_h}^{N_{\text{Tot}}} \binom{N_{\text{Tot}}}{k} p_h^k (1 - p_h)^{N_{\text{Tot}}-k}, \qquad (5)$$

where p_h is the individual "heads" probability.

Tifft then computes the Bernoulli probability for the Virgo data as

$$\text{Prob}(N_h \geq 8 | N_{\text{Tot}} = 14, 1/8) \approx 9 \times 10^{-5}. \qquad (6)$$

We have one objection to this computation: Since a peak was counted as "heads" if it came up in any of the combinations of W with V, the probability that a given peak would appear in one of the P intervals about a $P(N)$ is actually *greater* than 1/8. This difference can be very significant. If, for example, $p_h = 1/4$,

$$\text{Prob}(N_h \geq 8 | N_{\text{Tot}} = 14, 1/4) \approx 0.01 . \qquad (7)$$

A second objection would be that the total number of peaks (N_{Tot}) in the range $20 \leq P \leq 100$ km s^{-1} is itself a random variable. However, it seems true that one can recast the problem by considering all random events for which the total number of peaks is the fixed N_{Tot}. The analysis then proceeds as usual for Bernoulli trials.

To evaluate the significance of the periodicities in the Perseus sample, Tifft looked for power peaks with $w \geq 4$ which fell in the period range of 17 to 250 km s^{-1}. The restrictions on W, S/N, and F/W were as stated above.

As was done with the Virgo data, the number of matches within a limiting peak/period range was counted, and Bernoulli statistics were applied. This 53-galaxy sample had 9 peaks within 0.004 (instead of the 0.005 in Virgo) of unity and a total of 27 power peaks above 4. The probability of 9 or more "heads" with $p_h = 1/10$ in 27 trials is 0.0009 by by the Bernoulli statistic. Also, a 28-galaxy sample having 4 heads out of 9 trials was stated as unlikely at the 0.008 level. A further refinement of the data, defined by raising the cutoff period to 36 km s^{-1} to reduce noise, yielded 6 hits out of 10, which was stated as being unlikely at the 0.0002 level. These results were stated as consistent with the Virgo results.

The third sample, the TCF sample, consisted of galaxies from the Tifft & Cocke (1988) precision-redshift study. These are mostly local-group galaxies. The sample included only galaxies for which there were also measurements in the Fisher-Tully (1981) catalogue. Tifft (1996) restricted the sample further to include galaxies with $170 \leq W \leq 250$ km s^{-1}, a criterion for which there is a precedence in Cocke & Tifft (1996). This restriction defines a subsample of 92 galaxies, as stated above. Tifft did not do a Bernoulli test on this subsample, but we have verified his statements about sample size and spectral peaks.

To examine the accuracy of the Bernoulli-trials statistics in the above three subsamples, we employ Monte Carlo simulations.

4. Monte Carlo Simulations of the Occurrence of Spectral Peaks

In view of our doubts about the accuracy of the Bernoulli-trials statistic discussed in the previous section, we calculated the probabilities of getting N_h heads for the three data sets by using standard Monte Carlo techniques. This procedure also avoids the errors involved in assuming the approximation e^{-w} for the probability of getting a power level of w or greater at a given point in the spectrum. For a discussion of these errors see Newman, Haynes, and Terzian (1989, 1994).

To do the simulations we added normally distributed random variables to each velocity data point in the data sets. For Perseus, we used the 53-galaxy subsample; and for TCF, the 92-galaxy subsample. For Virgo, we arbitrarily picked the 46-galaxy subsample defined by $60 \leq W \leq 100$ and $+500 \leq V \leq 3500$ km s^{-1}. The standard deviation of the random variables was taken to be 1000 km s^{-1}, a value which preserves the general shapes of the velocity distributions but insures sufficient randomization. We generated the random numbers by means of the routines ran1 and gasdev from Press et al (1992) and added them to the CBR-corrected redshifts.

The results are given in Table 1, where we list the probabilities for each N_h, with $\alpha = .004$ for Perseus and TCF, and $\alpha = .005$ for Virgo. As before, a "heads" is having a power peak with $w \geq 4$ in the given window. These windows are defined, as in the previous section, to give $p_h = 1/8$ for Virgo and $p_h = 1/10$ for Perseus and TCF.

The errors in the probabilities may be computed from the fact that each of the simulations was done with a total of 10^4 independent runs. Thus, for

TABLE 1.   Monte Carlo Probabilities for N_h "heads", defined as peaks with $w \geq 4$ in the given windows

Subsample:	Perseus	Virgo	TCF
α:	.004	.005	.004
Range (km s^{-1}):	$16 - 250$	$20 - 100$	$7 - 20$
N_h	Prob	Prob	Prob
0	0.0599	0.3867	0.1820
1	0.1699	0.3752	0.3270
2	0.2533	0.1750	0.2796
3	0.2335	0.0495	0.1468
4	0.1538	0.0112	0.0504
5	0.0846	0.0021	0.0124
6	0.0294	0.0003	0.0016
7	0.0012	0	0.0001
8	0.0033	0	0.0001
9	0.0008	0	0
10	0.0003	0	0

example, $N_h = 5$ for Perseus has a total number of heads of 846. The error in this number is approximately $\sqrt{846} \approx 30$, and thus the table entry has an associated error of ± 0.0030.

We compare these probabilities to those computed by Tifft (1996) using the Bernoulli-trials formula. The results are listed in Table 2. The "Monte Carlo" line was calculated by summing all the probabilities in Table 1 for entries with $N_h \geq$ the appropriate number of "heads". For Virgo, we choose the subsample $60 \leq W \leq 100$ km s^{-1}, for which there are 2 peaks in the 1/8-bins out of a total of 4 peaks, and its Bernoulli entry is correspondingly Prob$(2|4, 1/8) = 0.079$. Tifft (1996) did not assign Bernoulli probabilities to the TCF subsample, but there are 3 peaks in the 1/10-bins in the range $7 - 20$ km s^{-1}, out of a total of 7, giving a Bernoulli entry of Prob$(3|7, 1/10) = 0.026$.

We see that for the Perseus data the Monte Carlo simulation gives a result that is quite close to that of the Bernoulli approximation. However, our Monte Carlo result for Virgo is a factor of 3 larger than *our* Bernoulli result, and is much larger than Tifft's figure of 9×10^{-5}, which he computed from the Bernoulli formula as Prob$(8|14, 1/8)$. As stated in the previous section, his computation seems to have been misapplied in this case. For TCF the Monte Carlo result is a factor of 8 greater than the Bernoulli result.

TABLE 2.   Probabilities for $\geq N_h$ peaks
with $w \geq 4$.

Subsample:	Perseus	Virgo	TCF
Heads	9	2	3
Trials	27	4	7
Bernoulli	0.00087	0.079	0.026
Monte Carlo	0.0011	0.23	0.21
M. C. 3	0.0029	0.50	0.46

Thus we conclude that one may not use the Bernoulli approximation even for a rough guide. The Monte Carlo calculation is certainly more nearly correct.

We finish this section by discussing the results of a supplementary Monte Carlo calculation (listed as "M. C. 3" in Table 2) that simulates a researcher who is investigating a data set but is unsure of how to break the set up into subsamples. He wishes to apply the Monte Carlo test defined above. For the first test, he uses all of a previously defined subsample. He then breaks the subsample in half and applies the test separately to the two halves. He reports only the most favorable of the three results.

Of course, such a result is less significant than reported. To gauge the effect of this sort of bias on the part of the experimenter, we performed this simulation on the three subsamples discussed above. The overall effect is of course to make "heads" more probable. This is shown in the bottom line of Table 2. Apparently, if the result is reported as "highly significant", as in the Perseus subsample with the reported probability of 0.0011, the real probability is almost a factor of 3 higher. Therefore, results which are said to be highly significant (say, ≤ 0.01) may still be truly significant even if the researcher has biased them in this way. When statistical procedures and data samples are first being explored, it is very difficult to avoid such biases. Unfortunately, this situation is exacerbated by the complex nature of the RSP's.

5. A Simpler Statistic Based on the Unbinned Occurrence of High Power Levels

The statistical tests discussed in the previous section are complicated by the fact that one must choose an interval within which a spectral power peak might occur. It seems more straightforward to inquire about the occurrence of high power levels exactly at the velocities $P(N)$ given by equation (1).

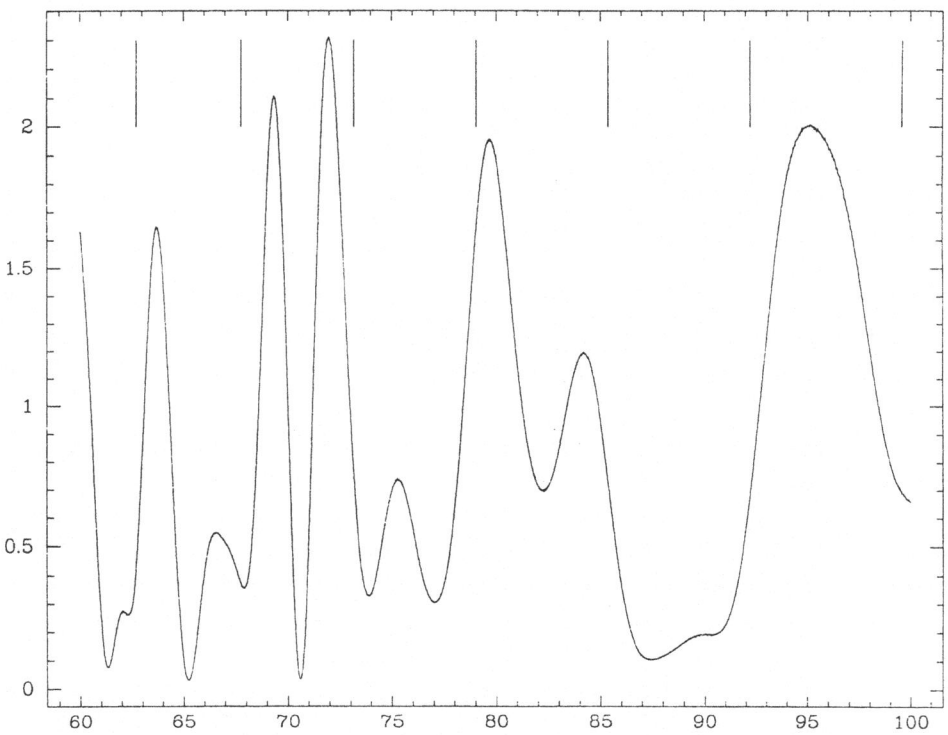

Figure 1.   Power spectrum for the 92-galaxy TCF sample for the period range $60 \leq P \leq 100$ km s^{-1}. The vertical lines are the Lehto-Tifft periods.

If the power levels at these velocity points are statistically independent, then the Bernoulli model is appropriate. If not, a Monte Carlo simulation is more to the point. But even so, if the power peaks in a particular period range are as far apart as the distance between adjacent L-T numbers, the test becomes weak.

To judge where this might happen, we plot part of the TCF power spectrum in Figure 1. In the top panel, covering the range $60 \leq P \leq 100$ km s^{-1}, we see that there are two or three fluctuations in power between the individual L-T velocities, so it is likely that the power levels there are statistically independent of each other. The Lehto-Tifft velocities are marked on the top of the graph. This is not likely to be the case at higher velocities, where the power fluctuations are broader. To test this idea further, we compared the results of the Bernoulli formula with those of simulations, for our three subsamples. The results of these simulations were very nearly identical with those given by the Bernoulli formula. Nevertheless, we feel that this test should not be applied for ranges of P where the L-T numbers are as close together as the spectral peaks, since even strong periodicities may be pulled away from their true locations in the spectrum.

To see where this might occur, note that the argument of the exponential in equation (3) can be written as $2\pi i V_{Q_j}/P = 2\pi i (V_{Q_j}/\Delta V)(\Delta V/P)$, where ΔV is the total range of the redshifts in the sample. It is known (Lake & Roeder 1972) that if the argument of the exponential is written $2\pi n x_j$, where the x_j are uniformly distributed over the interval $(0,1)$, then the values of w_n are statistically independent of each other for n an integer. Now, the redshifts are not uniformly distributed over the interval ΔV, but one should still be able to say that the intervals of statistical independence are roughly given by $\Delta V/P =$ integer. Therefore the correlation length in P should be given by $\Delta(\Delta V/P) \approx 1$, or

$$\Delta P \approx \frac{P^2}{\Delta V}. \tag{8}$$

The Perseus data extend from 5,000 km s^{-1} to about 20,000 km s^{-1}. Setting $\Delta V = 1.5 \times 10^4$ in the above equation, one can show that ΔP becomes equal to the spacing between the L-T numbers at and above about 1000 km s^{-1}. So for these data, this problem does not arise.

The Virgo and TCF data both have ranges $\Delta V \approx 3000$ km s^{-1}, and the above equation shows that the equality of spacing occurs at about $p \approx 100$ km s^{-1}. This is confirmed by Figure 1. Therefore in what follows we restrict the range of P to be less than 100 km s^{-1} for these two data sets.

The differences between the Bernoulli test that we propose here and the previous one are that the definition of "trials" and "heads" are different. In the present test, a trial is "check the spectral power level at a Lehto-Tifft velocity." "Heads" is simply "a power level $w \geq w_0$ at a Lehto-Tifft velocity," where w_0 is a level specified a priori. In the previous test, a trial was "find the locations of peaks having $w \geq w_0$" ($w_0 = 4$), while "heads" was "the location is within a bin centered on one of the Lehto-Tifft velocities."

The "heads" probabilities are therefore different between the two tests, and our test assumes the exponential law $p_h = e^{-w_0}$. In any case, this test is quite different from that used by Tifft. Inevitably, our version of the test misses some periodicities that Tifft's version would have found.

In a preliminary version of this research, we picked $w_0 = 4$, to facilitate the comparison with Tifft (1996). Tifft, however, pointed out later that our new Bernoulli test seems appropriate only for samples that one might expect to show a diversity of periods. The Perseus sample is one such, since it encompasses a very large redshift range – up to about 20,000 km s^{-1} –

and takes its membership from several galaxy clusters. The Virgo and TCF samples, however, were optimized for homogeneity: The Virgo data are all from the Virgo Cluster, and the TCF are all local dwarfs.

We have therefore done two tests, one for $w_0 = 4$, appropriate for a sample expected to have several peaks of low power, and one for $w_0 = 7$, for samples expected to have a few high-power peaks.

Table 3 shows the results of the Bernoulli trials (eqn. (5) for the Perseus data for the power level $w \geq 4$ ($w_0 = 4$), for which we assume that the "heads" probability is given by $p_h = e^{-4} \approx 0.0183$. We test this subsample for the following ranges in Lehto-Tifft velocities: $7 - 17(N_{Tot} = 12, 17 - 250(N_{Tot} = 34)$, and $7 - 250(N_{Tot} = 46)$, all in km s^{-1}.

TABLE 3. Bernoulli Probabilities for the Perseus data for $\geq N_h$ heads, defined as occurrences of $w \geq 4$ at the given velocities, with $p_h = e^{-4}$.

Range (km s^{-1})	$7 - 17$ $N_{Tot} = 12$	$17 - 250$ $N_{Tot} = 34$	$7 - 250$ $N_{Tot} = 46$
Perseus	$N_h = 0$ Prob $= 1.0$	$N_h = 4$ Prob $= 0.0034$	$N_h = 4$ Prob $= 0.010$

Table 4 shows the results of the Bernoulli trials (eqn. (5) for the Virgo and TCF data for the power level $w \geq 4$ ($w_0 = 4$). As stated above, we test these subsamples for a more restricted range in Lehto-Tifft velocities: $7 - 17(N_{Tot} = 12), 17 - 100(N_{Tot} = 23)$, and $7 - 100(N_{Tot} = 35)$, all in km s^{-1}.

TABLE 4. Bernoulli Probabilities for the Virgo and TCF data for $\geq N_h$ heads, defined as occurrences of $w \geq 4$ at the given velocities, with $p_h = e^{-4}$.

Range (km s^{-1})	$7 - 17$ $N_{Tot} = 12$	$17 - 100$ $N_{Tot} = 23$	$7 - 100$ $N_{Tot} = 35$
Virgo	$N_h = 0$ Prob $= 1.0$	$N_h = 4$ Prob $= 0.00075$	$N_h = 4$ Prob $= 0.0037$
TCF	$N_h = 2$ Prob $= 0.020$	$N_h = 1$ Prob $= 0.35$	$N_h = 3$ Prob $= 0.024$

We then did simulations to test both the validity of the approximation $p_h \approx e^{-4}$ and the assumption that the trials are statistically independent.

They were carried out as stated in the previous section: we used the routines **ran1** and **gasdev** from Press et al (1992) to create normally distributed random variables with variances of 1000 km s^{-1}. These were added to the CBR-corrected redshift data, and histograms were compiled of the number of times for which $w \geq 4$ at the Lehto-Tifft velocities. There were 10^4 such experiments for each subsamaple.

As stated above, the results of these simulations were very nearly identical to those given by the Bernoulli formula, within the expected errors of the simulation technique. Thus the exponential approximation for the occurrence of $w \geq 4$ and the assumption of statistical independence of the trials seem justified, in this case.

Note that this Bernoulli statistic scores very highly for the Perseus data and Virgo data, except in the range $7 - 17$ km s^{-1}, where there are no "heads" at all. The TCF data, on the other hand, show significance in this range, but have only 1 "heads" in the $17 - 100$ range. For the entire range (7-100), the TCF data score as quite significant, at the 97.6% level.

Table 5 shows the results of the same test carried out for $w_0 = 7$ for the Virgo and TCF data. Otherwise, the parameters defining the test are the same as for Tables 3 and 4. However, as pointed out by Newman et al (1994), the exponential law for the distribution of the power w may break down for $w \geq 4$. Thus we repeated the simulations described above in connection with Table 3 and found that equation (5) overestimates the true probabilities by about 20%. (That is, the law $p_h = e^{-w}$ underestimates the statistical significance by 20%.) Thus we list the simulation results in Table 5. The Perseus data show no "heads" at all for this test, so no table is necessary for them.

TABLE 5. Monte Carlo calculations for the Virgo and TCF data for $\geq N_h$ heads, defined as occurrences of $w \geq 7$ at the given velocities.

Range (km s^{-1})	$7 - 17$ $N_{\text{Tot}} = 12$	$17 - 100$ $N_{\text{Tot}} = 23$	$7 - 100$ $N_{\text{Tot}} = 35$
Virgo	$N_h = 0$ Prob = 1.0	$N_h = 1$ Prob = 0.017	$N_h = 1$ Prob = 0.026
TCF	$N_h = 0$ Prob = 1.0	$N_h = 1$ Prob = 0.018	$N_h = 1$ Prob = 0.025

We see that the power spectrum of the Perseus data has no Lehto-Tifft numbers at all where the power is ≥ 7, whereas the Virgo and TCF subsamples have one peak each. This is in accord with the idea that inhomogeneous data sets (e. g., Perseus) generally have lower values of the power at the Lehto-Tifft numbers, but more L-T numbers where the power is still elevated. In contrast, homogeneous subsamples have fewer such L-T numbers singled out, but the powers there tend to be higher. Table 5 shows that the corresponding statistical significance in the (homogeneous) Virgo and TCF subsamples is substantial, at the 98% and 97.5% levels.

Tifft has remarked that the L-T numbers that are in fact present do not seem to be random. Referring to the parameters in equation (2), there are apparently preferred values of the integer T. The pure cube-roots are among these ($T = 0, 3, 6$). In fact, a significant majority of the L-T numbers found in the power spectra computed for Tables 3 and 4 correspond to the velocities 73.2, 36.6, and 18.3 km s^{-1}, for which $T = 0$. We list these velocities and the corresponding power values below in Table 5. Note that 4 out of the 11 peaks listed are for $T = 0$, and all of the highest peaks are of this type. If the T-values were random, the expected number for which $T = 0$ would be $11/9 \approx 1.1$. Thus it might be well to refine the test presented in this section by placing different emphases on different values of T.

TABLE 6. T-values for power peaks with $w \geq 4$.

Perseus			Virgo			TCF		
$P(n)$	Power	T	$P(n)$	Power	T	$P(n)$	Power	T
18.30	5.98	0	23.05	4.12	6	7.84	4.09	2
85.38	4.75	7	36.60	8.13	0	16.94	4.73	1
116.18	4.79	3	49.80	5.74	5	18.30	7.64	0
215.1	5.09	4	73.19	4.29	0			

6. Conclusions

Our analysis of Tifft's methods shows that his conclusions about the statistical significance of the periodicities are justified even though his statistical methods are occasionally questionable. The connection between the periodicities in the data and those predicted by equation (2) seems to be real. The Perseus data show high significance in the Monte Carlo analysis presented in Table 2, and all three data sets show significance when submitted to the new Bernoulli tests, the results of which are presented in Tables 3, 4, and 5.

It seems necessary, however, to use different power levels in the new test, depending on the homogeneity of the sample in question. For inhomogeneous samples (e. g., Perseus), $w_0 = 4$ seems reasonable; whereas for more nearly homogeneous samples (Virgo and TCF), $w_0 = 7$ might recommend itself.

Unfortunately, the physical meaning of RSP remains obscure, in spite of the connection with Lehto's empirical formula (Lehto 1990).

We thank W. G. Tifft for his comments.

References

Cocke, W. J., & Tifft, W. G.: 1996, *Ap&SS* **239**, 35
Fisher, J. R., & Tully, R. B.: 1981, *ApJS* **47**, 139
Giovannelli, R., & Haynes, M.: 1985, *AJ* **90**, 2445
Giovannelli, R., & Haynes, M.: 1989, *AJ* **97**, 633
Guthrie, B. N. G., & Napier, W. N.: 1991, *MNRAS* **253**, 533
Hoffmann, G. L. it et al.: 1987, *ApJS* **63**, 247
Lake, R. G., & Roeder, R. C.: 1972, *JRAS Canada* **66**, 111
Lehto, A.: 1990, *Chinese J. Phys.* **28**, 215
Newman, W. I., Haynes, M. P., & Terzian, Y.: 1989, *ApJ* **344**, 111
Newman, W. I., Haynes, M. P., & Terzian, Y.: 1994, *ApJ* **431**, 147
Press, W. H. it et al.: 1992, *Numerical Recipes,* 2nd edition (Cambridge)
Smoot, G. F. it et al.: 1992, *ApJ* **396**, L1
Tifft, W. G.: 1976, *ApJ* **206**, 38
Tifft, W. G.: 1991, *ApJ* **382**, 396
Tifft, W. G.: 1996, *ApJ* **468**, 491
Tifft, W. G., & Cocke, W. J.: 1984, *ApJ* **287**, 492
Tifft, W. G., & Cocke, W. J.: 1988, *ApJS* **67**, 1
Yu, J. T., & Peebles, P. J. E.: 1969, *ApJ* **158**, 103

ZOOMING IN ON THE REDSHIFT PROBLEM

P.A. STURROCK
Center for Space Science and Astrophysics
Stanford University
Stanford, CA 94305

Abstract. Scientific inference offers a way to help organize and clarify our thinking about controversial areas of science such as the redshift problem. Scientists typically devote considerable effort to evaluating the probability that data relevant to a controversial area may be due to the "null hypothesis" (i.e. that there is no new phenomenon). However, it is usually not clear whether a small probability for the null hypothesis can be interpreted as a high probability for some other hypothesis, if only for the reason that the alternative hypothesis may not be specified, and it is not clear how strong a case is required to establish the new hypothesis. Thinking about such topics can be clarified by a simple procedure based on the methods of scientific inference. This procedure is referred to as "ZOOM" for "Zero-Order Organizing Model." This article proposes a ZOOM for the redshift problem, and presents the results of a preliminary trial.

1. Introduction

This article deals with two redshift-related controversies that are associated with the names of Halton Arp and William Tifft. Arp (1987, 1997) has argued that some objects have redshifts that cannot be explained purely in terms of the known processes of Doppler shift, gravitational redshift, and cosmological redshift. Tifft (1995, 1997) has presented evidence to support the claim that the redshift distributions of some classes of objects, when analyzed in a certain way, prove to be periodic. (See also Napier and Guthrie 1993.)

The strength of the evidence for such claims is usually computed in a way that is familiar in astrophysics and other branches of science. One assumes that there is no such effect, and then calculates the probability of

Astrophysics and Space Science **244**:159-166,1996.
© 1996 *Kluwer Academic Publishers.*

finding patterns that have been extracted from the data. One problem with this approach is that evidence against the null hypothesis is not equivalent to evidence for the proposed alternative hypothesis. Another difficulty is that there is no way to decide how much evidence is required to substantiate the considered hypothesis. Questions such as these may be clarified considerably by using the procedures of scientific inference (Sturrock 1973, 1994).

Section 2 presents some of the basic concepts of scientific inference that will be required in this article. Section 3 discusses, in more detail, the concept of priors or prior probabilities. It is argued that it is sometimes useful to regard prior probabilities as dependent upon more basic assumptions. This leads to the introduction of a ZOOM, or Zero-Order Organizing Model. A simple ZOOM is proposed for the redshift problem.

A worksheet summarizing questions in the ZOOM were distributed to participants at the TIME conference held in Tucson in April 1996. Fourteen of the participants completed the worksheet. An analysis of these data is presented in Section 4. Concluding remarks are presented in Section 5.

2. Scientific Inference

Scientific inference – and, one might argue, all of science – rests upon Bayes' theorem. (See, for instance, Good 1950, Jeffreys 1931, Sturrock 1973, 1994.) This theorem may be expressed as follows:

$$P(H|NI) = \frac{P(N|HI)}{P(N|I)} P(H|I) \tag{1}$$

The symbols in this equation have the following interpretation: H is an hypothesis under consideration; I represents initial information; and N represents a new item of information. $P(H|I)$ is the prior probability, i.e. the probability to be assigned to hypothesis H on the basis only of the initial information I. $P(H|NI)$ is the post probability, i.e. the probability to be assigned to H on the basis of both the initial information I and the new information N; $P(N|I)$ is an assessment of the probability that N will be true, based only on the initial information I; and $P(N|HI)$ is an assessment of the probability that N will be true, based on both the initial information I and the hypothesis H.

It is useful to introduce the notation

$$\Omega(A|B) = \frac{P(A|B)}{P(\bar{A}|B)} \tag{2}$$

for the odds on A, based on information B. In this equation, $\bar{A}$ signifies "not A". Another concept that will be used is that of "log-odds", defined by

$$\Lambda(A|B) = \log_{10}[\Omega(A|B)]. \tag{3}$$

An alternative is to introduce the notation

$$\Delta(A|B) = 10 \log_{10}[\Omega(A|B)], \tag{4}$$

that is equivalent to measuring the log-odds in "db." For instance, 1 db is equivalent to $\Lambda = 0.1$, $\Omega \approx 1.26$, and $P \approx 0.56$; 10 db is equivalent to $\Lambda = 1$, $\Omega = 10$, and $P \approx 0.91$; and -20 db is equivalent to $\Lambda = -2$, $\Omega = 0.01$, and $P \approx 0.01$.

We see, from equation (1), some of the limitations of the usual approach to problems such as we are now discussing. To compute the probability that a certain pattern would arise by chance is equivalent to calculating $P(N|Z)$, where we now introduce the symbol Z to represent the uncontroversial "baseline" or "zero-base" information. However, in order to evaluate the post-probability, we need to evaluate also $P(N|HZ)$, the probability that the pattern would arise if the considered hypothesis were true, and also $P(H|Z)$, the prior probability of the hypothesis. Studies of the redshift controversies typically involve estimates of $P(N|Z)$, but typically do not involve consideration of $P(H|Z)$ and $P(N|HZ)$. The quantities $P(H|Z)$, known as the "prior probabilities," are crucial. In the minds of most scientists, these quantities would be small if the topic is controversial, and very small if the proposal is heretical. But how can these quantities be estimated, and how do we deal with the fact that these quantities represent vague, individual, subjective estimates, not precise, agreed, objective measurements? This question is taken up in the next section.

3. A ZOOM for the Redshift Problems

Our goal is to arrive at estimates of the prior probabilities $P(A|Z)$, $P(C|Z)$. It would be possible simply to ask astrophysicists to give their personal assessments of these quantities, and then find a way to summarize the distribution of values that are given. (For reasons set out in Sturrock (1994), we will where necessary summarize a distribution of probability estimates by means of the mean and the standard deviation of the corresponding log-odds estimates.) However, most scientists have little experience of making such estimates, so some guidance could be helpful. Giving some structure to the thinking involved in arriving at these estimates will also make it easier for us to understand what trains of thought lead to differences in these estimates. Furthermore, by regarding the final prior probabilities as the output from two or three levels of decision-making, we can estimate an average and standard deviation of the log-odds for each decision, and so obtain average values – and, if we wish, standard errors of the means

– for $\Lambda(A|Z)$ and $\Lambda(C|Z)$. These estimates can of course be converted into estimates of $P(A|Z)$ and $P(C|Z)$. It appears, from admittedly limited experiments with this procedure, that we refer to as a "ZOOM" for "Zero-Order Organizing Model," that such a structure leads to a closer consensus in the final judgements than would have been obtained if there had been no structure.

It is first necessary to specify the hypotheses under discussion. The following propositions are, for reasons to be developed, regarded as belonging to "Level 3."

Level 3.

Proposition A: Some astronomical objects have "Anomalous" redshift contributions, in addition to gravitational, Doppler, and normal cosmological contributions.

Proposition C: Some astronomical objects have redshift distributions that exhibit periodicities ("Cycles").

As explained above, we propose to view these prior probabilities as depending upon more basic assumptions. These assumptions will be organized in two levels, Level 1 and Level 2.

Level 1.

This concerns the following pair of propositions, that use terms defined and used by Kuhn (1962):

Proposition E: There is extraordinary physics yet to be discovered.
By "extraordinary" is meant a truly revolutionary development similar to the development of quantum mechanics.

Proposition O: Present-day physics, or "ordinary physics", is essentially complete.
This proposition is equivalent to $\bar{E}$, the negative of E, and asserts that there are no more extraordinary developments to be discovered.

We now require each investigator to estimate $P(E|Z)$ and $P(O|Z)$. Since E and O form a complete and mutually exclusive set, we require that

$$P(E|IZ) + P(O|IZ) = 1. \tag{5}$$

Alternatively, we could ask for $\Omega(E|Z)$ or $\Lambda(E|Z)$, the odds or log-odds, on E. Of course, the information that is considered – explicitly or implicitly – will vary from investigator to investigator, so that, here and elsewhere, we will have a distribution of estimates to deal with. It would therefore be a more precise representation if we were to replace Z by Z_α, where α designates different investigators, but – for simplicity – we avoid this formality.

In considering Proposition E, it is convenient to consider some more specific possibilities. We here consider explicitly only the following three possibilities in the next level.

Level 2.

Proposition X: New particles and/or fields are yet to be discovered.

Proposition M: The space-time manifold is more complex than is now assumed.

Proposition Q: There is some other extraordinary development in physics other than X or M) yet to be discovered.

It is then necessary for each investigator to evaluate the contingent probabilities $P(X|EZ)$, $P(M|EZ)$, and $P(Q|EZ)$. We note that, E being given, X, M, and Q form a complete set, but they are not mutually exclusive and in what follows we regard them as independent of each other.

Finally, at Level 3 (the level of the redshift hypotheses), it is necessary to evaluate further contingent probabilities. Concerning hypothesis A, it is necessary to estimate $P(A|OZ)$, the probability that A is true if there is no extraordinary physics yet to be discovered, and $P(A|EZ)$, the probability that A is true if there is extraordinary physics yet to be discovered. Clearly, the former does not involve consideration of the Level-2 propositions. On the other hand, we will need to consider the Level-2 propositions in order to arrive at $P(A|EZ)$ and $P(C|EZ)$.

We first note that

$$P(A|EZ) = P(A, X \& or M \& or Q|EZ), \qquad (6)$$

since, if E is true, at least one of X, M, and Q must be true. Hence we may evaluate $P(A|EZ)$ in terms of $P(AX|EZ)$, $P(AM|EZ)$ and $P(AQ|EZ)$, by repeated application of the rule

$$P(A \& or B|C) = P(A|C) + P(B|C) - P(A|C)P(B|C). \qquad (7)$$

Finally, one may evaluate $P(A|Z)$ by noting that

$$P(A|Z) = P(A|EZ)P(E|Z) + P(A|OZ)P(O|Z), \qquad (8)$$

with a similar equation for $P(C|Z)$.

4. Results of Survey

Of the participants in the TIME workshop held in Tucson in April 1996, fourteen completed the ZOOM worksheet. These were analyzed according to the process proposed in Section 5 of Sturrock (1994). In terms of log-odds, the results may be summarized as shown in Table 1.

TABLE 1. Summary of probability estimates for the redshift ZOOM, expressed as mean and standard error of the mean of the log-odds, all estimates rounded to one decimal place.

	$\Lambda(E\|Z) = 2.1 \pm 0.4$	
	$\Lambda(O\|Z) = -2.1 \pm 0.4$	
$\Lambda(A\|OZ) = -1.4 \pm 0.4$		$\Lambda(C\|OZ) = -1.5 \pm 0.3$
	$\Lambda(X\|EZ) = 1.1 \pm 0.5$	
	$\Lambda(M\|EZ) = 1.2 \pm 0.6$	
	$\Lambda(Q\|EZ) = 2.0 \pm 0.4$	
$\Lambda(A\|XZ) = -0.8 \pm 0.5$		$\Lambda(C\|XZ) = -1.2 \pm 0.4$
$\Lambda(A\|MZ) = -0.2 \pm 0.5$		$\Lambda(C\|MZ) = -0.3 \pm 0.5$
$\Lambda(A\|QZ) = 0.6 \pm 0.5$		$\Lambda(C\|QZ) = 0.2 \pm 0.6$

It is convenient to introduce the expression

$$P(A, [QE]) = P(A|QEZ)P(Q|EZ)P(EZ), \text{ etc.,} \tag{9}$$

to give an estimate of the contribution to the final estimate of $P(A|Z)$ that comes from the "channel" QE, etc. The results, computed from the mean values of the log-odds shown in Table 1, are shown in Table 2.

TABLE 2. Contributions to the final probabilities from the possible "channels." Estimates given to only one significant figure.

$P(A, [O]) = 0.0003$	$P(C, [O]) = 0.0002$
$P(A, [XE]) = 0.1$	$P(C, [XE]) = 0.06$
$P(A, [ME]) = 0.4$	$P(A, [ME]) = 0.3$
$P(A, [XE]) = 0.8$	$P(A, [XE]) = 0.6$
$P(A, [E]) = 0.9$	$P(C, [E]) = 0.7$

We see from Table 2 that, if one considers only the possibility of conventional physics, the propositions A and C are considered quite unlikely, with probabilities of only 0.0003 and 0.0002, respectively. However, if one considers the possibility of extraordinary physics, the same propositions become quite likely. If it is assumed that E is correct, i.e. that there is extraordinary physics to be discovered, the probability of A becomes 0.1, 0.4 or 0.8, and the probability of C becomes 0.06, 0.3 or 0.6, if one assumes X, M, or Q, respectively, to be correct. Clearly the most significant "avenue" for explaining anomalous redshifts or redshift periodicity comes from assuming there is extraordinary physics to be discovered, that development being neither new particles and fields, nor new space-time structure, but something else that is unspecified and is probably unspecifiable.

If one combines the X, M, and Q "channels", one arrives at the estimates $P(A|EZ) = 0.9$ and $P(C|EZ) = 0.7$. These are changed insignificantly by considering that O is correct (no new physics), leading to the final values $\Lambda(A|Z) = 0.9$ and $\Lambda(C|Z) = 0.4$. However, the standard error of the mean is of order 0.8 in both cases, yielding the following ranges of probabilities: $P(A|Z) = 0.6$ to 0.98, and $P(C|Z) = 0.3$ to 0.94.

An alternative procedure is to carry through the above calculations for each respondent, and then take the mean of the log-odds on A and C. This leads to the final values $\Lambda(A|Z) = -0.2\pm0.2$ and $\Lambda(C|Z) = -0.5\pm0.3$, that can be translated into the following ranges of probabilities: $P(A|Z) = 0.3$ to 0.5, and $P(C|Z) = 0.1$ to 0.4. We see that the difference between the two sets of results is within the estimated errors.

5. Discussion

The author found the results of this exercise to be surprisingly "liberal." When respondents think their way through the three steps of the ZOOM, they seem to be much more likely to accept the possibility of anomalous redshifts or redshift periodicities than would have been the case if they had been asked to jump straight to the prior probabilities of A and C. The difference probably arises in the following way.

If one is asked to estimate the probability of A or C (or some equally strange hypothesis), one will instinctively base the estimate on current knowledge. That is to say, the estimates will really represent what we call $P(A|OZ)$ and $P(C|OZ)$. On the other hand, if one is first asked to consider the possibility that present-day physics is incomplete, and that there are further great discoveries to be made, the chain of thought is completely different. Once one considers that our knowledge of physics may be very limited, and that there may be further truly extraordinary discoveries to be made, it becomes much more difficult to remain narrow-minded concerning strange hypotheses.

Of course, one is not used to making such decisions in normal everyday science. In a sense, they are not scientific decisions, since they cannot be based on known data or known theory. On the other hand, such decisions are essential for the purposes of scientific inference. One may perhaps refer to them as "meta-scientific."

It may well prove that respondents of this questionaire, drawn from participants in the TIME Workshop, were not representative of most physicists or most astrophysicists. It may be that they are, on average, more "liberal" in their scientific outlook than most scientists would be. This is a hypothesis that can be checked experimentally, by giving the same questionaire to a group of scientists drawn randomly from the physics and astrophysics professions. It is hoped that it will be possible to carry out this test in the near future.

References

Arp, H. C.: 1987, *Quasars, Redshifts and Controversies*, (Berkeley, CA: Interstellar Media)

Arp, H. C.: 1997, *Astroph. & Sp. Sci.*, Proc. of this conference

Good, I. J.: 1950, *Probability and The Weighing of Evidence*, (London: Griffin)

Jeffreys, H.: 1931, *Scientific Inference*, (Cambridge University Press)

Kuhn, T. S.: 1962, *The Structure of Scientific Revolutions*, (Chicago: University of Chicago Press)

Napier, W. M. and Guthrie, B. N. G.: 1993, *Progress in New Cosmologies – Beyond the Big Bang* Eds. Arp, J. C., Keys, C. R., and Rudnicki, K., (New York, Plenum), 29

Sturrock, P. A.: 1973, *ApJ*, **182**, 569

Sturrock, P. A.: 1994, *J. Sci. Exp.*, **8**, 491

Tifft, W. G.: 1995, *Astroph. & Sp. Sci.*, **227**, 25

Tifft, W. G.: 1997, *Astroph. & Sp. Sci.*, Proc. of this conference

NEW APPROACHES TO COSMOLOGY
GRAVITATION AND TIME IN GENERAL RELATIVITY

DAY 2
Pima Community College, West

Figure 1. Sketch art courtesy Janet A. Tifft

TWO UNIVERSES

G. BURBIDGE
*Center for Astrophysics and Space Sciences and
Dept. of Physics
University of California, San Diego
La Jolla, California 92093-0424*

Abstract. The community of astrophysicists see the universe in two different ways. Most of them believe that the evidence points to a hot big bang universe. The minority, largely represented at this meeting, believe that if proper weight is given to all of the observational evidence, rather than only a part of it, a very different model of the universe is indicated. Here I summarize that part of the evidence ignored by the majority, which shows (a) that not all redshifts are due to expansion, and (b) that galaxies and other coherent objects probably did not form from the condensation of diffuse gas.

1. Introduction

We all live in one universe, but we see it through two very different pairs of spectacles. Thus, we are really talking about two universes. I shall call them A and B. Almost everyone at this meeting sees it one way (B) and the bulk of the astronomers who attend almost any other meeting on cosmology and extragalactic astronomy (say the Princeton meeting held in June 1996) see it the other way (A).

What are the major differences between the two positions?

There is one overriding effect which separates the two positions. Observational results which are not easily explained by conventional ideas are disregarded or claimed not to be correct by those in the majority position (A), but they are taken seriously by those in the minority (B).

If this occurred on a small scale, it would be considered natural, since many early observational or experimental results in physical science are initially questioned if theory has not already predicted them. However, in

Astrophysics and Space Science **244:**169-176,1996.

the present situation we are dealing with something on a much larger scale which has built up over some thirty years or more.

2. Universe A

Since the fundamental work of Hubble and Humason on galaxies which started in the 1920s, and which was based on the earlier observations made by V. M. Slipher, the redshifts were interpreted using Friedmann and Lemaitre's solutions to Einstein's equations as showing that we live in an expanding universe. From about 1930 on, this view was generally accepted.

While it was speculated at early times by a minority that the redshift might not be due to the expansion of the space-time metric, it took more than 60 years for an observational test to be carried out which showed that the bulk of the shift for normal galaxies is an expansion shift and is not due to other causes. Time reversal of such a universe would lead to contraction to an exceedingly small volume. Thus the concept which has become widely held is that the universe has expanded from a very small volume "*the primeval atom*" of Lemaitre and, the "Big Bang" in the common parlance today.

From the 1930s on this model was generally accepted but no physics was put in. Starting in the 1940s attempts were made to understand how all of the chemical elements could have been made from fundamental particles at such an early stage. Gamow, and Alpher and Herman showed that deuterium and helium could be built in an early universe, but it was impossible to build elements beyond mass 5 (which is not stable). Starting in 1946, Hoyle proposed that the heavier elements were built in the interior of stars, and in a series of investigations culminating in 1957 Cameron, and Burbidge, Burbidge, Fowler and Hoyle (1957) showed that all of the elements beyond helium, together with some helium, were built in stellar interiors (we are neglecting here Li, Be and B, whose origin is probably in cosmic rays, etc.).

In the course of their studies of the physics of the early universe, Gamow and his colleagues realized that with an expanding cloud of protons, neutrons, electrons, positrons and neutrinos, there would also be a hot ball of radiation which would cool as a black body as the universe expanded. They estimated that its temperature would be 5-10° K at the present epoch, but they did not consider the possibility of detecting it.

Had the connection been made, it would have been clear that the observations of McKellar (1941) and his colleagues on the interstellar CN, CH, and CH^+ molecules already suggested that a microwave background flux must be present with an intensity such that $T \leq 3°$ K. Also it had become clear that to attain the observed helium abundance from hydrogen burning in stars, it was required that there be a radiation field produced either by galaxies at an earlier stage in their evolution, or in an early universe (cf Bondi, Gold and Hoyle 1955; Burbidge 1958), and that the temperature of this radiation if it were transformed into black body radiation would be about 2.7° K, though this was not specifically stated by either Bondi et.al. or Burbidge.

In the early 1960s, Robert Dicke with the Princeton group traveled again along the road laid out by Gamow et.al., but went further and started looking for the black body radiation, on the assumption that there was a big bang. Also calculations of the D, He^3 and He^4 produced in a big bang were made (Hoyle and Tayler 1964, Peebles 1966, Wagoner, Fowler and Hoyle 1967). With the discovery by Penzias and Wilson (1965) of the microwave background radiation, and confirmation of its black body form (cf COBE 1990) it was concluded that we have a strong observational basis for the belief in the hot big bang model.

Galaxies in this scheme must have formed from the collapse of higher than average density fluctuations which are invoked ad hoc in this model. Thus all of the phenomena involving discrete objects – galaxies, and QSOs, and anything else that is found, must be attributed to the evolution of the density fluctuations as a function of time and space.

The discovery of phenomena which imply that not everything can be traced back to evolution and gravitational interaction, means that this picture is either incomplete or just plain wrong.

Above all, it is necessary to argue that apart from the small effect due to peculiar motions of galaxies ≤ 300 km s^{-1}, and the random motions expected in groups and clusters, the whole of the observed redshifts are expansion shifts. The belief that all of the groups and clusters have characteristic ages corresponding to the age of the universe leads to a general belief that for all such systems the virial theorm holds, and it is this argument which forms one of the observational bases for the belief in the large scale existence of dark matter.

To summarize, believers in category A (most astronomers) require that

(i) Almost all of the redshifts of all extragalactic objects are due to the expansion of the universe. Their distances can be determined from the redshifts.

(ii) The universe began in a hot big bang which already contained the seeds of galaxies.

(iii) All groups and clusters are bound so large amounts of dark matter are present.

3. The Universe B

There have been a number of observational discoveries, many of which are being discussed at this meeting, which if accepted contradict (i) a part of (ii) and (iii). It is these which force us toward viewpoint B. We discuss the evidence under several headings.

3.1. EXPANSION PHENOMENA IN GALAXIES AND IN GROUPS AND CLUSTERS

Forty years ago Ambartsumian (1958 and other references) by analogy with expanding associations of O and B stars argued that the evidence was growing that there were expanding associations of galaxies. He was particularly intrigued by early work on compact groups like Stephan's Quintet, Seyfert's Sextet, and VV 172 each of which has one member with a highly discrepant redshift. We now know that out of 100 compact groups cataloged by Hickson nearly 40% have one member with a discrepant redshift greater than 1000 km s^{-1} from the mean of the others. Many attempts have been made to explain this phenomenon as due to statistical accidents – but they fail.

This suggests that some galaxies have quite significant intrinsic redshift components, or that very high speeds of ejection are present in some cases (both discrepant redshifts and blueshifts are present).

Even if the galaxy with a discrepant redshift is ignored the remaining group is a problem for the conventional view. This is because the groups are so small that the crossing times are $\sim 2 \times 10^8$ years and the ratio of galaxy diameters to size of the group is such that there would be many inelastic collisions if the group were as old as 10^{10} years; i.e. the group members should not exist as separate dynamical systems.

In some cases, the X-ray flux from hot gas shows that much dark matter must be present, but that does not solve the dynamical or timescale problems. In large clusters also it is well known that the kinetic energy of the visible matter is much greater than the potential energy and this is usually interpreted, applying the virial expression, as evidence of the presence of dark matter. In clusters which from their form and regularity are clearly relaxed, this is appropriate but in many systems, e.g. the Virgo cluster and the Hercules cluster. it is clear that they are far from being relaxed.

There are two ways of interpreting these results. The first is to argue with Ambartzumian that these systems are all coming apart. This is the case if we attribute the whole of this redshift dispersion in the group or cluster to the Doppler effect. The consequences of this is to conclude that the systems are much younger than $\sim 2/3H_0$. Thus the conventional view of galaxy formation cannot be correct.

The alternative explanation is to suppose that part of the differential redshift is due to an intrinsic component so the actual velocity dispersion which determines the crossing times and the amount of dark matter (through the virial theorem) is significantly reduced.

3.2. ANOMALOUS REDSHIFTS IN QSOS AND RELATED OBJECTS

Since about 1967 extensive evidence has been found which shows that many QSOs with large redshifts are physically associated with galaxies with low redshifts. This work has been extensively described by Arp and his collaborators and by Burbidge et.al. in a series of papers (cf Arp 1987, Burbidge 1996, Hoyle and Burbidge 1996, Burbidge et.al. 1990 for many references). The evidence involves a few cases where there are luminous connections between low redshift galaxies and high redshift QSOs, many statistical samples, and many geometrical configurations which strongly suggest that QSOs are ejected from the galaxies. If z_0 is the observed redshift, z_c is the cosmological redshift, z_d is the Doppler (velocity) component of the redshift and z_i is the intrinsic redshift component

$$(1 + z_0) = (1 + z_c)(1 + z_d)(1 + z_i).$$

For nearby galaxies with $z_c \leq 0.02$, and close QSOs, $z_0 \approx z_i$ but there are some associations involving fainter galaxies where $z_c \geq 0.2$.

There is also a correlation between the QSO-galaxy angular separation and the distance of the galaxy, of the form $\theta d \cong$ constant. (Burbidge, Strittmatter and O'Dell 1972; Burbidge et.al. 1990). Also some QSOs have redshifts approximately equal to the companion galaxies. Thus sometimes $z_i \sim 0$.

The existence of finite z_i in many QSOs must be due to something intrinsic to the QSO. It means that all distances derived from the Hubble relation for QSOs are highly suspect. The intrinsic redshift cannot be a Doppler shift (cf Burbidge and Burbidge 1967) because we see no blueshifts – which would dominate. Thus there is hard evidence of intrinsic redshifts in a special class of objects which are dominated by a non-thermal process: z_i can range from 0 to ~ 3 at least.

3.3. EJECTION OF COHERENT OBJECTS

The geometric configurations in many galaxies and ejected QSOs and other systems (e.g. NGC 4258 and the QSOs aligned across it) – (cf Burbidge 1995), and M84 lying in exactly the same position angle as the non-thermal jet coming from the center of M87 (Wade 1960), show clearly that coherent objects with masses up to those of galaxies can be ejected from parent galaxies.

Of course this suggests that galaxies and related objects have a very different origin from that believed by those who subscribe to A.

3.4. QUANTIZED REDSHIFTS

The existence of quantized redshifts as an observational fact is now well established. It has been discussed extensively here.

(a) Normal Galaxies

Starting more than 20 years ago Tifft (1976) showed that the differential redshifts of galaxies in the Coma cluster showed distinct periodicities with a value of $\Delta cz \sim 72$ km s^{-1}. These results were confirmed by Weedman and over the years the phenomenon has been found in pairs of galaxies (Tifft and Cocke 1989) and in the redshift differences between satellite galaxies and the central galaxies in small groups (cf Arp and Sulentic 1985). Tifft (1995, 1996) has extended the work to the global scale, and Guthrie and Napier (1996) have confirmed that these quantized effects, with $\Delta cz \approx 37$ km s^{-1} are found in accurate observed redshifts of normal galaxies within the local supercluster.

In addition to this, there is growing evidence from pencil beam surveys of faint galaxies that a periodicity with a large value $\Delta cz = 12800$ km s^{-1} is also present (Broadhurst et.al. 1990). Arp (1987) has also shown that in a few cases intrinsic redshift components with $c\Delta z$ up to ~ 10000 km s^{-1} are present.

(b) Quasi-Stellar Objects and Related Objects

In 1968, it was first shown that the redshifts of these objects form quantized sheets with $\Delta z = 0.061$. Peaks could be easily seen in $n\Delta cz$ from $n = 1$ to $n = 10$ (Burbidge 1968). With time the effects have strengthened. With more than 700 objects with $z < 0.2$, the same effect has been seen. Ninety-four of the objects have z between 0.055 and 0.065 (Burbidge and Hewitt 1990). Burbidge and O'Dell (1972) and later Duari et.al. (1992) carried out statistical tests on the samples and showed that the strong periodicity is real – the exact value of Δz_0 of 0.0565 and its significance is increased when the redshifts are transformed to the galactocentric frame. A second period $\Delta z = 0.0128$ is also found at high significance. In addition to these effects peaks in the wider redshift distribution have been known since the early days. These peaks come at $z = 0.3$, 0.60, 0.96, 1.41 and 1.95. It was shown by Karlsson (1977) that they can be fitted by the periodic formula $\Delta \log(1 + z) = 0.223$. There has been much debate about the peaks but when selection effects are taken out the peaks at 0.3, 0.6, 1.41 and 1.95 in particular are unquestioningly present.

All of the observed phenomena discussed above must be taken into account when we try to determine what sort of cosmological model is viable, and what pattern of formation, and evolution has been followed by galaxies.

4. Summary

Those who accept all of the observational data – those in category B, find it hard to accept the simplistic viewpoint espoused in A. The main stumbling block comes when we try to interpret the redshift simply as a distance indicator. For normal galaxies it can still be argued that the bulk of the redshift is due to expansion, together with a small term involving quantized effects which are intrinsic. There is no accepted theory of the quantized effect though Tifft and Cocke (Tifft, Cocke and DeVito 1996, Tifft 1997) are exploring various possibilities.

The redshift phenomena involving QSOs and related objects suggest:

(1) That many QSOs are not at the distance derived from the redshifts.

(2) That they are ejected from galaxies.

(3) That coherent objects in general are ejected from galactic nuclei.

(4) For Redshift components which are intrinsic, probably the masses of the fundamental particles are not the same as those in our own Galaxy.

It is possible, but not certain that these results point to a cosmology very different from that espoused in A. It certainly requires a change in the conventional approach to galaxy formation and evolution starting from a very dense phase in the universe.

How long will it be possible for the community to believe in A, and by ignoring the observational basis for B, treat it as irrelevant? This depends on the sociology of science and not on theory and observation. As long as astronomers are rewarded for following the herd, and punished for behaving independently, we are in trouble.

References

Ambartsumian.: 1958, *Solvay Conf. Reports*, ed. R. Stoops, (Bruxelles)

Arp, H.: 1987, *Quasars, Redshifts and Controversies*, (Interstellar Medium, Berkeley)

Arp, H. and Sulentic, J. W.: 1985, *ApJ*, **291**, 88

Bondi, H., Gold T. and Hoyle, F: 1955, *Observatory*, **75**, 80

Broadhurst, T. J., Ellis, R. S., Koo, D. C. and Szalay, A. S.: 1990, *Nature*, **343**, 726

Burbidge, E. M.: 1995, *A&Ap*, **298**, L1

Burbidge, G.: 1958, *PASP*, **70**, 83

Burbidge, G.: 1968, *ApJ*, **154**, L41

Burbidge, G.: 1996, *A&Ap*, **309**, 9

Burbidge, G. and Burbidge, E. M.: 1967, *Quasi Stellar Objects*, (W. H. Freeman, San Francisco)

Burbidge, G., Burbidge, E. M., Fowler, W. and Hoyle, F.: 1957, *Rev. Mod. Phys.*, **29**, 547

Burbidge, G. and Hewitt, A.: 1990, *ApJ*, **359**, L33

Burbidge, G., Hewett, A., Narlikar, J. V. and Das Gupta, P.: 1990, *ApJS*, **74**, 675
Burbidge, G. and O'Dell, S. L.: 1972, *ApJ*, **178**, 583
COBE; Mather, J. C., Cheng, E. S., Eplee, R. E., et.al.: 1990, *ApJ*, **354**, L37
Duari, D., Das Gupta. P. and Narlikar, J. V.: 1992, *ApJ*, **384**, 35
Guthrie, B. N, G., and Napier, W. N.: 1996, *A&Ap*, **310**, 353
Hoyle, F. and Burbidge, G.: 1996, *A&Ap*, **309**, 335
Hoyle, F. and Tayler R.: 1964, *Nature*, **203**, 1108
Karlsson K. G.: 1977, *A&Ap*, **58**, 237
McKellar A.: 1941, *Dom. Ap. Obs. Publ.*
Peebles, P. J. E.: 1966, *ApJ*, **146**, 542
Penzias, A. and Wilson R.: 1965, *ApJ*, **142**, 419
Tifft, W. G.: 1976, *ApJ*, **206**, 38
Tifft, W. G.: 1995, *Ap&SS*, **227**, 25
Tifft, W. G.: 1996, *ApJ*, **468**, 491
Tifft, W. G.: 1997, *Ap&SS*, these Time Conf. proceedings
Tifft, W. G., and Cocke, W. J.: 1989, *ApJ*, **336**, 128
Tifft, W. G., Cocke, W. J., DeVito, C. L.: 1996, *Ap&SS*, **238**, 247
Wagoner, R., Fowler, W. and Hoyle, F.: 1967, *ApJ*, **148**, 3
Wade C. M.: 1960, *Observatory*, **80**, 235

ANOMALOUS REDSHIFTS
AND THE VARIABLE MASS HYPOTHESIS

JAYANT V. NARLIKAR
Inter-University Centre for Astronomy and Astrophysics
Post Bag 4
Ganeshkhind, Pune 411 007, India

Abstract. There are several observations of extragalactic objects that do not appear to be consistent with the cosmological hypothesis that their redshifts arise from the expansion of the universe. These phenomena are looked at in a spacetime framework that is wider in its scope than general relativity. This framework directly incorporates the Machian notion of inertia and is conformally invariant. The consequence of this approach is that the mass of a particle may not stay constant. Two alternative viewpoints are presented to explain how large redshifts could arise from emission of radiation by particles of low masses.

1. Introduction

The velocity distance relation first announced by Hubble (1929) set the theme for the present mainstream of cosmological models. These models have the universe expanding, i.e., its typical distance scale S, separating two extragalactic objects, *increases* with the cosmic epoch t. If a typical extragalactic object, say, a galaxy G emitted light at epoch t, which is received by us today at epoch t_0, the object would exhibit a redshift z given by

$$1 + z = \frac{S(t_0)}{S(t_1)}. \tag{1}$$

Thus, if $S(t)$ has been steadily expanding, the ratio $1 + z$ will be larger, the farther back in time (t_1) we go into. Since this would also increase $(t_0 - t_1)$ and hence the distance D of the object, we have a relation of the type

Astrophysics and Space Science **244:**177-186,1996.

$$z = f(D), \tag{2}$$

with $f(D)$ increasing with D. The form of $f(D)$ is determined by the specific model chosen. For small D, this relation takes the linear form

$$z = \frac{DH}{c}, \tag{3}$$

where H is the Hubble constant, and c the speed of light.

This result does appear to hold, in the form (3) for nearby galaxies and (2) for distant ones. The latter has several sources of errors and uncertainties and so we cannot as yet fix the form of $f(D)$ with any degree of confidence. For first ranked cluster members, however, $f(D)$ does seem to provide a good fit with modest scatter (Kristian, et al 1978).

This has generated a confidence that the rule (2) applies to all extra-galactic redshifts. This paradigm is often called the *cosmological hypothesis*. Nevertheless, there are, by now several claims by observers and theoreticians that there are situations where this paradigm does not apply. Redshifts of such objects are often referred to as *anomalous redshifts*, i.e., redshifts that don't fit into the cosmological hypothesis. We begin with a brief review of the field (*see* for details Arp 1987, Narlikar 1989).

2. Examples of Anomalous Redshifts

2.1. THE REDSHIFT MAGNITUDE RELATION FOR QSOS :

Astronomers estimate distances by using apparent magnitudes. The method works provided they are looking at a class of objects which are standard candles, i.e., objects of a fixed absolute luminosity. This seems to be the case for galaxies of elliptical type that dominate a cluster, which is the reason why the relation (2) gets verified in a redshift (z) – magnitude (m) diagram. For the quasi-stellar objects, however, the $(z - m)$ diagram is a typical scatter diagram. This had been first pointed by Hoyle and Burbidge (1966) three decades ago when there were only about 100 QSOs known. Today with more than 7000 QSOs plotted on the $z - m$ diagram there is no trend discernable : certainly, there is no prima-facie correlation between m and z as predicted by the cosmological hypothesis.

2.2. QUASAR-GALAXY ASSOCIATION :

There are examples of pairs of quasars and galaxies separated by small angular deviation on the sky. Given the magnitude of the quasar we can estimate the surface density of such (or brighter) quasars on the sky. From these data we may estimate the probability of a galaxy being found within the observed angular separation purely by chance. If the probability is low (say $< 10^{-2}$) we may consider such association real. Burbidge et. al (1990) have compiled cases of such associations in which the members' redshifts do not match. Clearly if (2) holds, two objects in physical proximity of each other should have the same redshifts. If large redshift QSOs are in close proximity to bright galaxies (of low redshift) clearly the relation (2) breaks down.

2.3. CLOSE PAIRS OF QSOS :

As QSOs are rare objects (compared to galaxies) the chance of finding two QSOs with different cosmological redshifts projected within, say 60 arcsec. of each other is very small. By finding several such pairs Burbidge, Narlikar and Hewitt (1985) highlighted this anomaly.

2.4. GALAXY-GALAXY ASSOCIATION WITH CONNECTION :

Arp (1987) has pictures of pairs of galaxies in which typically a large galaxy is connected by a filament to a smaller companion. Unless the connection is fortuitous, the main and companion galaxies should show very little difference in redshift. The observed pairs, by contrast show redshift differences Δz of the order of $c\Delta z \geq 5000$ km s^{-1}. These velocity differences are too high to be explained away as velocity dispersion in a bound system. Thus the anomaly appears significant. Further, in almost all cases the companion galaxy has excess redshift whereas in a dynamical model one would expect Δz to be negative as well as positive.

2.5. COMPACT GROUPS :

Burbidge and Burbidge (1961) had highlighted the case of the discrepant redshift in the Stefan's Quintest. A few years ago Hickson (1982) compiled data on compact groups of galaxies. If these groups are real dynamically bound systems their internal velocity differences should not exceed $\sim 1000 - 2000$ km s^{-1}. Sulentic (1988) has analyzed the data and finds that a substantial fraction contain members with discrepant redshifts.

2.6. SPECIAL CONFIGURATIONS :

In addition there are several special alignments of QSOs as well as of QSOs with galaxies and of extraordinary concentrations of QSOs near galaxies (Narlikar 1989) to suggest their physical proximity. Yet the redshift differences are such that one cannot reconcile them with the cosmological hypothesis.

2.7. PERIODICITIES :

Recently Duari et. al (1992) have carried out several statistical analyses of QSO redshifts and they find that the period $\Delta z = 0.056$ occurs with a large degree of significance. This confirms an early result of Burbidge (1968) for same 70 QSOs (in the sample examined by Duari et. al there were ~ 30 times as many QSOs!).

It is also known from other studies (Karlsson 1977, Depaquit et. al 1985) that a periodicity of large amplitude is also present in the QSO redshifts, given by

$$\Delta \log(1 + z) = \text{constant} = 0.089. \tag{4}$$

For galaxy samples Tifft (1988; and references therein), Napier (1996; and references therein) have been reporting very significant periodicities of the form $c\Delta z \approx 37.5$ km s^{-1}. All these results are clearly beyond the scope of the cosmological hypothesis.

3. The Variable Mass Hypothesis

It is always argued by the conventional supporters of the cosmological hypothesis that the data described in §2 are not a serious threat to the cosmological hypothesis because of one or more of the following reasons :

a) There are subtle selection effects that are not taken into account,

b) Probabilities for observed configurations are computed a-posteriori and hence they don't mean much,

c) Effects like gravitational lensing can explain dense concentrations,

d) The observed connections are not real.

These issues, pros and cons of the cosmological hypothesis and alternative explanations that go beyond the cosmological hypothesis have been discussed by Narlikar (1989). My purpose here is to accept the reality of at least some of the anomalous effects and look for an explanation. Here I will talk of a redshift that arises from variability of masses of elementary particles in a Machian theory of gravity. Such a theory was proposed by Hoyle and Narlikar (1964, 1966) and its basic features are as follows.

Mach's principle broadly states that the inertia of matter arises from other matter in the universe. To put the statement in a mathematical form Hoyle and Narlikar (op. cit.) assumed that the spacetime geometry is Riemannian with metric

$$ds^2 = g_{ik}dx^i dx^k \tag{5}$$

for coordinates $x^i [i = 0, 1, 2, 3; x^0$ timelike, signature $(+ - - -)]$.

Now imagine particles of matter labeled $a, b, c, \ldots$ with x_a^i the coordinates of the ath particle, whose worldline will be denoted by Γ_a. Then the 'mass-function' $m(X)$ at a world point X is defined as the contribution to inertia at X, of all particles $a, b, \ldots$ etc. :

$$m(X) = \sum_a m^{(a)}(X) \tag{6}$$

where

$$m^{(a)}(X) = \int_{\Gamma_a} G(X, A) ds_a. \tag{7}$$

Here the inertia at X due to particle a is communicated by the propagator $G(X, A)$ which satisfies a conformally invariant wave equation. The simplest form of such an equation is

$$\Box m^{(a)} + \frac{1}{6} R m^{(a)} = N^{(a)} \tag{8}$$

where $\Box$ is the wave operator, R the scalar curvature and $N^{(a)}$ the number density function of particle a at point X.

The dynamical equations of this theory are derived from the variation of a simple action :

$$\sum_a \int m_a(A) ds_a, \tag{9}$$

where

$$m_a(A) = \sum_{b \neq a} m^{(b)}(A). \tag{10}$$

The action (9) may be varied with respect to g_{ik} to get the field equations and with respect to particle worldlines to get the equations of motions. The former gives, in the many particle approximation

$$\frac{1}{2} m^2 (R_{ik} - \frac{1}{2} g_{ik} R) = -3 T_{ik} + m \{ g_{ik} \Box m - m_{;ik} \} + 2 \{ m_{,i} m_{,k} - \frac{1}{4} g_{ik} m^{,l} m_{,l} \}. \tag{11}$$

These equations allow us to talk of a variable inertial mass. Since the equations are conformally invariant, we may be able to choose a conformal frame in which $m = \text{constant}$. In such a frame (11) becomes

$$R_{ik} - \frac{1}{2} g_{ik} R = -\frac{6}{m^2} T_{ik}, \tag{12}$$

identical with those of general relativity if we identify

$$\frac{8\pi G}{c^4} = \frac{6}{m^2},\tag{13}$$

G being the Newtonian gravitational constant.

However, this transformation breaks down if we choose part of spacetime which has $m = 0$. Indeed, one can show that the spacetime singularities of general relativity are due to the 'forcing' of equations (11) into the more compact form (12) even when $m = 0$ hypersurfaces exist (Kembhavi 1978). It is at such hypersurfaces that relativistic singularity is found. As we shall see later, one can avoid referring to the equations (12) and their singular solutions and instead use the nonsingular equations (11).

This is when we encounter a new interpretation for redshift that applies equally well to the regular as well as the anomalous situations.

4. A Flat Spacetime Solution

We illustrate this statement with the flat spacetime solution of the equations (11). It can be easily verified that the solution of these equations is given by the Minkowski metric

$$ds^2 = c^2 dt^2 - dr^2 - r^2(d\theta^2 + \sin^2 d\phi^2),\tag{14}$$

with the mass function

$$m = at^2, \qquad a = \text{constant};\tag{15}$$

the number density of particles being constant in the comoving reference frame (r, θ, ϕ).

We have here a flat spacetime cosmology in which light waves travel without spectral shift. How then do we explain redshift? Consider a galaxy G at a given radial coordinate r, the observer being at $r = 0$. A light ray leaving the galaxy at $t_0 - r/c$ reaches the observer at time t_0. Since the masses of all subatomic particles scale as t^2, the emitted wavelengths go as $m^{-1} \propto t^{-2}$. Hence we get the factor

$$1 + z = \frac{t_0^2}{\left(t_0 - \frac{r}{c}\right)^2}\tag{16}$$

as the ratio of the wavelength *actually emitted* by the galaxy to the wavelength emitted in the laboratory of the observer. As such the observed cosmological redshift is the consequence of the systematic increase in particle masses with the t-epoch.

This solution is observationally *no different* from the Einstein de Sitter model of standard relativistic cosmology because we can effect a conformal transformation that makes the mass function constant by choosing a conformal function $\propto t^2$. Thus, writing

$$ds_R \propto t^2 ds \tag{17}$$

the line element in the *relativistic frame* ds_R^2 becomes the familiar Einstein de Sitter line element if we make the coordinate transformation

$$t \propto \tau^{1/3}, \qquad t_0 = 3\tau_0. \tag{18}$$

The present value of the Hubble constant in the model is $H_0 = 2/t_0$.

Notice that in a well behaved conformal transformation the conformal function should not vanish or become infinite. Here we have to pay the price of choosing a conformal function that vanishes at $t = 0$: for in the relativistic frame the $\tau = 0$, $t = 0$ hypersurface has the (big bang) singularity.

The flat spacetime cosmology admits anomalous redshifts in a natural way, as was shown by Narlikar (1977), hereafter *Paper I*. Suppose the zero mass hypersurface has a kink as shown in Figure 1. The worldline of a QSO, Q (say) intersects it at an epoch $t_1 > 0$. As shown in Paper I, the particle mass function in Q starts ticking from *this epoch*. Thus at an epoch $t > t_1$ it will be $\propto (t - t_1)^2$. The interpretation of this result is simple; the particle receives all inertial contributions of $1/r$ type from a past light cone extending from t to t_1.

In Figure 1 we see a QSO, Q, and a galaxy, G, both close neighbours but the worldline of Q passes through the kink while that of G does not. For particles in G the mass function is $\propto t^2$ at epoch t. If both Q and G are at a distance r from the observer, formula (16) gives the respective redshifts as

$$1 + z_Q = \frac{t_0^2}{(t_0 - r/c - t_1)^2}, \qquad 1 + z_G = \frac{t_0^2}{(t_0 - r/c)^2}. \tag{19}$$

So we have $z_Q > z_G$ and an anomalous redshift for the QSO! Narlikar and Das (1980), hereafter *Paper II*, considered such pairs.

As illustrated in Figure 1, the worldlines of Q and G continue on both sides of the zero mass hypersurface. However, the appearance of $m = 0$ corresponds in the relativistic frame to the spacetime singularity, thus giving an incomplete (and erroneous) view of a universe 'beginning' at $\tau = 0$. In practice we may interpret the Figure 1 as describing a QSO ejected from the neighbour galaxy. *Paper II* had given a detailed dynamical study of such pairs.

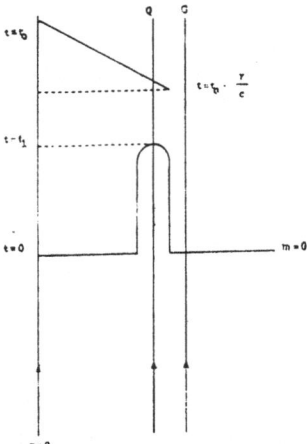

Figure 1. Spacetime diagram showing the worldlines of a QSO Q and a galaxy G crossing the zero mass hypersurface. The latter crosses the hypersurface at $t = 0$ while the former crosses it at $t = t_1 > 0$. The hypersurface has a kink which raises it from the generic value $t = 0$ to a local value $t = t_1$.

Stability : How can a static, matter-filled universe remain stable? Would it not collapse as Einstein (and even earlier Newton) found? The answer is that stability is guaranteed by the mass dependent terms on the right hand side of (11). Small perturbations of the flat Minkowski spacetime would lead to small oscillations about the line element (14) rather than to a collapse.

Quantized redshifts : Redshifts which arise from a difference in age, however, could solve the quantization problem in a natural way. Creation processes which produce galaxies at different times must originate at a zero mass surface. Close to the zero mass surface the classical action is very small and hence physics is dictated by quantum considerations. Thus one could argue that the material that emerges from the zero mass surface, emerges within a quantum mechanical realm and may do so in discrete bursts spaced at discrete intervals instead of continuously. This could lead to a quantized distribution of redshift intervals.

For example, consider a small variation of t_1 in Eq. (19), which leads to a variation of z :

$$\frac{\Delta z_Q}{1 + z_Q} = \frac{2\Delta t_1}{t_0 - r/c - t_1}. \tag{20}$$

Thus a small difference in the epoch of creation would lead to a small difference in the observed redshift. For the nearby samples considered by Tifft (op. cit.) the redshifts are small. Thus we set $z_Q \approx 0$ and neglect r/c and t_1 in (20) in comparison with t_0. Thus we get

$$\Delta z_Q \cong \frac{2\Delta t_1}{t_0} = H_0 \Delta t_0 \qquad (21)$$

This tells us that quantized steps of $c\Delta z_Q = 37.5$ km s^{-1} arise from spacings in the epochs of creation, of magnitude $(8000)^{-1}$ of the Hubble time scale H_0^{-1}.

Although this is at the moment only a crude suggestion, the alternative of trying to explain the observed quantization in a velocity-only universe seems quite daunting.

This explanation must assume that there are no other 'contaminating' redshift contributions such as the Doppler or gravitational ones which would spoil the observed exact periodicity. This remains a serious difficulty of the present explanation as, indeed of any other explanation of this effect.

5. Alternative View of Burbidge and Hoyle

As mentioned above the crucial element of the idea is that *new matter appears with anomalously high redshift*. Hoyle and Burbidge (1996) on the other hand have argued that *the anomalously redshifted matter must be very old*. The interpretation is based on the quasi-steady state cosmology (QSSC) of Hoyle, Burbidge and Narlikar (1993, 1994 a,b, 1995). I give a brief description here of how the Burbidge-Hoyle scheme operates.

The QSSC has no beginning and no end on the time axis and its scale factor is given by

$$S(t) = e^{t/P}\left\{1 + \alpha \cos \frac{2\pi t}{Q}\right\}, \qquad (22)$$

where P, Q are time scales while α is a dimensionless parameter with $|\alpha| < 1$. We will take $\alpha > 0$. Typically $P \approx 20Q$, $Q \approx 40 - 50$ Gyr. The universe, in this model, has a long-term (P) trend of expansion superposed with alternative cycles of contraction and expansion (with period Q). The dynamics are controlled by (and in turn control) the matter creation process going on near collapsed massive objects. These occur predominantly near minima of S.

Now consider a species of particles created at a typical minimum given by

$$T = T_r = -\left(r + \frac{1}{2}\right)Q \qquad (23)$$

where r is an integer > 0. The most recent minimum corresponds to $r = 0$.

Now the particles created at $t = t_r$ will acquire the mass contribution from all existing particles, but not from particles created at $t > t_r$. Thus if we observe the universe from the present epoch which lies in the cycle beginning at T_0, we will see particles created at $T_1, T_2, T_3, \ldots$, but with masses in decreasing order. The masses turn out to be in a geometric series with common ratio (of decrease from one term to next) of exp $(-Q/P)$. Consequently, the redshifts of objects made of these particles will systematically increase in a geometric series with $(1 + z)$ rising at each term by a fraction exp (Q/P). This interpretation thus has the advantage of having $\Delta \log(1 + z) = $ constant in a natural way.

6. Conclusion

It is too early to comment on the merit or disadvantage of either of the interpretations. What is needed are further observations to decide whether the anomalously redshifted matter is systematically younger or the other way round, compared to standard matter.

References

Arp, H. : 1987, *Quasars, Redshifts and Controversies*, Interstellar Media, Berkeley
Depaquit, S., Pecker, J.-C., Vigier, J.-P. : 1985, *Astron. Nach.* **306**, I, 7
Duari, D., Das Gupta, P. and Narlikar, J. V. : 1992, *ApJ.* **384**, 35
Burbidge, E. M., Burbidge, G. R. : 1961, *ApJ.* **134**, 244
Burbidge, G. : 1968, *ApJ.* **154**, L41
Burbidge, G., Hewitt, A., Narlikar, J. V. and Das Gupta, P. : 1990, *ApJ. Supp.* **74**, 675
Burbidge, G., Narlikar, J.V. and Hewitt, A. : 1985, *Nature*, **317**, 413
Hickson, P. : 1982, *ApJ.* **255**, 382
Hoyle, F. and Burbidge, G. : 1966, *Nature* **210**, 1346
Hoyle, F. and Burbidge, G. : 1996, *A&A.* (to be published)
Hoyle, F. Burbidge, G. and Narlikar, J. V. : 1993, *ApJ.* **410**, 437
Hoyle, F. Burbidge, G. and Narlikar, J. V. : 1994a, *M.N.R.A.S.* **267**, 1007
Hoyle, F. Burbidge, G. and Narlikar, J. V. : 1994b, *A&A.* **289**, 729
Hoyle, F. Burbidge, G. and Narlikar, J. V. : 1995, *Proc. Roy. Soc. London* **A448**, 191
Hoyle, F. and Narlikar, J. V. : 1964, *Proc. Roy. Soc. London* **A282**, 191
Hoyle, F. and Narlikar, J. V. : 1966, *Proc. Roy. Soc. London* **A290**, 143
Hubble, E. : 1929, *Proc. Nat. Acad. Sci.* (USA) **15**, 168
Karlsson, K. G. : 1977, *A&A.* **58**, 237
Kembhavi, A. K : 1978, *M.N.R.A.S.* **185**, 807
Kristian, J., Sandage, A. and Westphal, J. A. : 1978, *ApJ.* **221**, 383
Napier, W. : 1996, in *Proceedings of This Conference*
Narlikar, J. V. : 1977, *Ann. Phys.* **107**, 325
Narlikar, J. V. : 1989, *Space Sci. Rev.* **50**, 523
Narlikar, J. V. and Das, P. K. : 1980, *ApJ.* **240**, 401
Sulentic, J. : 1988 in *New Ideas in Astronomy*, Eds. F. Bertola, J.W. Sulentic and B.F. Madore, Cambridge University Press, Cambridge, p.123
Tifft, W.G. : 1988 in *New Ideas in Astronomy*, Eds. F. Bertola, J.W. Sulentic and B.F. Madore, Cambridge University Press, Cambridge, p.173

THREE-DIMENSIONAL QUANTIZED TIME IN COSMOLOGY

W. G. TIFFT

Steward Observatory
University of Arizona
Tucson, Arizona 85721

Abstract. Starting from a model of 3-d time in units of the Planck energy, it is possible to model fundamental particles and forces. Masses are associated with 3-d volumes of time; forces are related to 4-d space-time structures from which the fine structure constant can be derived. Fundamental particles may then be assembled into larger objects, up to galaxies, within which special relativity is satisfied. The component parts of an object retain a common quantized temporal structure which appears to link the spatially distributed parts together. The flow of time is associated with a flow of the common temporal structure within a general 3-d temporal space. Each galaxy evolves along a 1-d timeline such that within a given galaxy standard 4-d space-time physics is satisfied. The model deviates from ordinary physics by associating different galaxies with independent timelines within a general 3-d temporal space. These timelines diverge from a common origin and can have different flow rates for different classes of objects. The common origin is consistent with standard cosmology. The radius of temporal space replaces the standard radius of curvature in describing redshifts seen when photons transfer between objects on different timelines. Redshift quantization, discordant redshifts, and other observed cosmological phenomena are natural consequences of this type of model.

1. Introduction

In this paper we explore concepts, from fundamental particles to the cosmic scale, representing mass-energy in terms of quantized three-dimensional

Astrophysics and Space Science **244:**187-210,1996.
© 1996 *Kluwer Academic Publishers.*

time. Tifft (1996a,b) suggests that the redshift contains periodicities

$$P = c2^{-\frac{9D+T}{9}}. \tag{1}$$

This relationship arises as a modification of Lehto's (1990) equation associating permitted quantized energies with the Planck energy,

$$E = E_o 2^{-\frac{3D+M}{3}}. \tag{2}$$

The fractional root reduces the multidimensional temporal representation to the form observed in ordinary 3-d 'sigma-space'. We will begin at the fundamental particle level and extend the 3-d temporal concept to encompass a full 3-d temporal space. This 'tau-space' contains particles, the conventional building blocks of more complex 'objects' up to galaxies. The model permits redshift quantization, offers new ways to explain cosmological observations, and appears to relate to masses and forces operating at the fundamental particle level. It is locally compatible with relativity.

2. Leptons and Baryons

Figure 1 illustrates one schematic concept. Tau-space is presumed to contain a set quantized 'instants'. They define discrete structures which we perceive as particles. More realistically we may have an energy field where the structure represents boundary conditions. Scales associate with the Planck energy through doubling sequences inherent in equation (2). These permitted 'vacuum' energies will be called 'Gamma' energies. We presume a 'flow' within tau-space; objects move along 'flowlines'. As the flow progresses quantized structures evolve in discrete steps although the flow itself can be smooth. We associate one coordinate with the flow direction and call this the radial coordinate. Lateral coordinates distinguish independent flowlines. Figure 1 schematically depicts a region of tau-space which is manifest in sigma-space as a point-like particle. The particle has a mass energy defined by the associated tau-space structure. We associate the 1-d tau-space flow with the passage of cosmic time in sigma-space. Location in sigma-space is arbitrary; sigma-space is a continuous space with a continuous flow of time. From the sigma-space viewpoint a tau-space flowline is a cosmic timeline.

A philosopher will see this as a 'hidden variables' formulation connecting quantum (tau) and dynamical (sigma) behavior. A physicist will see a 'lepton', a pointlike particle with mass, and other properties, inherent in the tau-space structure. Functionally we can describe our particle as

$$\psi(x, y, z, t_c, f(\tau_1), g(\tau_2), h(\tau_3)). \tag{3}$$

f, g and h, describes the 3-d tau-space structure.

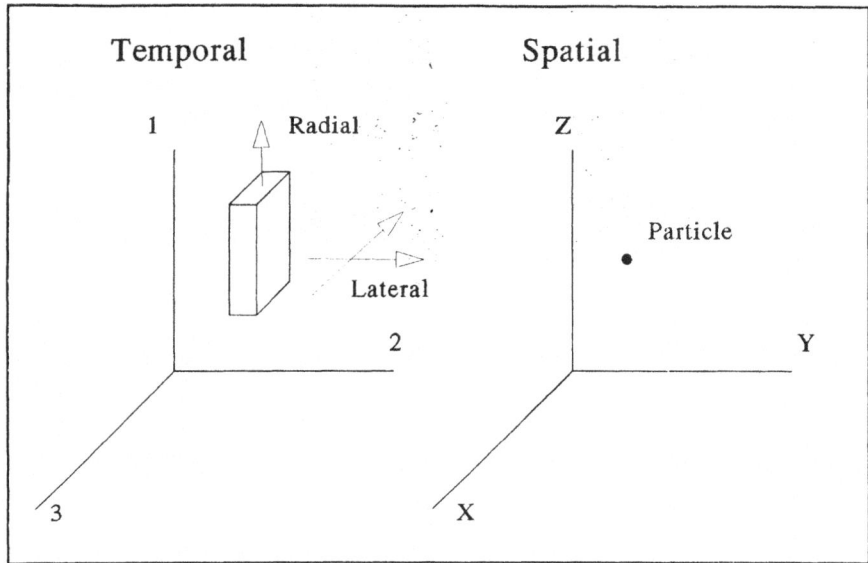

Figure 1. Representation of a particle in space as a 3-d quantized construct in a 3-d temporal space. The temporal construct has three characteristic scales. One 'radial' dimension is associated with the flow of time along a 'timeline'; 'lateral' dimensions distinguish timelines. The quantized temporal 'volume' defines the mass-energy of the particle.

Given a point particle with an associated tau structure we can restrict the model to yield appropriate statistics and satisfy special relativity. A key step is to assume that the tau region is the same for all particles in a specific object, say a galaxy or an appropriate part of one. The particles have the same clock, the 3-d tau structure; locally observable space experiences a common 4-d space-time structure. With this restriction there is no apparent reason why consistency with standard 4-d space-time physics cannot be obtained. Our particle is a fermion; such particles, defined by a common tau region, require unique spatial locations. Other particle properties can be associated with tau properties such as rotation. Spatial motion in conjunction with the passage of time and the quantized tau structure may be related to momentum quantization. Formal relationships between tau and sigma-space remain to be developed, but empirical agreements, starting with particle mass fits, suggests that a consistent model is possible. The extent, repeatability, and symmetries in the fits goes far beyond what is conventionally dismissed as 'numerology'.

Lehto's (1990) original model associated the properties of the electron with a specific power of two in equation 2. The electron-positron pair energy corresponds very closely with the energy at level (3D+M)=224. This number is not arbitrary; 224 and other key powers involved can be written as a sum of three numbers which are themselves powers of two, 224=32+64+128. The three dimensional 'volume' product for the electron-

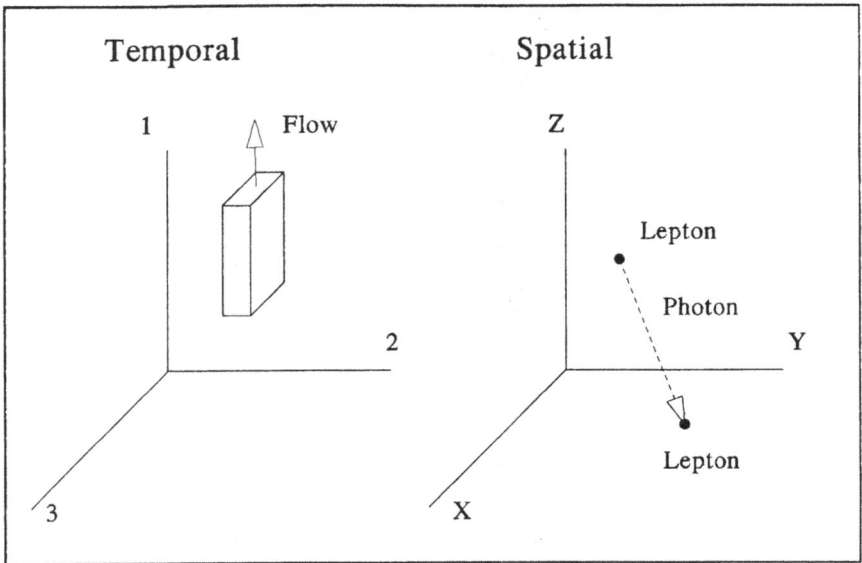

Figure 2. Leptons which belong to one 'object' are modeled as point particles in space with a common 3-d temporal structure. The common quantized structure links the particles regardless of spatial location; it defines the lepton mass and the electromagnetic force interaction strength. Photon exchange occurs through continuous space where particles are individually distinguished and experience a continuous flow of time.

positron pair contains the factor $2^{-(l+m+n)}$ where l, m and n are powers of two. We will call l, m and n the 'dimensions' of the tau-structure. More precisely they are three doublings which we associate with the electron. For reasons which are not known the properties of fundamental particles and forces seem to be intimately associated with such numbers. The minimum structure with the same dimensional character as the electron-positron pair is the (1,2,4) triad.

Figure 2 illustrates a pair of leptons within one object. There is a single tau-space structure and two points in sigma-space. The electric force operates between such particles. The strength of the force, characterized by the fine structure constant, can be quite logically and accurately derived from the tau structure. The actual effect of the force depends upon particle separation, a $1/r^2$ term from sigma-space. In conventional physics the electric force is an exchange force involving the photon. In this model there is an alternative. Given two particles, each defined by equation (3), the common tau link means that the information needed to define a force could be known without an exchange. This has major implications if we separate tau regions. The internal properties of a locally linked region do not automatically apply between separate regions. Conventional 4-d space-time physics may be only locally applicable.

The next challenge is to construct a baryon. Baryons are spatially structured particles which must also have the common clock we associate with leptons. The standard three quark construct for baryons suggests a relationship to the three tau axes, but details relating to quarks is not required here. It is necessary to introduce the strong and weak forces and the associated mesons and gauge bosons. A key to modeling the proton comes from the fact that its mass-energy is *not* matched by one of the Gamma energies given by equation (2). Lehto (1995) found that it could be represented in a second energy sequence, the 'Pi' sequence, scaled by $\pi^{-1/2}$ from Gamma values. Tifft and Lehto (1996) subsequently found that the mass energies of the other ordinary low energy stable particles can be matched using these two sequences. This includes not only the baryons, but the mesons and gauge bosons.

There are two ways to represent a particle mass-energy. It may match an energy level in the Pi or Gamma sequence, and/or it may match an energy difference, a transition, between the sequences. The low energy particles stable against strong decay are matched both ways. Allowing for symmetries present an accidental fit seems improbable. Decay processes that involve the weak force appear to involve a transition between the sequences. We associate the weak force with a such a sequence change. Excited states or resonances are not represented by levels or transitions. These involve strong decay which we associate with excitation of particle structures rather sequence energies. Figure 3 is an energy level diagram for ordinary baryons. They all either occupy or connect to level 64.00.

Two energy sequences, connected by the geometric constant $\sqrt{\pi}$, suggests that a basic structural change occurs during a transition between the sequences. One sequence may be determined by radial, the other by angular effects. The electron-positron pair was described earlier using a (32,64,128) triad of powers of two. The proton involves the cubic triad (64,64,64). The product of the two sets of 'dimensions' is the same. The gauge bosons can also be associated with a structure involving the power 64. Figure 4 schematically illustrates leptons and baryons out of which objects with common tau-space structures can be constructed.

3. Bosons and Forces

Leptons and baryons are 3-d masses. Their representation, equation 2, involves cube roots; the exponent of two ends in .00, .33 or .67 which characterizes 3-d structures. The electron falls at Gamma 75.67, the proton (and neutron) at Pi 64.00. Fits rarely deviate by more than 1% of the rest mass. Particles associated with the forces, the bosons, should be and are different. A force involves energy acting over distance. This introduces an additional dimension; 4-d structures should be involved. A 1-d observable vector can

Baryon Energy Levels

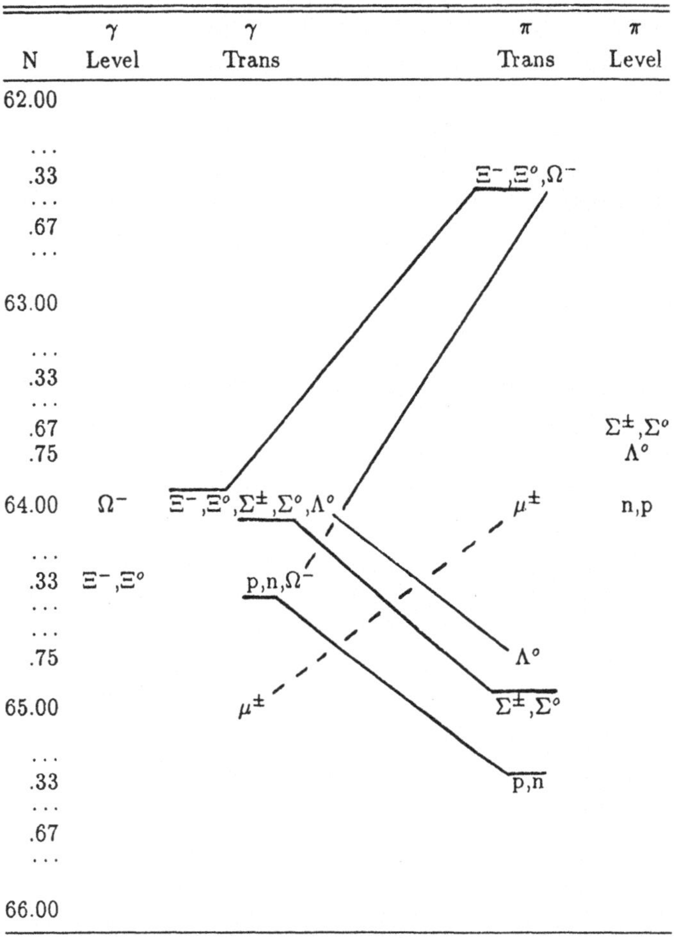

N	γ Level	γ Trans	π Trans	π Level
62.00				
...				
.33			Ξ^-,Ξ^o,Ω^-	
...				
.67				
...				
63.00				
...				
.33				
...				
.67				$\Sigma^\pm,\Sigma^o$
.75				Λ^o
64.00	Ω^-	$\Xi^-,\Xi^o,\Sigma^\pm,\Sigma^o,\Lambda^o$	$\mu^\pm$	n,p
...				
.33	Ξ^-,Ξ^o	p,n,Ω^-		
...				
...				
.75			Λ^o	
65.00		$\mu^\pm$	$\Sigma^\pm,\Sigma^o$	
...				
.33			p,n	
...				
.67				
...				
66.00				

Figure 3. The common baryon mass-energies are represented as quantized levels, and/or transitions between two sets of levels (Gamma and Pi), related to the Planck energy. Gamma levels are negative cube or forth roots of 2, times the Planck energy; Pi levels are scaled by $\pi^{-1/2}$. N is the root associated with a particular level. Standard symbols indicate where mass-energies match levels and are connected where they match energy differences (transitions). The common baryons, and the muon, occupy or connect to level 64.00; except for the Λ^o particle only 3-d levels, which characterize mass-energy, are involved.

be obtained using a forth-root reduction. Assuming appropriate scaling the equivalent of equation (2) for bosons is

$$E = E_0 2^{-\frac{4D+F}{4}}. \tag{4}$$

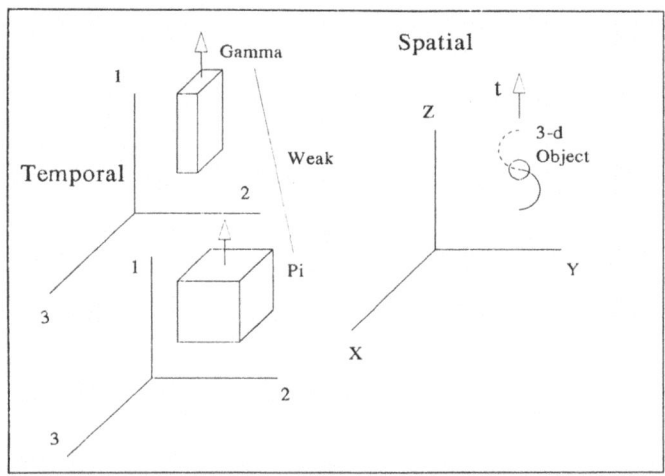

Figure 4.   An extended object in continuous space is represented by collections of fundamental particles which retain a common quantized temporal structure. The object experiences only standard 4-d space time; its 1-d timeline arises from the flow of its common temporal structure in 3-d temporal space. Structural changes within quantized temporal assemblies appears to occur during transitions between the Gamma and Pi energy sequences. This change is associated with the weak force; the strong force appears to involve excitation of the individual structures. Objects may involve associated temporal structures connected by dynamics through space, as indicated by the solid and dashed parts in the spatial frame.

Bosons should associate in some way with levels or transitions for exponents with fractional parts .75, .50, .25, and .00. Allowing for ambiguity at .00 this is what is found. The distinction between cube and fourth root fits is one clear indication that the fits are not arbitrary. Specific fourth roots (F values) can in fact be predicted.

Figure 5 illustrates the 4-d concept of bosons. A common tau region remains; all particles within an appropriate part of an object share a common 1-d timeline. A spread in sigma-space provides a fourth dimension. No unique spatial location is occupied consistent with Bose-Einstein statistics. Figure 6 illustrates some level and transition fits associated with common mesons. Figure 7 shows transitions near D=57 which match the gauge bosons. Table 1 lists fits for most common particles. Transition fits for bosons connect 3-d to 4-d levels and the low energy level fits are also uniquely 4-d, F=3. F shifts to 1 and 0 at the high energy end. The common mesons have 4-d properties; leptons and baryons match 3-d. The appearance of the gauge bosons near level 57 is also consistent with expectations as noted later.

The fine structure constant, α, is defined as 2π times the ratio of a force

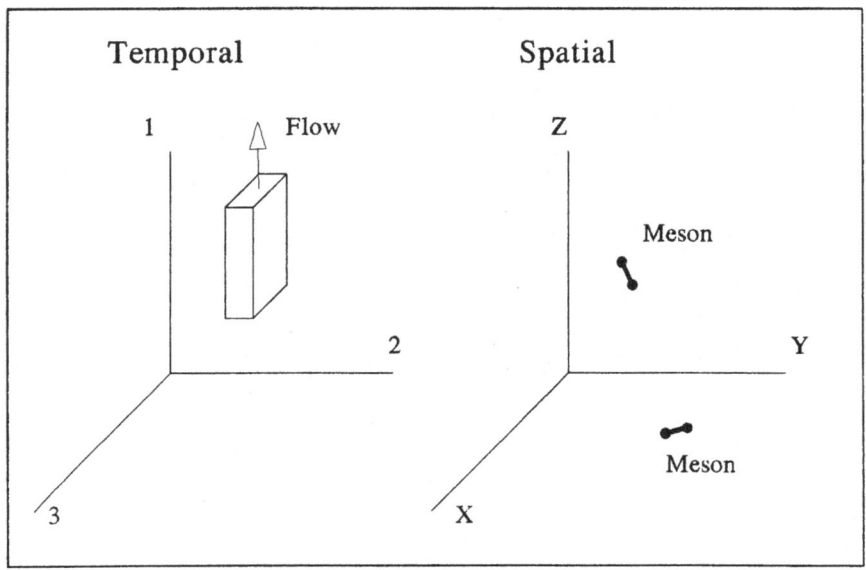

Figure 5. Bosons, associated with forces, are modeled as 3-d temporal structures (representing mass-energy) as are the pure mass-energy fermions which make up matter. In addition a related spatial interval is involved since force involves distributed energy; bosons have a 4-d character. Interaction strengths of the forces, characterized by the fine structure constants, is determined by the 4-d structure.

interaction strength to a reference strength hc. The interaction strength has units of energy times distance. In a 4-d representation of forces a 3-d volume-energy, $V_0 2^{l+m+n}$, combines with one spacelike dimension, $r_0 2^i$, to give a 4-d interaction strength $V_0 r_0 2^{i+l+m+n}$. r_0 can be viewed as a transit time consistent with tau scales. The reference interaction strength, $V_0 r_0$, cancels in the ratio leaving only power of 2 scaling terms. The 4-d ratio reduces to 1-d in space as $2^{-(i+l+n+m)/4}$; the fine structure constant is

$$\alpha = 2\pi 2^{-(i+l+n+m)/4}. \tag{5}$$

The choice of l, m and n seems arbitrary, but for several reasons the minimum (1,2,4) triad scaled like the electron-positron pair appears to be appropriate. It is the minimal such structure, it generates observed values of α, and it uniquely predicts observed F values associated with mesons. Table 2 gives values of α and/or α^{-1} for a series of i values stepped in powers of 2. For $i = 32$, a displacement consistent with the minimum doubling dimension of the electron, we find $\alpha^{-1} = 137.045$ very close to the value associated with the electric force.

At smaller intervals α assumes a value close to one, consistent with the strong force. The F values are especially interesting. At scales above $i = 2$ the F value is 3 which is what is observed for low energy mesons. At higher energy both observed and predicted F values shift to 1 and then

TABLE 1. Particle Masses

Particle	Mass (MeV)	Mass (MeV)	Level	Dev (MeV)	Transition	Dev (MeV)
p^+, n	938.3	939.6	$\pi_{64.00}$	1.0, 2.3	$\gamma_{64.33} - \pi_{65.33}$	$-8.3, -7.0$
Λ^o	$\cdots$	1115.6	$\pi_{63.75}$	1.0	$\gamma_{64.00} - \pi_{64.75}$	11.6
$\Sigma^{\pm}, \Sigma^o$	1189.4	1192.5	$\pi_{63.67}$	8.5, 11.6	$\gamma_{64.00} - \pi_{65.00}$	$-3.2, -0.2$
	1197.3			16.5		4.7
Ξ^-, Ξ^o	1321.3	1314.9	$\gamma_{64.33}$	$2.8, -3.6$	$\pi_{62.33} - \gamma_{64.00}$	7.0, 0.5
Ω^-	1672.4	$\cdots$	$\gamma_{64.00}$	11.2	$\pi_{62.33} - \gamma_{64.33}$	15.4
$\pi^{\pm}, \pi^o$	139.6	135.0	$\pi_{66.75}$	0.2, -4.4	$\gamma_{66.00} - \pi_{65.75}$	2.9, -1.7
$K^{\pm}, K^o$	493.7	497.7	$\gamma_{65.75}$	$-0.2, 3.8$	$\pi_{63.33} - \gamma_{64.75}$	$-6.4, -2.3$
η^o	$\cdots$	548.8	$\pi_{64.75}$	-8.5	$\gamma_{64.00} - \pi_{63.75}$	2.1
$F^{\pm}$	1971.0	$\cdots$	$\gamma_{63.75}$	-4.6	$\pi_{61.33} - \gamma_{62.75}$	-29.1
$D^{\pm}, D^o$	1869.3	1864.6	$\pi_{63.00}$	$-5.2, -10$	$\gamma_{62.25} - \pi_{62.00}$	30.6, 25.9
$B^{\pm}, B^o$	5270.8	5274.2	$\gamma_{62.33}$	$-3.4, 0.0$	$\pi_{60.33} - \gamma_{62.00}$	13.3, 16.7
Z^o	$\cdots$	92600	None	$\cdots$	$\gamma_{57.00} - \pi_{57.00}$	$-71*$
$W^{\pm}$	80410	$\cdots$	None	$\cdots$	$\gamma_{58.00} - \pi_{59.25}$	$-689**$
$e^{\pm}$	0.511	$\cdots$	$\gamma_{75.67}$	0.000	$\pi_{73.67} - \gamma_{75.33}$	0.002
$\mu^{\pm}$	105.7	$\cdots$	$\gamma_{68.00}$	1.8	$\pi_{64.00} - \gamma_{65.00}$	-1.0

$* = \pm 1700$ $** = \pm 1500$

TABLE 2. Fine Structure Constants

i	$\frac{(i+l+m+n-4)}{4}$	F	α^{-1}	α	Force
1	1.00	0	0.637	$1.57 = \frac{\pi}{2}$	Strong(B):
2	1.25	1	0.777	1.32	Strong(D):
4	1.75	3	1.071	0.934	Strong:
8	2.75	3	2.141	0.467	$\cdots$
16	4.75	3	8.565	0.117	$\cdots$
32	8.75	3	137.045	0.0073	Electro
64	16.75	3	$\cdots$	$2.85 \times 10^{-5?}$	Weak:
128	32.75	3	$\cdots$	4.35×10^{-10}	$\cdots$
256	64.75	3	$\cdots$	1.01×10^{-19}	$\cdots$
512	128.75	3	$\cdots$	5.49×10^{-39}	Gravity:

0. Higher energy quark flavors can be associated with progressively shorter scales. This pattern of F values is a specific property of the (1,2,4) triad

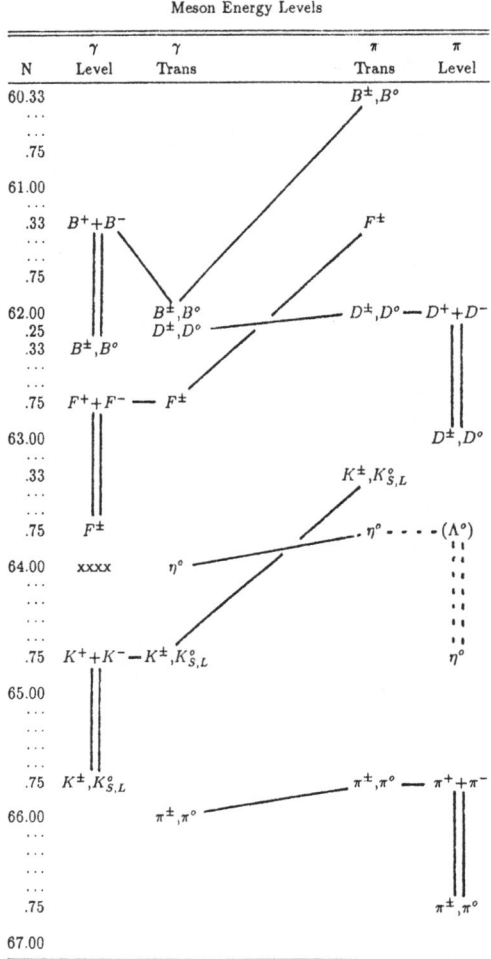

Figure 6. A representation of meson mass-energies as quantized levels, and/or transitions, related to the Planck energy. See Fig. 3 for a description of the energy scales. Symbols indicate where mass-energies match levels; the symbols are connected in the center where mass-energies match differences between levels. Double lines connect particles and particle-antiparticle pairs. Mesons show a pattern alternating between the sequences. They form doubling series within each sequence with distortions at the high energy end. Common mesons occupy or connect to 4-d levels which characterize forces. The Λ° baryon is anomalous; it fits a slot which would normally be filled with a meson (dashed lines). xxxx at N = 64.00 marks the level associated with common baryons.

and suggests that the structure is indeed basic in determining interaction strengths.

The 64 scale, consistent with baryon dimensions, returns an α value consistent with the weak force. For larger i values α drops precipitously but matches the gravitational interaction strength at $i = 512$. The gravita-

Gauge Boson Energy Levels

Level	γ Level Energy	Tran Energy	γ Neut Trans	γ Chg Trans	π Chg Trans	π Neut Trans
56.00						...
...						
...						
...						
...						
...						
57.00	212640.8	92671.1	Z^o ——————			Z^o
...						
...						
...						
...						
...						
58.00	106320.4	81099.9		$W^\pm$		...
...						
...						
...						
...						
...						
59.00 .25	53160.2				$W^\pm$	...
...						
...						
...						
...						
60.00						...

Figure 7. A representation of the gauge boson mass-energies as quantized transitions between levels related to the Planck energy. See Fig. 3 for a description of the energy scales. Symbols are connected where mass-energies match energy differences between levels. Level and transition energies are shown. A 4-d association for the Z boson is ambiguous, .00 can be 3-d or 4-d; the W particle does connect to a 4-d level at 59.25.

tional constant does enter into defining the Planck scale. Although some i values are skipped, the F correspondences and α matches suggest a possible unified picture of the forces. If the model is at all correct a model of the forces and particle masses may be possible which involves properties of time common to all particles within an object. Objects may be bound through their common tau structure and mediating particles may not be required. Transmission speeds in excess of c are possible and whether any force at all connects independent tau-space structures, becomes a possibility. There is no obvious inconsistency with known particle properties or standard 4-d space-time models within a single object. Bosons, for example, will be present as an effect rather than a cause.

The gauge bosons provide an example. We do not need to embezzle energy, it arises from tau-space restructuring. Because transitions are fast the gauge bosons have very short lifetimes and ranges. They do not fit energy levels where they could reside for longer periods. Such energy shifts could be the equivalent of 'virtual' particles. Since there is always a specific energy change there is always a specific particle.

We previously noted a relationship between the powers of two connecting the electron-positron pair and proton. There is a similar connection with the gauge boson energy level. The weak force appears to involve (64,64,64) type structures dispersed over scales comparable to the minimum scale of the electron, 32. The force could therefore be expected to associate with a doubling level $(32+64+64+64)/4 = 56$. This is in fact the transition level of a Z^o particle-antiparticle pair. The Z^o particle matches the Gamma to Pi transition at level 57.00. The .00 level is ambiguous as to dimension but the W particle involves .25, associated with 4-d.

One fourth root which does not occur in observed fits or α estimates is F=2. Lehto (1990), noted that the Gamma sequence energy associated with level 224/2=112 corresponds to the energy of the 21 cm transition, the electron spin flip energy in the ground state of the proton-electron configuration. The electron and proton, the only two completely stable 3-d particles, occupy the lowest populated 3-d levels in the two energy sequences.

4. Objects

Objects consist of associated tau elements. For baryons the structure involves the strong or color force which we associate with tau intervals at or near $i = 4$. If gravity can be associated at the $i = 512$ scale we may be able to construct stars and ultimately galaxies. Stars internally balance gravity with electrical forces; galaxies balance gravity with dynamics. Any balance in the temporal model presumes that characteristic time (tau) intervals are maintained to provide phase closure in a quantum mechanical system. An exchange particle concept illustrates this. If the spacing between instants is a precise multiple of wavelengths associated with underlying energy field frequencies, a force is transmitted. If the spacing is incorrect exchange particles misconnect; they pass neighboring instants in their past or future. Within or between quantized regions precise spacings are required for force linkage.

Two tau regions drifting towards one another may experience a *periodic* force as they move in and out of phase. They can merge if nothing keeps them apart. In a gravitational collapse angular momentum prevents a direct merger. The centripetal force in sigma-space may balance gravitation of tau-space. Temporal dipoles provide a possible model for galaxies; timelines, laterally distinct in tau-space, are stabilized by angular momentum in sigma-space. This fits the original view of galaxies as 'redshift dipoles' or combinations of 'states' (Tifft, 1976), and is a first step in redshift quantization. Beyond such 'molecular structure' galaxies may be high order independent particles. Large scale geometry may be determined by the nature of tau-space.

Laterally distinct regions of quantized tau-space may have no significant force linkage but they should readily see one another through continuous sigma-space. Photons, and particles, are sigma-space entities which move freely and should interact between timelines. Ordinary relativistic rules may not apply. A photon emitted and absorbed within its own tau region has identical initial and terminal tau values. No 'time' has passed for the photon; relativity applies within a common tau region. Initial and terminal tau values are not the same for a photon which changes timelines, time does pass for these photons. Lateral transit time between timelines falls outside of relativity.

Figure 8 illustrates linkage in a quantized space to the left and a continuous space to the right. Cosmic time progresses smoothly upward. A hypothetical exchange particle may link or miss adjacent instants in the quantized frame; a force requires linkage. The continuous case has a standard photon light cone. A photon will reach any object whose timeline it crosses. There is a restriction on the timing of emission and absorption, but photons from *some* point in the past history of an object are always available. Although cosmic time (tau flows) are continuous the quantized tau state of an object must evolve discretely. The tau state of a photon must update for such evolution; the photon is part of its parent object in tau-space. The spatial particle travels. Upon receipt within the same object there is no net tau difference; receipt within a different object will show lateral, and possibly radial, differences. These quantized differences convey information which appears to set photon frequencies. We see quantized redshifts.

Knowledge of photon structure is not required here, but it can be readily accommodated in the temporal model. Particles with mass have a 3-d tau structure; the radial tau dimension exists for particles with mass since temporal volume relates to mass. A 2-d structure has zero volume. The photon is consistent with being a massless 2-d energy packet entirely in lateral tau-space through which it can link timelines. It experiences no effect of the radial coordinate associated with the passage of time. A second 2-d construct can be formed combining radial and lateral coordinates. Such a pattern may fit the neutrino since it is involved in 3-d to 4-d transitions associated with particle decay.

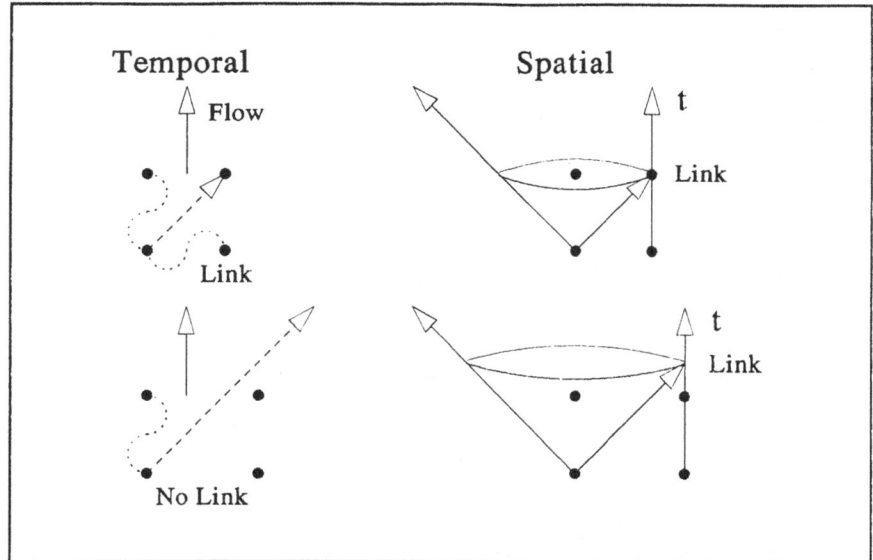

Figure 8.   Phase and force linkage in temporal space contrasted with object visibility in ordinary space. The model presumes that phased temporal quantum intervals are linked by a force appropriate to the spacing and associated fine structure constant involved. The dashed arrow symbolizes a classical exchange particle. The linkage breaks down if the spacing is incorrect; no force is transmitted. Particle exchange need not occur in temporal space; particles are a property of ordinary space. In ordinary continuous space, where photons do move between objects, we have standard lightcones; objects are visible regardless of force linkage.

5. Timeline Structure

We have invoked laterally distinct timelines to distinguish independent galaxies. We now suggest a specific tau-space model to construct a toy cosmology. We suggest a singular 'big-bang' event in *time* – something like a particle pair formation. Time, unlike space, can have a unique origin point. The result of this tau-space event (the origin of cosmic timelines) can be modeled as a simple spherical temporal expansion, the tau flow. Figure 9 shows individual unbound timelines radial about an origin. Within local sigma-space regions there is no obvious difference between such a model and the usual big-bang as far as early universe effects are concerned. The large scale picture is different, tau-space timeline structure determines the cosmological properties.

Sigma-space observers, within individual synchronized regions on individual timelines, will readily see galaxies on other timelines, and will recognize that cosmological space is curved. 3-d sigma-space should appear to be curved into a spherical geometry set by the radius of tau-space. The geometry should be well described by the ordinary Robertson-Walker metric, with the curvature term replaced by the tau-space radius – the age of the

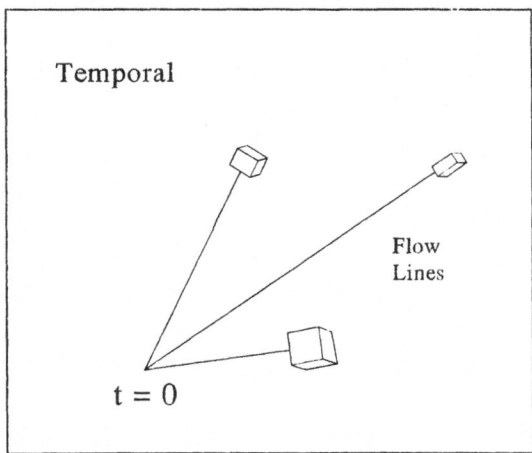

Figure 9. Individual flowlines diverging from a common origin in 3-d temporal space. Individual objects experience standard 4-d space time physics associated with its common temporal structure on a single flow line. The complete 3-d structure of diverging timelines determines larger scale physics and cosmology. Different flow rates, associated with different structures (hence mass or morphology), can be manifest in space as different Hubble constants for different classes of galaxies.

Universe. The curvature is not set by gravitation which may be largely limited to synchronized regions. If the expansion is well advanced local space should look very flat; unless flows become turbulent sigma-space should approach a flat space asymptotically. The spherical model also gives a possible rationale for the cube-root rule in equation (2). The doubling of a volume increases the radius by the cube root of two.

The above geometry has important observational consequences. Unbound timelines diverge. The photon transit time between unbound timelines increases with cosmic time; from the sigma-space viewpoint space is expanding. There is a difference from conventional expansion, however; not all rates need be the same. 'Lightweight' regions may expand faster than 'heavyweight' regions, as suggested in Figure 9. The Hubble 'constant' may be a function of galaxy type. The quantity that varies here is the 'radius' of tau-space, not cosmic time which we take to be an absolute time for all tau-space. The situation resembles an explosion in ordinary space where equipartition of energy gives different radial motion to fragments with different mass. The Hubble constant, measured using galaxies less massive than our own, may seem too large.

Timeline divergence may serve another function, it may tend to pull objects apart laterally. This may not be important at the particle level where quantized internal tau properties enter, but it could be a limiting factor for galaxies. In the bound timeline model mentioned earlier spatial angular momentum balances a tau-space quantum linkage. Divergence stress could be manifest as expansion in the plane of rotation. Divergence increases with distance while gravitation drops. A mechanism may exist which can generate flat rotation curves and slowly expanding galaxies.

If we associate the direction of the angular momentum vector of a galaxy with the radial tau direction, as implied in the force balance model, we may relate the CBR and galactic center transformations for the Sun. Within small deviations, consistent with local random motion of the Sun, the galactic radial (disk expansion) 'motion' components should be and are the same $(-36, -31)$ km s^{-1}. The tangential components are equal and opposite $(+232, -243)$ km s^{-1} as though one was a reflex of the other. The Z component is zero for the galactic transformation but $+275$ for the CBR transformation which could reflect the actual radial tau flow. The CBR Z transformation component is equal and opposite to the sum of the other components which could reflect a compensation or conservation rule.

If the underlying tau structure of a galaxy is constrained within a quantum pattern, the structure cannot yield smoothly to divergence stress; it must decay stepwise as shown in Figure 10. Lateral stress could be what drives the observed doubling processes. Such a stepwise decay seems to be required to produce redshift quantization. Tau flows, and cosmic time, can be continuous but the tau response must be quantized. At a given cosmic 'radius' in tau-space a galaxy occupies a specific temporal state. It must change that temporal state in discrete steps. A photon emitted and received within the same galaxy sees no change since its initial and final tau coordinates are synchronized. When photons connect different galaxies there will be a discrete difference in the lateral, and perhaps radial, tau values. This quantized difference must determine the apparent energy of photons; we see quantized redshifts. No transit time distortion occurs within an object since the tau properties of its photons are synchronized; photons from all parts are observed simultaneously by us at our specific 'now' location. The amount and timing of discrete changes which produces redshift quantization must depend upon the type of galaxy involved since redshift periods and phases are observed to depend upon galaxy type. The lateral timeline spacing, and quantized changes within the parent object during the photon transit time, must fix the redshift; this difference is detectable only in comparisons between objects.

If the tau flow is continuous, the passage of cosmic time and the real expansion of sigma-space can be continuous. Such a real spatial growth should, and possibly does, produce a small continuous classical redshift effect added to the discrete effect. The effect will be nonlinear since timeline spacings are growing during photon transit times. This could be the origin of the continuous 'cosmological' correction applied to observed redshifts

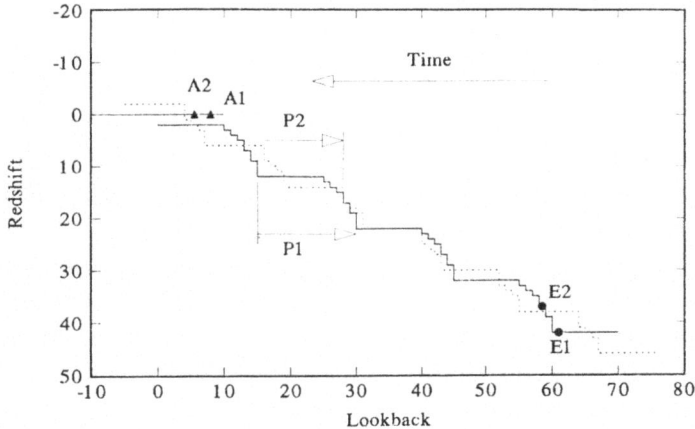

Figure 10.   Temporal structures associated with galaxies are presumed to evolve in discrete steps as objects flow outward from a common temporal origin. The evolution involves 3-d volume doubling which is detected in space as the cube-root and ninth-root periods. The quantized redshift arises when discrete temporal differences are detected for photons arriving from different timelines. Variability can occur if galaxies change state between observing epochs. Different types of objects evolve in different characteristic periodic patterns as shown. Long plateaus of constant redshift may make the Universe look more structured than it really is. The axis scales are schematic only.

prior to a periodicity analysis.

$$z_{corr} = 4[(1 + z)^{1/4} - 1] + \cdots. \tag{6}$$

This correction was derived (Tifft 1991b, Cocke & Tifft 1989) by assuming that quantum intervals varied in proportion to the square root of the time dependent Hubble constant. Using the classical relationship between H(t), q_0 and z, one can derive equation (6). Observationally, when q_0 is set equal to 1/2 observed redshift periods correspond very closely with equation (1). The relationship of quantum intervals to $H(t)^{-1/2}$ is consistent with the temporal model. The Hubble constant should be a measure of the time rate of change of temporal volumes. Since volumes depend upon t^3, H should vary as t^2, and intervals could show an $H(t)^{-1/2}$ dependence arising from the classical spatial expansion.

6. The Redshift

To illustrate cosmological effects it is instructive to calculate redshifts for some timeline geometries and different rates of tau-space expansion. In standard cosmology the redshift is given by Lemaitre's relationship; the redshift is determined by the ratio of the scale of the universe at photon absorption compared to the scale at emission. For a radius of curvature, R, and the dimensionless redshift $z = V/c$

$$\frac{R_a}{R_e} = 1 + z. \tag{7}$$

$(1+z)$ is the frequency ratio ν_e/ν_a. In the temporal model R is the tau-space radius of the photon's parent object at emission or absorption. Equation (7) should apply to any given object since R is proportional to time and time ratios are inverse frequency ratios. Nonlinearities, deceleration effects, may be different in the temporal model. The temporal model actually assumes that the redshift involves ratios of discrete states which evolve stepwise in response to changes in R as illustrated in Figure 10. To first order we ignore the steps and use a continuous variable R. Equation (6) apparently removes any nonlinearities due to growth during transit times. The following discussion is not dependent upon quantization effects.

We can assume that at emission all photons have standard properties; fundamental particles and energies need not evolve once assembled in the early universe. We assume it is the evolution of larger tau structures, objects, that generates the redshift. It acts like a response to changes in the energy density within an object, but no actual model exists. To allow for different rates of tau flow the radius of an object in tau-space R can be written as

$$R = st_c. \tag{8}$$

t_c is a continuous cosmic time and s is a scale factor; $s = 1$ for our Galaxy.

Several conditions must be met for transmission of photons between objects. We see at any instant those photons which arrive at our sigma-space location at a common cosmic time, $t_a = $ 'now'. Each photon in the set which comes from a different timeline must have left its timeline (been emitted), at a cosmic time, t_e, such that the transit time between the timelines is equal to the cosmic time difference between emission and absorption (now). If all regions in tau-space flow at the same rate there is a simple relationship between lookback time, z and ordinary distance, lookback time is proportional to z and distance. If regions of tau-space flow out at different rates then the quantities are *not* so simply related.

The tau radius of an object when a photon is absorbed is related to its radius at emission as

$$R_a = R_e + s \times t_t, \tag{9}$$

where t_t is the transit time between timelines. Using equation (7) the observed redshift z is given by

$$\frac{R_a}{R_e} = 1 + z = 1 + \frac{st_t}{R_e} \qquad z = \frac{st_t}{R_e}. \tag{10}$$

We can replace the transit time with d/c where d is the distance the photon travels in sigma-space. Using this and equation (8) at t_e we can summarize several useful forms for z,

$$z = \frac{st_t}{R_e} = \frac{sd}{cR_e} = \frac{d}{ct_e}.$$ (11)

Unbound divergent timelines, as shown in Figure 9, can be characterized by separation angles, θ, measured at the origin instant. Ignoring the nonlinearity due to the growth of R during photon transit, the distance between lines is $d \approx R_e\theta$. From equation (11)

$$z = \frac{s\theta}{c}.$$ (12)

We see that z increases with distance (θ), is proportional to s, and is independent of time if s, c and θ are constants. Equation (12) has profound implications. Objects with different s values which are close together in sigma-space will have progressively discordant redshifts as the distance from an observer is increased. The difference scales with θ so as seen from one another or from very nearby the effect vanishes. This is a simple case of objects with different H values seen from the same distance.

If flow rates depend upon mass, then dwarf and late spiral galaxies may have larger flow rates than giant ellipticals. Given an aggregate of galaxies the rapidly flowing objects will spread, leaving a core containing the massive objects. Viewed from a distance we will see radial morphological sorting. The effect applies to any aggregates, including superclusters where the effect is observed but where dynamics is not applicable. The effect is independent of cluster or group richness, again as observed for radial elliptical/spiral ratios in groups and clusters. In addition to radial morphological sorting, redshift gradients should be present. The excess redshift of dwarf companions noted by Arp, the higher redshifts of outlying spirals in various clusters, and redshift-morphology correlation within redshift-magnitude bands in the Coma cluster (Tifft 1974a,b) are all predictable effects.

The effects of separation of objects in time produces interesting effects relating to lookback times. The condition that photon transit time must equal the cosmic time interval between emission and absorption requires

$$t_a - t_e \approx \theta R_e.$$ (13)

Using equation (8) to replace R_e and solving for t_e yields

$$t_e \approx \frac{t_a}{1 + \theta s}.$$ (14)

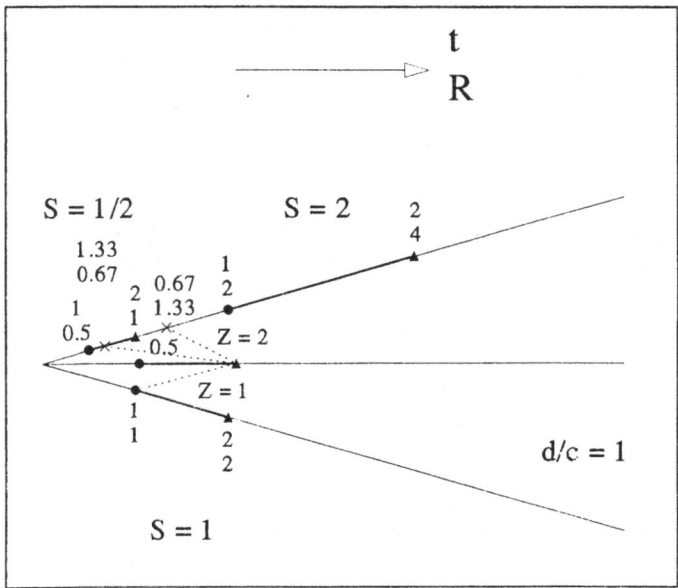

Figure 11.   The geometry of diverging timelines with differing rates of expansion, s, in temporal space. An observer, on the central timeline with s=1, simultaneously receives photons from three galaxies with s=1, 1/2 and 2 on timelines the same angular distance from the observer. Filled circles show the radius, R (lower number in vertical pair), at cosmic time t=1 (upper number). The radius at t=2 is shown with filled triangles; the scale is set (transit time = d/c = 1) so for s=1, photons are received at t=2. Since the lateral spacing of objects at t=1 depends upon s, photons seen at t=2 must have been emitted at different cosmic times and radii, shown by x symbols. Lookback times and redshifts are different for different classes of objects on equivalent timelines.

The photons received (at t_a) leave their parent objects at different times determined by s values. Lookback is greater for spirals and dwarves than for ellipticals at a given θ. The discrepancy increases with distance but vanishes for nearby objects as θ approaches zero.

Figure 11 shows why this happens using three diverging timelines. The central line represents our Galaxy with $s = 1$. The lower timeline contains a galaxy which also has $s = 1$. The upper timeline, at the same θ, has galaxies with $s = 2$ and $s = 1/2$. Pairs of numbers along timetracks give the cosmic time (upper) and tau-space radius (lower) at points along the timelines. Each object is plotted at cosmic time 1 and 2; three photons are received on the central track at $t = 2$, $R = 2$. The scale is defined so a photon emitted at $t_e = R_e = 1$ on an $s = 1$ track will reach the $s = 1$ observer at $t_a = 2$; the transit time is 1 and at absorption the parent object has $R_a = 2$. Equation (7) shows that $z = 1$ is observed. Although there are dashed lines connecting points these do not denote photon tracks. The photon is a sigma-space entity which travels through sigma-space. In tau-space it is viewed as fixed in its parent structure where it evolves in step with the object.

On the upper timeline the galaxy with $s = 1/2$ is closer to the central timeline at $t = 1$; a photon emitted at that time will reach the observer before $t = 2$. The photon which will connect at $t = 2$ must be emitted later so additional timeline divergence nullifies the reduced transit time $t_a - t_e$. The opposite occurs for the $s = 2$ object, the photon must leave before $t = 1$. Table 3 collects information for photons received at $t_a = 1$ and 2. Some consequences of the redshift and lookback differences are quite interesting.

TABLE 3.   Examples Using Divergent Timelines

t_a	θ	s	t_e	R_e	R_a	z
2	1	2	2/3	4/3	4	2
		1	1	1	2	1
		1/2	4/3	2/3	1	1/2
	1/2	2	1	2	4	1
1	1	2	1/3	2/3	2	2
		1	1/2	1/2	1	1
		1/2	2/3	4/3	1/2	1/2
	1/2	2	1/2	1	2	1

If a low mass quasar and a massive companion galaxy are observed from a distance, we see the quasar and the galaxy at different cosmic times. The galaxy, being close to the quasar, will see the quasar at a later stage in the quasar's evolution. Any effect the quasar activity we currently see had on the galaxy will have occurred long before the galaxy reached the stage we are viewing from a distance. Interactions between different types of neighboring objects will be desynchronized from what distant observers see.

Another interesting consequence relates to the apparent luminosities and evolution stages seen for various classes of galaxies. In a general deep survey we should be able to look much further back in the history of spirals than we can for ellipticals. Deep surveys may find strong evolution in spirals but more normal looking ellipticals. They may not be being surveyed to the same cosmic epoch. A more subtle effect concerns luminosities. At equal z spirals may be closer than ellipticals. Spirals should appear to brighten with respect to ellipticals at the same redshift as z is increased simply due to the inverse square law. The galaxies will be assigned to the wrong portion of the luminosity function distorting both counts and morphological comparisons. Dwarfs will be progressively interpreted as spirals. Such effects occur naturally in the temporal model.

A second simple timeline geometry involves parallel timelines. It could apply to loosely bound groups or clusters if larger scale forces are not completely unlinked. Timeline separations and transit times are constant. Equation (11) gives

$$z = \frac{d}{ct_e} = \frac{\text{Const}}{t_e}. \tag{15}$$

We see that z increases with distance as usual. The quantized redshift is a lookback effect independent of the physical expansion of sigma-space. Any spatial change may add a small correction; there should also be a small nonlinearity due to a lookback decrease in t_e as d increases. z should increase slightly faster than d and contribute to or explain the correction given in equation (6).

A second consequence of parallel timelines is that the redshift is no longer dependent on s. In loosely linked systems there could be a tendency to suppress discordant redshifts if diverging lines can be pulled toward parallel configurations. The outer fringes of clusters or even individual galaxies would still be expected to diverge and show an excess redshift. Such an effect has been reported in disk galaxies (Tifft 1991a).

A third consequence of equation (11) is that the redshift is now time dependent; as the universe ages t_e increases so z should decrease. This is the trend observed in redshift variation. Figure 12 shows parallel timelines with a layout similar to Figure 11. At time 2, for the indicated separation, z is 1 independent of s. By time 2.5 z has dropped to 0.67. Figure 13 examines a special case of such decay. The central line is the observer and the upper line is an external $s = 1$ object. On the lower timeline we allow 'new' matter to be created, by which we mean resetting R to zero at a given t. We get a rapid decay from a very large z, reminiscent of the excess redshift decay proposed in Machian models. Discordant redshifts need not be limited to quasars. If near the origin time a dispersion in R values occurred, we would see this as excess redshift scatter or discordant objects within groups well after a spectacular initial event. This might account for small discordant effects and enhanced velocity dispersions in groups or clusters conventionally ascribed to dark matter.

In summary, we have examined a model based upon two coexisting 3-d spaces, one of time, one of space. Quantum physics resides in tau-space and conventional dynamics operates in sigma-space. Although there as yet exists no formal mathematical framework linking these spaces, there is a wealth of empirical consistency with observations. This includes properties ranging from the masses and forces at the fundamental particle level

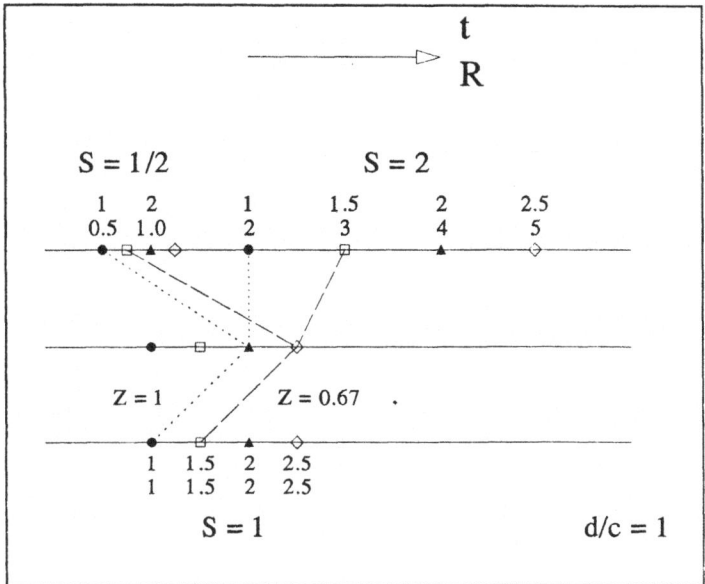

Figure 12. The geometry of parallel timelines with differing rates of expansion, s, in temporal space. An observer, on the central timeline with s=1, simultaneously receives photons from three galaxies with s=1, 1/2 and 2 on timelines the same distance from the observer. Filled circles show the radius, R (lower number in vertical pair), at cosmic time t=1 (upper number). The radius at t=2 is shown with filled triangles, and other symbols denote other times; the scale is set (transit time = d/c = 1) so photons are received at t=2 (dotted lines). Since the lateral spacing of objects is constant, photons seen at t=2 have identical emission times but different radii. Lookback times and redshifts are the same for different classes of objects on equivalent timelines, but redshifts decrease with time (dashed line case) as R continues to grow.

through redshift quantization to cosmological effects on the largest scale. Time as a 3-dimensional quantity appears to be a promising subject for investigation.

7. Acknowledgements

The author is indebted to John Cocke who was instrumental in the derivation of equation (6) and the pursuit of the CBR connection. The author is especially indebted to Ari Lehto for his work in developing 3-d temporal concepts relating to the description of fundamental particles. Carl DeVito and Anthony Pitucco also contributed ideas which assisted in development of the current 3-d temporal model.

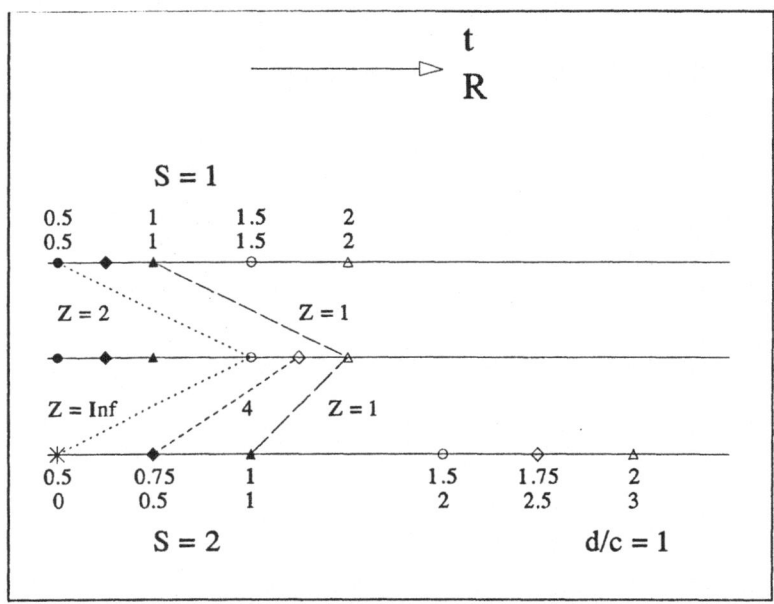

Figure 13.   The geometry of parallel timelines with differing rates of expansion, s, in temporal space. An observer, on the central timeline with s=1, simultaneously receives photons from galaxies with s=1, and 2 on timelines the same distance from the observer. This diagram is similar to Figure 12, but now 'new' material is created; the s=2 object starts with R=0 at t=0.5. The redshift of the object is initially infinite, and it rapidly decays. Refer to Figure 12 for a description of this type of diagram.

References

Cocke, W. J., and Tifft, W. G.: 1989, *ApJ*, **346**, 613

Lehto, A.: 1990, *Chinese J. Phys.*, **28**, 215

Lehto, A.: 1995, personal communication

Tifft, W. G.: 1974a, *ApJ*, **188**, 221

Tifft, W. G.: 1974b, **Formation and Dynamics of Galaxies**, J. Shakeshaft ed., (Reidel)

Tifft, W. G.: 1976, *ApJ*, **206**, 38

Tifft, W. G.: 1991a, *ApJ*, **368**, 105

Tifft, W. G.: 1991b, *ApJ*, **382**, 396

Tifft, W. G.: 1996a, *ApJ*, in press (Sept. 10)

Tifft, W. G.: 1996b, this conference

Tifft, W. G., and Cocke, W. J.: 1989, *ApJ*, **336**, 128

Tifft, W. G.,& Lehto, A.: 1996, in preparation

THE STRESS-ENERGY TENSOR
AND THE DEFLECTION OF LIGHT
IN 6-DIMENSIONAL GENERAL RELATIVITY

W. J. COCKE
Steward Observatory
University of Arizona
Tucson, AZ 85721, USA

Abstract.

We find the stress-energy tensor of a perfect fluid in the 6-dimensional spacetime proposed by Cole. Using the weak-field Newtonian approximation of general relativity gives a constant of proportionality in Einstein's field equations that differs by a factor of 4/6 from the usual one and shows that Cole's extension of the Schwarzschild metric to 6 dimensions is not valid for a gravitating mass of "ordinary" matter. A subsequent evaluation of the deflection of starlight for the 6-d spacetime gives a result that is 4/6 of the 4-d result. We conclude that if spacetime is 6-dimensional, one must find a different way to deal with gravity.

1. Introduction

There have been many attempts to embellish special and general relativity by extending the number of dimensions of spacetime to more than 4. These enclude Kaluza-Klein projective geometries (Kalligas, Wesson, and Everitt, 1995) and the compact higher dimensions contemplated by certain versions of string theory.

Cole (1978, 1985) has investigated an extension of special relativity to 6 dimensions. Three of these dimensions would be time-like, thus providing a pleasing symmetry to the 3 space-like dimensions. Cole (1980) has also devised an extension of the Schwarzschild metric to 6 dimensions, based on a preliminary version of the weak-field approximation in 6-d general relativity. He did this without regard to the possible form of the stress-

Astrophysics and Space Science **244**:211-218,1996.

energy tensor and without considering a possible change in the coupling constant in Einstein's field equations.

Lehto (1990) has also proposed such a symmetric spacetime, based on certain empirical facts that seem to call for period doubling in 6 dimensions. There are interesting connections between his work and the redshift periodicities observed by Tifft (1996a, 1996b) and discussed by Cocke and Tifft (1996). Following this, Tifft, Cocke, and DeVito (1996) discuss cosmology in a 6-d spacetime. Thus one is led from this point of view to investigate gravitation theories in 6 dimensions.

In Section 3 of this paper, we derive the form of the stress-energy tensor of a perfect fluid in 6-d spacetime. This leads to a self-consistent Newtonian approximation which requires a change in the coupling constant in Einstein's field equations, from the usual $8\pi G$ to $(4/6)8\pi G = 16\pi G/3$, G being the Newtonian gravitational constant.

In Section 5, we use the resultant theory to find a value for the deflection of starlight.

2. The Extension to 6 Dimensions

As stated above, Cole (1978, 1985) has made an interesting extension of spacetime to 6 dimensions, three of which are time-like. In this version of special relativity, the proper time of a particle may be written

$$d\tau^2 = \eta_{\mu\nu}dx^\mu dx^\nu = -dx^2 - dy^2 - dz^2 + dt_1^2 + dt_2^2 + dt_3^2, \qquad (1)$$

where the $\{dt_a\}$ are the time-like coordinates. We use the convention that Greek indices go from 1 to 6, Latin indices a, b, and c go from 1 to 3, and Latin indices i, j, and k go from 4 to 6.

Cole (1980) introduced a 6-d version of general relativity, in which he treated the weak-field Newtonian limit and calculated expressions for the precession of perihelia and the deflection of starlight. The proper time is of course generalized as

$$d\tau^2 = g_{\mu\nu}dx^\mu dx^\nu = (\eta_{\mu\nu} + \gamma_{\mu\nu})dx^\mu dx^\nu, \qquad (2)$$

where $|\gamma_{\mu\nu}| \ll 1$ in the weak-field limit.

Cole assumed that $\gamma_{\mu\nu} = 0$ if $\mu \neq \nu$, and $\gamma_{44} = \gamma_{55} = \gamma_{66} \equiv \gamma$. He then found a corresponding strong-field expression for the extension of the Schwarzschild metric and obtained values for the perihelion precession and starlight deflection which disagree with the observations. We show below, however, that the above assumptions about the metric are too restrictive.

3. The Stress-Energy Tensor in 6-D Spacetime

In this section we use the Boltzmann distribution function to define the stress-energy tensor $T^{\mu\nu}$ for a collisionless system of particles. It is then easy to show that the proper conservation law is satisfied, and we use this definition as a guide for defining $T^{\mu\nu}$ for a 6-d perfect fluid. As usual, we write (Misner, Thorne, and Wheeler, 1973)

$$T^{\mu\nu} \equiv \frac{1}{\sqrt{-g}} \int d^6 P \, P^\mu P^\nu \frac{f(x,P)}{m_0}, \tag{3}$$

where $P^\mu \equiv m_0 dx^\mu/d\tau$ is the 6-momentum of the particles, $f(x,P)$ is the Boltzmann distribution function for the particles in the 12-dimensional phase space (x,P), and g is the determinant of $g_{\mu\nu}$. m_0 is the rest mass of the particles, and we denote the 6-velocity as $dx^\mu/d\tau \equiv U^\mu$. If the particles have a discrete distribution of rest masses, $f(x,P)$ must contain δ-functions so that the particles inhabit the proper mass shells in momentum space. Otherwise, the rest mass distribution could be arbitrary.

To show the appropriate conservation law for the general case where the particles are subject to a possible non-gravitational force F^μ, we write the geodesic equation in terms of the proper time τ of the particles as

$$\frac{dP^\nu}{d\tau} = -\frac{1}{m_0}\Gamma^\nu_{\alpha\beta}P^\alpha P^\beta + F^\nu. \tag{4}$$

One may then take $\partial/\partial x^\nu \equiv {}_{,\nu}$ of equation (3) and use the identity $\partial(-g)^{-1/2}/\partial x^\nu = -(-g)^{-1/2}\Gamma^\alpha_{\nu\alpha}$ to find

$$T^{\mu\nu}_{,\nu} = \frac{1}{\sqrt{-g}} \int d^6 P \, P^\mu P^\nu (f_{,\nu} - \Gamma^\alpha_{\nu\alpha}f)m_0^{-1}. \tag{5}$$

But the collisionless Boltzmann equation is

$$U^\nu f_{,\nu} = \frac{1}{m_0}P^\nu f_{,\nu} = -\frac{\partial}{\partial P^\nu}\left(\frac{dP^\nu}{d\tau}f\right) = 0. \tag{6}$$

Using this relation and equation (4) and integrating equation (5) by parts yields

$$T^{\mu\nu}_{,\nu} = -\Gamma^\mu_{\alpha\beta}T^{\alpha\beta} - \Gamma^\alpha_{\nu\alpha}T^{\mu\nu} + <F^\mu>, \tag{7}$$

where $<F^\mu>$ is the non-gravitational force averaged over momentum space. Equation (7) is obviously $T^{\mu\nu}_{;\nu} = <F^\mu>$, showing that the covariant divergence of stress-energy tensor of the particles is properly coupled to the non-gravitational force.

Fluids are, by definition, collision-dominated; but collisions *per se* play no role in defining $T^{\mu\nu}$. We therefore conclude that equation (3) is an adequate model on which to base the definition of the corresponding $T^{\mu\nu}$ for the perfect fluid.

4. The Perfect Fluid in 6-D Spacetime

In ordinary spacetime, the stress-energy tensor for the perfect fluid is completely characterized by its eigenvalues and eigenvectors: Its four eigenvectors can be chosen to be mutually perpendicular, and the three space-like ones have the same eigenvalues $-p$, where p is the pressure. These three eigenvectors are therefore degenerate. The time-like eigenvector U^μ is the fluid velocity and has the eigenvalue ϵ, the mass-energy density. The tensor itself is then written $T^{\mu\nu} = (p+\epsilon)U^\mu U^\nu - pg^{\mu\nu}$. The space-like eigenvectors $V_{(a)}^\mu$, $a = 1, 2, 3$, are any three mutually perpendicular unit vectors that are at the same time perpendicular to U^μ. The set $\{V_{(a)}^\mu, U^\mu\}$ form an *orthonormal tetrad* (Synge, 1960) characterizing an observer moving with the fluid.

The extension of the perfect fluid to 6-D spacetime is straightforward. There are still three degenerate space-like eigenvectors all having minus the pressure $-p$ as eigenvalues. But there are now *three* time-like eigenvectors $U_{(i)}^\mu$, $i = 4, 5, 6$, corresponding to the three time dimensions. The eigenvalues ϵ_i are, roughly speaking, mass-energy densities projected onto the three time axes, and they are not necessarily equal. One can easily verify that the stress-energy tensor becomes

$$T^{\mu\nu} = \sum_{i=4}^{6}(p + \epsilon_i)U_{(i)}^\mu U_{(i)}^\nu - pg^{\mu\nu}. \tag{8}$$

The six unit eigenvectors form what one might term a "hexad," representing three observers all carrying the same space-like triad, each at rest with respect to the fluid, but moving in mutually perpendicular directions in temporal space.

It is perhaps surprising that the pressure p is "blind" to the direction of motion of the particles in temporal space. We can use equation (3) to check this conclusion by going into a frame of reference that is comoving with the fluid; i. e., one in which, at the event of interest, $U_{(j)}^\mu = \delta_j^\mu$ and where also $g_{\mu\nu} \approx \eta_{\mu\nu}$. Equation (3) then implies, for the xx-component,

$$T^{11} = p = \int d^6 P(P^1)^2 f(x, P)m_0^{-1}. \tag{9}$$

This is the average of the square of the quantity $\sqrt{m_0}dx^1/d\tau$ and contains no direct reference to the motion of the particles in temporal space. Mixed components such as T^{12} would vanish if $f(x, P)$ were an even function of

P^a, $a = 1, 2, 3$, as is the case for an ordinary 4-d fluid. In that case, we see immediately that $T^{a\nu} = 0$ for all $\mu \neq a$, consonant with equation (8) evaluated in a comoving frame of reference.

The fact that, in the comoving frame, $T^{ij} = 0$ for $i \neq j$ ($i, j = 4, 5, 6$) is more difficult to understand. The difficulty is as follows: In an ordinary 4-d fluid all particles have $P^4 > 0$, and therefore $f(x, P)$ cannot be an even function of P^4. The same holds for such a fluid imbedded in our 6-d spacetime, in which case the particles are all said to have the same time-track. But then $P^5 = P^6 = 0$ for all the particles, and still $T^{ij} = 0$ for $i \neq j$.

For the more general case, where the particles do not all have the same time-track, the transformation to the comoving frame has diagonalized the submatrix T^{ij} by an orthogonal rotation in temporal space, so that in fact it is possible (although not necessary) that $f(x, P)$ be an even function of the P^i in this coordinate system.

5. The Weak-Field Newtonian Approximation in 6-D Spacetime

In a paper on gravitation in 6-d spacetime, Cole (1980) used the weak-field approximation under the assumption that $g_{44} = g_{55} = g_{66} = 1 + \gamma$, where $|\gamma| \ll 1$, and $g_{ij} = 0$ for $i \neq j$. He used these assumptions to argue that the extension of the Schwarzschild solution for strong fields should also satisfy $g_{44} = g_{55} = g_{66}$. He did not, however, develop the field equations beyond the vacuum equations $R_{\mu\nu} = 0$. We show below that it is necessary to do this and that in fact the assumption $g_{44} = g_{55} = g_{66}$ is too restrictive.

We now work out the weak-field approximation and show that the field equations become $R^{\mu\nu} - \frac{1}{2}g^{\mu\nu}R = (16\pi/3)GT^{\mu\nu}$, where G is the gravitational constant and $G^{\mu\nu} \equiv R^{\mu\nu} - \frac{1}{2}g^{\mu\nu}R$ is the Einstein tensor. Note the factor $16\pi G/3$ which appears as the coupling constant instead of the usual $8\pi G$.

The development of the weak-field approximation proceeds as in the 4-d case, except that the transformed weak-field tensor $\bar{\gamma}_{\mu\nu}$ is defined via $\gamma_{\mu\nu} = \bar{\gamma}_{\mu\nu} - \frac{1}{4}\eta_{\mu\nu}\bar{\gamma}$, with $\bar{\gamma} \equiv \eta^{\mu\nu}\bar{\gamma}_{\mu\nu}$. The difference here is the factor of $\frac{1}{4}$ instead of the $\frac{1}{2}$ in the 4-d case.

One can check to see that in 6 dimensions the weak-field expression for the Einstein tensor, together with the usual gauge condition $\bar{\gamma}^{\mu\nu}_{,\nu} = 0$ gives $G_{\mu\nu} \approx -\frac{1}{2}\Box\bar{\gamma}_{\mu\nu}$, where $\Box$ is the 6-dimensional Laplacian, $\Box f \equiv \eta^{\alpha\beta}f_{,\alpha,\beta}$.

The Newtonian assumption is that $p \ll \epsilon_i$ for some $i = 4, 5, 6$, and that the space velocities are all small, so that $U^{\mu}_{(i)} \approx \delta^{\mu}_i$. The stress-energy tensor (equation (8)) thus becomes $T^{\mu\nu} \approx \text{diag}(0, 0, 0, \epsilon_4, \epsilon_5, \epsilon_6)$

To verify the factor $16\pi/3$, it is expedient to write the field equations as $G_{\mu\nu} = c_0 T_{\mu\nu} \approx -\frac{1}{2}\Box\bar{\gamma}_{\mu\nu}$. We want to recover Poisson's equation in the limiting case of ordinary matter ($\epsilon_5 = \epsilon_6 = 0$) as $\nabla^2\phi_4 = 4\pi G\epsilon_4$, where ϕ_4 is the Newtonian potential due to ordinary matter. This gives the relation

$$\bar{\gamma}_{44} = (c_0/2\pi G)\phi_4, \tag{10}$$

Similar expressions then arise for ϕ_5 and ϕ_6. One also finds

$$\bar{\gamma} \equiv \eta^{\mu\nu}\bar{\gamma}_{\mu\nu} = \frac{c_0}{2\pi G}(\phi_4 + \phi_5 + \phi_6). \tag{11}$$

The geodesic equations for a particle moving with 6-velocity U^μ are as usual $dU^\mu/d\tau = -\Gamma^\mu_{\alpha\beta}U^\alpha U^\beta$. For nonrelativistic velocities and $U^5 = U^6 = 0$ we obtain $dU^a/d\tau \approx -\Gamma^a_{44} \approx -\frac{1}{2}\gamma_{44,a}$, for $a = 1, 2, 3$. Equating this to $-\phi_{4,a}$ and using equations (10) and (11), one can then show that

$$c_0 = \frac{16\pi G}{3}. \tag{12}$$

Since $\bar{\gamma}_{a\mu} \approx 0$ for $a = 1, 2, 3$, we find from equations (11) and (12) that

$$\gamma_{11} = \gamma_{22} = \gamma_{33} = \frac{1}{4}\bar{\gamma} = \frac{2}{3}(\phi_4 + \phi_5 + \phi_6), \tag{13}$$

$$\gamma_{44} = 2\phi_4 - \frac{2}{3}(\phi_5 + \phi_6), \tag{14}$$

$$\gamma_{55} = 2\phi_5 - \frac{2}{3}(\phi_6 + \phi_4), \tag{15}$$

$$\gamma_{66} = 2\phi_6 - \frac{2}{3}(\phi_4 + \phi_5). \tag{16}$$

The off-diagonal terms all vanish.

The metric as a whole is thus

$$\begin{aligned} d\tau^2 &= [-1 + \frac{2}{3}(\phi_4 + \phi_5 + \phi_6)](dx^2 + dy^2 + dz^2) + \\ &+ [1 + 2\phi_4 - \frac{2}{3}(\phi_5 + \phi_6)](dx^4)^2 + \\ &+ [1 + 2\phi_5 - \frac{2}{3}(\phi_6 + \phi_4)](dx^5)^2 + \\ &+ [1 + 2\phi_6 - \frac{2}{3}(\phi_4 + \phi_5)](dx^6)^2. \end{aligned}$$

We see that the condition $g_{44} = g_{55} = g_{66}$ (Cole, 1980) is not satisfied.

6. The Deflection of Starlight

The deflection of starlight around a spherically symmetric mass M is easily done in the weak-field approximation. Let the 6-momentum of the photon be $P^\mu \equiv dx^\mu/d\lambda$, where λ is a geodesic parameter. Set $P^\mu = P^\mu_{(0)} + \delta P^\mu$, where $P^\mu_{(0)}$ is its constant, unperturbed momentum (energy), and δP^μ is the change in P^μ due to the presence of the mass M.

We assume that the photon's unperturbed motion is in the x^1-direction and that it is an "ordinary" photon (its time-track is x^4). Then $P^\mu_{(0)} = E_0(\delta^\mu_1 + \delta^\mu_4)$, where E_0 is the unperturbed photon energy. The deflection angle θ is very small, and if the perturbing mass is in the $x^1 - x^2$ plane, θ is found by evaluating the ratio

$$\theta \approx \left.\frac{P^2}{P^1}\right|_{x \to \infty} \approx \left.\frac{\delta P^2}{E_0}\right|_{x \to \infty}, \tag{17}$$

where we call $x^1 \equiv x$ and $x^2 \equiv y$.

Writing the geodesic equation as

$$\frac{dP^\mu}{d\lambda} = -\Gamma^\mu_{\alpha\beta} P^\alpha P^\beta$$

and using the usual expression for $\Gamma^\mu_{\alpha\beta}$ in terms of the γ's, we find to first order of smallness

$$\frac{dP^2}{d\lambda} = \frac{dx}{d\lambda}\frac{dP^2}{dx} \approx E_0 \frac{dP^2}{dx} \approx -E_0^2(\Gamma^2_{11} + 2\Gamma^2_{14} + \Gamma^2_{44}) = -\frac{E_0^2}{2}\frac{\partial}{\partial y}(\gamma_{11} + \gamma_{44}).$$

This equation is exactly the same as in the 4-d case; and since it is linear in P^2, we can make the comparison with the 4-d result by examining the corresponding expressions for $\gamma_{11} + \gamma_{44}$. In 4-dimensions, it is well known that

$$\gamma_{11} + \gamma_{44} = 4\phi. \quad \text{(4 dimensions)}$$

In the 6-d case, however, we refer to equations (13) and (14) to find that

$$\gamma_{11} + \gamma_{44} = \frac{8}{3}\phi_4, \quad \text{(6 dimensions)}$$

ϕ_5 and ϕ_6 having cancelled out. If b is the impact parameter, the 4-d result is

$$\theta \approx \frac{4GM}{b},$$

and therefore the 6-d deflection is

$$\theta \approx \frac{8GM}{3b},$$

which is in serious conflict with observation.

7. Conclusions

We have seen that developing a self-consistent weak-field approximation in 6-d spacetime is possible only with some knowledge of the stress-energy tensor. For the Newtonian approximation, it suffices to know the form of this tensor for the perfect fluid [equation (8)].

This approximation may then be used to evaluate the coupling constant in the field equations. This results in the replacement of the 4-dimensional $8\pi G$ by its 6-dimensional analogue $(4/6) \times 8\pi G = 16\pi G/3$ [equation(12)].

The ensuing calculation of the deflection of starlight gives a result at variance with long-established experimental facts: For the sun, the observed value for a ray grazing the solar surface is about 1.7 arcsec. This agrees very well with the predictions of 4-d relativity. The 6-d prediction, however, is $4/6$ of this, or about 1.17 arcsec.

We conclude that if spacetime is actually 6-dimensional in the sense proposed by Cole (1980), one must incorporate gravitation in a different way.

References

Cocke, W. J. & Tifft, W. G.: 1996, *Astroph. & Space Sci.*, in press

Cole, E. A. B.: 1978, *Nuovo Cimento B*, **44**, 157

Cole, E. A. B.: 1980, *Nuovo Cimento B*, **55**, 269

Cole, E. A. B.: 1985, *Nuovo Cimento B*, **85**, 105

Kalligas, D., Wesson, P. S., and Everitt, C. W. F.: 1995, *Astroph. J.*, **439**, 548

Lehto, A.: 1990, *Chinese J. Phys.*, **28**, 215

Misner, C. W., Thorne, K. S., and Wheeler, J. A.: 1973, *Gravitation* (Freeman), pg. 585 ff.

Synge, J. L.: 1960, *Relativity: The General Theory* (North-Holland), pg. 8

Tifft, W. G.: 1996a, *Astroph. J.*, **468**, 491

Tifft, W. G.: 1996b, *Astroph. & Space Sci.*, this conference

Tifft, W. G., Cocke, W. J., and DeVito, C. L.: 1996, *Astroph. & Space Sci.*, in press

EINSTEIN'S GREATEST MISTAKE?

P. C. W. DAVIES
Department of Physics and Mathematical Physics
The University of Adelaide
Adelaide, Australia 5005

Abstract.
Recent astronomical observations have re-opened the old paradox that the universe appears to be younger than some of the objects within it. I suggest that a natural resolution of this paradox lies with the re-introduction of Einstein's cosmological constant, and that sentiment against this constant fails to take into account its connection with quantum field theory.

1. The Problem of the Age of the Universe

When Hubble first measured the rate of expansion of the universe, he was able to infer an estimate for the age of the universe, i.e. the time that has elapsed since the big bang. The figure came out at 1.8 billion years. This was clearly absurd if taken literally because the Earth was known from radioactive dating techniques to be 4.6 billion years old. At that time the mismatch was not regarded with much concern, because few astronomers took seriously the notion that the origin of the universe could be brought within the scope of science in a rigorous and quantitative way.

The "age problem" has waxed and waned over the years. In the 1950s it was one of the motivations for the steady state theory, which postulates that there was no cosmic origin. Since Hubble's initial work, the value of the Hubble constant has been revised several times, extending the age of the universe to 10 or 15 billion years. However, the ages of the oldest stars in globular clusters are estimated to be 14 or 15 billion years. This has not led to rejection of the big bang theory, though, because uncertainty persists over the value of the Hubble constant and other cosmological parameters.

Recently, however, the age problem has loomed large again, for several reasons. The first concerns the relationship between the inferred age and

Astrophysics and Space Science **244**:219-227,1996.

the density of matter in the universe. In a Friedmann universe with the critical density ($\Omega = 1$, and spatially flat, $k = 0$) the age is $T = 2/3H$, where H is Hubble's constant. For a low-density universe with $\Omega < 1$ and $k = -1$, T is larger for a given value of H (usually considered to lie in the range 40-80 km per second per megaparsec).

Conversely, the more matter there is, the shorter the age. If the luminous matter in the universe were all there is, Ω would be about one per cent of the critical value and T could be >14 billion years for plausible values of H. However, there is good evidence for substantial quantities of dark matter. Also, many cosmologists favour the inflationary universe scenario, which predicts $\Omega = 1$. Moreover, the spectrum of fluctuations from COBE is consistent with $k = 0$. In the standard big bang model (without a cosmological constant) these latter values lead to an unacceptably young universe unless H lies right at or below the low end of its observational range.

Another development is that the Hubble Space Telescope has recently returned values of H close to the *top* end of its observational range. If the HST measurements are combined with the inflationary universe scenario, T comes out at 8 to 10 billion years, i.e. several billion years less than the oldest stars.

This paradoxical state of affairs has led to a number of proposals, ranging from "fudge and fit" within the existing big bang paradigm, to rejection of the big bang theory altogether [1]. In this paper I shall argue that a natural resolution of the age problem exists within the framework of the big bang theory. This resolution involves re-introducing into the theory the cosmological constant Λ, the quantity that Einstein once rejected as "the biggest blunder" of his life.

2. The Cosmological Constant

The Einstein gravitational field equations including the cosmological term are (in units with $c = 1$):

$$R_{\mu\nu} - (1/2)Rg_{\mu\nu} + \Lambda g_{\mu\nu} = (8\pi G)T_{\mu\nu}. \qquad (1)$$

The first two terms in Eq. (1) can be regarded as representing normal gravitational attraction. The right hand side represents conventional cosmological matter, the source of the gravitational field. The interpretation of the cosmological term $\Lambda g_{\mu\nu}$ is that it represents a repulsion force that grows with distance. We shall see in §4 that such a force ameliorates the age of the universe problem.

Einstein introduced the cosmic repulsion term originally in order to model a static universe, but dropped it when he learned from Hubble about the cosmological expansion. Since then (c. 1931) Λ has been considered repulsive to most cosmologists on grounds of taste as well as action.

The Λ term in Eq. (1) is the simplest possible term for the left hand side of a tensor equation. Today, we can regard the gravitational field equations as a series expansion of higher and higher orders of derivatives of the spacetime curvature. Dropping the Λ term therefore amounts to setting the leading coefficient of this series equal to zero. It may indeed be zero, but we need some justification for asserting that. Observationally, one may only place the limit

$$\Lambda < 10^{-58} \text{ cm}^{-2}. \tag{2}$$

(I assume for now that Λ is non-negative.) However, as we shall see, this limit is within the range of values that permits a solution of the age problem.

Within the context of Einstein's original theory, the justification for setting $\Lambda = 0$ exactly is that the cosmological term was introduced for purely *ad hoc* reasons, i.e. solely to provide a counterbalancing repulsive force for a static universe model. It had no other manifestation. Therefore, once cosmological theory no longer had use for it, it could be discarded on the grounds of Occam's razor.

The situation is very different today, however, for the following reason. Suppose for the moment that the universe is devoid of normal matter, i.e. $T_{\mu\nu} = 0$. One may rearrange Eq. (1) more suggestively:

$$R_{\mu\nu} - (1/2)Rg_{\mu\nu} = -\Lambda g_{\mu\nu} = (T_{\mu\nu})_{\text{effective}}. \tag{3}$$

In this form, the cosmological term appears on the right hand side of the field equations as an effective source term for the gravitational field. In this role it represents an effective energy density ρ and pressure p given by

$$\rho = \Lambda/G \tag{4}$$

$$p = -\Lambda/G. \tag{5}$$

This form of invisible "matter" has a negative pressure equal to its energy density. It is this unusually large *negative* pressure that produces the repulsion, in accordance with the fact that pressure as well as energy is a source of gravitation in general relativity. (The Λ term is effectively the same as the C-field of Hoyle-Narlikar cosmology [1] and the inflaton field of the inflationary universe scenario [2].)

Because conventional matter is absent in Eq. (3), we can interpret Eqs. (4) and (5) as representing the energy density and pressure of otherwise empty space. A volume V of space has a mass $\rho V = \Lambda V/G$. In simple terms, Λ is a measure of the weight of empty space. Setting $\Lambda = 0$ then implies that empty space has zero weight. Why should empty space weigh anything other than zero?

First note that if space does have weight, it would not show up directly as a gravitational force in any laboratory experiment. Space is everywhere, so its weight does not pull (or push) in any priveliged direction. Only the weight of the universe as a whole would manifest itself – in the form of a cosmological repulsion (for $\Lambda > 0$).

The reason why empty space might have weight is to be sought in quantum field theory, a theory which did not exist when Einstein formulated his theory of gravitation. In simple terms, a field such as the electromagnetic field is quantised by decomposing it into an infinite set of simple harmonic oscillators. The ground state of the field corresponds to an absence of photons, and is called the vacuum state. In this state the expectation value of the energy, evaluated directly, is not zero, on account of the fact that each harmonic oscillator has a zero-point energy of $(1/2)h\nu$. The total energy of all the oscillators for the vacuum state, i.e. empty space, is then

$$(1/2)\Sigma h\nu \;\; \propto \int^{\infty} h\nu^3 d\nu \tag{6}$$

where the sum is over all the modes, and the cubed exponent in the integral arises from a density of states factor. The key point is that there is no upper limit to the summation and integration in (6) (no highest oscillator frequency exists for a continuous field). The sum is therefore formally divergent. This vacuum energy divergence has been known since the inception of quantum field theory, and is similar to the divergent mass-energy of the electron. So taken at face value, classical physics gives us no compelling reason to assume that Λ is non-zero, while quantum field theory suggests an infinite value for Λ!

The standard resolution of the vacuum energy divergence of quantum field theory is renormalization. In effect, one assumes a "bare" (or underlying) value for Λ that is large and negative, and then uses the fact that $-\infty + \infty$ can equal zero. (Actually, this procedure works only in flat spacetime. To do a proper renormalisation of Λ in a consistent metric theory of gravitation demands a much more elaborate theory [3].) However, $-\infty + \infty$ is mathematically ambiguous. It *can* equal zero, but it can also equal anything else. The theory cannot determine the final, observed value of Λ. Only direct observation can do that. Thus, the theory is no guide at all, and certainly does not require $\Lambda = 0$. It is entirely consistent with a nonzero value of Λ below the observational limit in Eq. (2).

Few physicists believe that infinite renormalisations constitute a completely satisfactory answer to the problems of divergences in quantum field theory. The hope is that some future theory, perhaps one involving unification of the fundamental forces (including gravitation) will lead to finite (though perhaps very large) quantities at all stages of the calculation, and possibly even permit the theorist to calculate the observed values of quantities such as Λ. Some theorists believe that superstrings [4] provide a good candidate for such a theory.

So far, I have argued that quantum field theory permits a nonzero Λ, but does not especially suggest it. However, the situation is dramatically transformed when the standard hot big bang theory is combined with theories that seek to unify the electroweak and strong nuclear forces. These quantum field theories, which involve gauge symmetry-breaking, all introduce a Λ term into the gravitational field equations as an inevitable byproduct of the unification process. It is appeal to this byproduct Λ that provides the repulsive force to drive inflation.

Can one still argue that the "real" (i.e. underlying) Λ plus the "byproduct" Λ taken together sum to zero, in accordance with Occam's razor? No, one cannot. Crucially, the value of Λ which emerges as a byproduct of the unification program is temperature-dependent. For this simple reason it cannot be zero at all temperatures.

It is conventional to argue, on rather flimsy grounds, that the final value of Λ is indeed zero when the universe cools down. In other words, the zero-temperature value of Λ is precisely zero. But this assumption involves a notorious sleight-of-hand (termed "the big fix" by Coleman [5]). The trick concerns the fact that there is not just one quantum field in nature, but many. Each has a formally divergent vacuum energy. In the context of unification theories, there may be several fundamental fields with nonzero values of Λ in the high-temperature phase of the very early universe. To get a final value of zero at zero temperature requires a conspiracy of cancellation between all these fields. It is possible that some deep symmetry of nature may be found to impose this felicitous cancellation (see §3), but none is so far known.

The problem is thrown into stark relief when account is taken of the numerical values associated with Λ in the unification theories. Purely on dimensional grounds, putting in the energy scale at which a characteristic symmetry would be broken, one predicts a vacuum energy at least 40 powers of ten larger than the value Einstein had in mind, and in some cases up to 120 powers of ten! So we must accept that at the end of the day, when the universe approaches zero temperature, the above-mentioned felicitous cancellations occur to an astonishing fidelity. Expressed differently, the observed value of Λ differs from its "natural" value by more, perhaps far more, than 40 powers of ten. This is referred to as "the cosmological constant problem", and has been described by Stephen Hawking as the greatest failure of dimensional analysis known [6]. The intractable nature of the cosmological constant problem has been stressed by Weinberg [7].

One may therefore conclude from a study of quantum field theory that, far from being "naturally" zero, it is actually deeply mysterious as to why Λ is in fact not "naturally" very much larger than the limit in Eq. (2).

3. Reasons Why Λ Is So Small

There have been several attempts to explain why the cosmological constant does not have a large observed value, and might yet turn out to be precisely zero as Einstein in the end claimed.

One type of explanation involves a reinterpretation of the nature of the underlying, or zero-temperature, Λ. Instead of treating it as a fundamental constant of nature, it could be regarded as a physical variable, a field in its own right, which has a quantum expectation value that depends on the quantum state of the universe. Coleman [5] has developed a theory of this type, in which the quantum state of the universe takes into account the nontrivial topology at the Planck scale that is familiar in the theory of quantum gravity. His claim is that the spacetime "foam" associated with this complex topology contributes to Λ in such a way as to produce values that are sharply peaked about $\Lambda = 0$. In other words, it is exceedingly probable that a measurement of Λ will produce a value very close to zero (much closer than the limit in Eq. (2)).

A more general argument has been used by Hawking [6] to claim that Λ should be precisely zero. Hawking compares the ratio of the observed upper limit of Λ to that deduced on dimensional arguments (e.g. 10^{-120}) with the ratio of the observed upper limit of the photon mass with a "natural" mass scale such as the Planck mass. The latter ratio is about 10^{-58}. Hawking then points out that when one is faced with such a small ratio as that of the photon mass, one seeks a symmetry of nature (in this case a gauge symmetry) that fixes the ratio to be zero, i.e. requires the photon rest mass to vanish. It would, he suggests, be odd for nature to throw up an exceedingly small, but still nonzero, ratio like 10^{-58}. He goes on to argue that if this reasoning is compelling in the case of the photon, where the ratio is "only" 10^{-58}, how much more compelling it is in the case of the cosmological constant, where the ratio might be as small as 10^{-120}. Therefore, Hawking says, we have every reason to expect nature to possess a deep symmetry that compels Λ to be precisely zero.

A counter to Hawking's argument is that, in the context of the fundamental constants, nature already throws up some well-known large and small number ratios. For example, the ratio of the strength of gravitation to the strength of electromagnetism in the hydrogen atom is about 10^{-40}. Perhaps it is no coincidence that $10^{-120} = (10^{-40})^3$. One can well imagine a future theory in which the values of the fundamental constants, instead of being determined empirically, emerge from the theory. Such a theory may link the constants as follows:

$$\Lambda_{\text{observed}}/\Lambda_{\text{Planck}} \approx (Gm_e m_p/e^2)^3 \tag{7}$$

where m_e and m_p are the masses of the electron and proton respectively and e is the fundamental unit of charge. This would place the actual value of Λ close to the observational limit in Eq. (2).

Another type of argument has been used by Unwin and myself [9], and by Weinberg [7] and Efstathiou [10]. It involves anthropic reasoning. Given that a variety of quantum fields will make large contributions to Λ, then the observed value of Λ will depend on the exact state of those fields. But the state may vary from one region of the universe to another. For example, if symmetry-breaking is involved, a symmetry may be broken in different ways in different spatial domains. The final value of Λ may be a purely random affair. In this scenario, the average value of Λ (averaged, that is, across the whole universe) may well be very large, but in a very small subset of domains felicitous cancellations may lead to extremely small values of Λ consistent with the limit in Eq. (2). In answer to the question of how we happen to be living in such an atypical region of the universe, the answer is that if Λ exceeded the limit in Eq. (2) by much, then the cosmic repulsion would be fierce enough to interfere with galaxy formation. An absence of galaxies would imply an absence of stars and hence an absence of biological organisms and observers. Similarly, if Λ took on a negative value much in excess of the numerical limit in Eq. (2), it would add positively to the gravitation of the universe, and bring about a collapse to a big crunch before life had a chance to evolve to the point of intelligent observers. Either way, it is no surprise that we find ourselves in a region of the universe with a cosmological constant as small as the limit in Eq. (2). We could not exist anywhere else.

Note, however, that the anthropic argument does not imply that Λ should be exactly zero, or even much smaller than the limit in Eq. (2). Given that life is consistent with a range of values, say

$$-10^{-57} \text{ cm}^{-2} < \Lambda < 10^{-57} \text{ cm}^{-2} \tag{8}$$

then the probability that the actual (absolute) value in our region of the universe is much smaller than the observational limit in Eq. (2) is very low. In other words, it is rather probable that the actual value is not much smaller than 10^{-58} cm^{-2} – the original Einstein value.

The conclusion, then, is that so long as the cosmological constant was based on classical gravitation theory, a nonzero value could be considered to be largely *ad hoc*. However, once the interpretation of Λ as a vacuum energy in quantum field theory is appreciated, there is no longer any compelling reason to put $\Lambda = 0$ exactly. The popularity of the $\Lambda = 0$ option can be traced to the high scientific esteem of Einstein and the fact that a nonzero Λ is no longer needed for his originally intended purpose. But one cannot argue, because of the particular historical sequence of events, that that rules out a nonzero Λ for *any* purposes. Pending a comprehensive theory of the constants of nature, and/or a fully unified theory of the fundamental forces, the onus is on those who claim that nature has chosen to make a free parameter precisely zero to justify this specific choice.

4. Solving the Age Problem

Once one admits the possibility of a value for Λ close to the observational limit in Eq. (2), then the way is open to solve the age of the universe problem I discussed in §1. In effect, the weight of space acts as surrogate dark matter, permitting a spatially flat universe (k = 0) without a critical ($\Omega = 1$) density for normal matter. But because the negative pressure associated with the cosmological term produces a repulsion, whereas normal matter is attractive, the Λ term acts to counter the decelerating effects of ordinary matter. If the universe decelerates more slowly, then it will have a greater age T for a given present value of the Hubble constant. (This is easy to see by analogy. Suppose two vehicles with their brakes applied pass an observer together at 20 mph, and the observer is told that each vehicle was moving at 100 mph when it started to brake. Then the vehicle which has the weaker brakes must have been travelling for longer.)

Cosmological models with nonzero Λ belong to the class of Friedmann models known as Eddington-Lemaitre, and are well-known. A careful analysis [11] shows that a big bang universe with T $\approx$ 15 billion years, H $\approx$ 75 and k = 0 is consistent with all the astronomical observations at this time. However, the model is falsifiable, because a nonzero Λ shows up in the statistics of gravitational lensing in cosmological surveys. It is likely that improved observations will settle the matter within a few years.

If, as I am suggesting here, Λ is nonzero, then Einstein's greatest mistake will turn out after all to have been his greatest triumph!

References

1. For example, see Narlikar in these proceedings
2. Linde, A.: 1990, *Inflation and Quantum Cosmology*, (Academic Press)
3. Birrell, N. D., and Davies, P. C. W.: 1982, *Quantum Fields in Curved Space*, (Cambridge University Press)
4. Brown, J. R., and Davies, P. C. W.: 1988, *Superstrings: A Theory of Everything*, (Cambridge University Press)

5. Coleman S.: 1988, *Nucl. Phys. B*, **310**, 643
6. Hawking, S. W.: 1983, *Phil. Trans. Roy. Soc. A*, **310**, 303
7. Weinberg, S.: 1992, *Dreams of a Final Theory*, (Random House)
8. Hawking, S. W.: 1984, *Physics Letts B*, **134**, 162
9. Davies, P. C. W. and Unwin, S. D.: 1988, *Proc. R. Soc. A*, **377**, 147
10. Efstathiou, G.: 1995, *M.N.R.A.S.*, **274**, L73
11. Efstathiou, G., Sutherland, W. and Maddox, S. J.: 1990, *Nature*, **348**, 705

TIME, SPACE AND COMPLEX GEOMETRY

R. PENROSE
Oxford University
Oxford, U.K.

Published by title only, no manuscript available.

DISCRETE SPATIAL SCALES IN A FRACTAL UNIVERSE

D. F. ROSCOE
School of Mathematics
Sheffield University
Sheffield, S3 7RH, UK.

Abstract. The work of this paper is based on work which has been described in a preliminary form in Roscoe (1995), and it applies the formalism developed there to the problem of deriving the cosmology for a universe which is in a state of gravitational equilibrium. It predicts that, in such a universe, material is distributed in a fractal fashion with fractal dimension *two* whilst redshifts necessarily occur in integer multiples of a basic unit and, given a certain model for light propagation, the measured magnitudes of peculiar velocities will increase in direct proportion to cosmological redshift.

The first of these predictions is strongly supported by the results of the most modern pencil-beam and wide-angle surveys, whilst the second conforms with the results of very recent rigorous analyses of accurately measured redshifts of nearby spiral galaxies and the third is in qualitative agreement with the very limited data available. The observational support for these predictions is described in detail in the text.

1. Introduction

The following work describes the application of the gravitation theory described in Roscoe (1995) to the problem of deriving a cosmology. This latter presentation is a preliminary and incomplete development of work now completed, and in preparation for publication elsewhere. Preprints are available on request. The underlying gravitation theory, which is predicated upon the idea of a discrete and finite model universe, is distinguished in the fact that, according to it, concepts of spatial and temporal measurement are undefined in the absence of mass - in this sense, it conforms to the strictest possible interpretation of Mach's Principle.

Astrophysics and Space Science **244**:231-248,1996.

There is evidence, discussed in §2, to suggest the real Universe is in a state of approximate thermodynamic equilibrium; this possible state is used to justify the cosmological principle that the model universe is in a state of exact gravitational equilibrium. The mass-distribution corresponding to this state is calculated in §3 and §4, and is found to be fractal with a fractal dimension of *two*. This mass-distribution prediction is very strongly supported by the results of several modern surveys, and this evidence is discussed in §5.

The discrete nature of material in the model universe is considered in §6, and is found to imply a discretization of distance scales which leads, in §7, to the conclusion that redshifts must increase in integer multiples of a basic unit; the evidence supporting this is discussed in §8. The discretization of distance scales occurs in such a way that spatial and temporal measurement scales in remote localities undergo systematic change, discussed in §9, which has implications for kinematics and the nature of light, discussed in §10 and §11 respectively. The predicted kinematics has implications for the apparent behaviour of the *peculiar velocities* of galaxies; these are discussed in §12 where it is shown how one consequence of the scale-change phenomenon is that the estimated magnitudes of peculiar velocities will appear to vary linearly with the cosmological redshift. The evidence supporting this conclusion is discussed in §13. The discussion of §3 also leads to the idea of a *material vacuum*, existing in the model universe, and the implications of this are briefly considered in §14.

The equations of motion, derived for a spherically symmetric distribution of material particles, with an isotropic velocity distribution, are given by

$$\ddot{\mathbf{r}} = -\frac{\partial V}{\partial r}\hat{\mathbf{r}},$$

where r is the position vector defined with respect to the global mass-centre,

$$V \equiv -\frac{1}{2} < \dot{\mathbf{r}}, \dot{\mathbf{r}} > = -\frac{r_0 \gamma A}{2} + \frac{B}{2A}\dot{\Phi}^2, \tag{1}$$

γ is the gravitational constant, r_0 is a constant defined below, and A, B are defined by

$$A \equiv \frac{M}{\Phi}, \quad B \equiv -\left(\frac{M}{2\Phi^2} - \frac{M'M'}{2\alpha_0 M}\right), \quad M' \equiv \frac{dM}{d\Phi}, \quad \Phi = \frac{1}{2} < \mathbf{r}, \mathbf{r} > . \tag{2}$$

The function M is the mass-distribution function, for which a broad admissable class is given by

$$M(r) = m_0 \left(\frac{r}{r_0}\right)^{\gamma_1}, \tag{3}$$

where m_0 has dimensions of *mass*, and r_0 is the radius of the volume containing mass m_0. It is to be noted from this expression that M/r^{γ_1} is a global constant, so that the particular choice of r_0 has no significance for (3). Finally, the defining relationship between time scales and distance scales is given by

$$dt^2 = \left(\frac{\Phi^2}{r_0 \gamma M^2} \right) g_{ij} dx^i dx^j. \tag{4}$$

whilst the metric tensor is given by

$$g_{ab} = A\delta_{ab} + Bx^a x^b. \tag{5}$$

It follows from (2), (4) and (5) that if $M = 0$, so that there is no mass, then concepts of time and distance are undefined.

The foregoing equations of motion can be identified with those given in Roscoe (1995) by making the substitution $M = \alpha U$. It is to be noted that the potential form of the equations is not given in this early development, nor is the interpretation of $M \equiv \alpha U$ as a mass-distribution made there. Preprints of the complete development are available on request.

2. A Simple Cosmological Principle

There is some evidence, briefly discussed below, which suggests the observable universe might be in a state of approximate thermodynamic equilibrium with respect to the various energy sources within it. If this is the case, it would follow that gravitational energy must be included as one of these sources; correspondingly, the most simple realistic cosmological principle applicable to the model universe is the condition that it is in a state of *gravitational equilibrium*. However, before the consequences of this most simple possible of cosmological principles are worked through, we shall consider some of the evidence supporting the argument that the cosmic ray flux, the cosmic background radiation and our own galaxy's starlight field are in thermal equilibrium.

One of the earliest (if not the earliest) predictions of a background temperature to space, and estimations thereof, is that of Guillaume (1896) who used Stefan's Law to calculate the equilibrium temperature, arising from stellar radiation, of an inert body placed in the interstellar space of contemporary understanding; this was equivalent to calculating the 'temperature of space', and the figure arrived at was $5.6^o K$. A similar black-body calculation was given by Eddington in 1926 (reprint 1988), and he arrived at the figure $3.18^o K$, calling it explicitly the 'temperature of interstellar space'.

By 1928 the work of Millikan and Cameron had shown that cosmic rays have an extragalactic origin and, subsequently, Regener (1933 or 1995 for an English translation) calculated the equilibrium temperature of an inert body (having the necessary dimensions to absorb cosmic rays) which is placed in a 'sea' of cosmic radiation, and found this to be $2.8°K$. Regener went on to argue that, because of the extragalactic origin of cosmic rays, and because of the (assumed) extreme weakness of starlight in inter-galactic space, then $2.8°K$ must be the 'temperature of intergalactic space'.

The earliest *Hot Big Bang* predictions for the existence of the CBR with a black-body spectrum were given by Alpher & Herman (1949) and Gamow (1953), and these authors estimated the 'temperature of space' variously in the range $5°K$ to $50°K$; After the observations of Penzias & Wilson (1965), we are now aware that the CBR does exist as an additional extragalactic energy field, with a temperature of $2.7°K$.

So, there are at least three independent sources of energy - galactic starlight, cosmic rays and the CBR - which have been used to estimate the 'temperature of space', giving answers which suggest that the three sources are in near thermodynamic equilibrium. In addition, Sciama (1971) has pointed out that the turbulent energy density of interstellar gases and the energy density of the interstellar magnetic field is similar to that of the aformentioned sources, and so the net picture is entirely consistent with the idea of a universe which is in an approximate thermodynamic equilibrium.

3. The Equilibrium Universe

If the model universe is in gravitational equilibrium, then the net gravitational force at every point within it is necessarily zero so that $\ddot{\mathbf{r}} = 0$ everywhere. Consequently, the potential is constant everywhere so that, by (1),

$$-\frac{r_0 \gamma A}{2} + \frac{B}{2A}\dot{\Phi}^2 = V_0, \tag{6}$$

where V_0 is the value of the constant potential. Using the definitions of A, B, Φ given at (2), this equation can be written as

$$-\frac{r_0 \gamma M}{r^2} - \frac{\dot{r}^2}{2}\left\{1 - \frac{1}{4\alpha_0}\left(\frac{r}{M}\frac{dM}{dr}\right)^2\right\} = V_0.$$

An easy means of solving this equation is arrived at as follows: The equation gives the form of $M(r)$ which is consistent with the constraint $\ddot{\mathbf{r}} = 0$ for all motions in the model universe. Of all possible trajectories of this type, there will be a subclass which pass directly through the centre-of-mass, and will therefore have zero angular momentum about this point. These *particular* trajectories satisfy $\dot{r} = constant$ where, because the speed of the particle concerned is arbitrary, then *constant* is arbitrary; consequently, these trajectories can be considered specified by $\dot{r}^2 = 2\lambda_1$, for arbitrary positive values of λ_1. The above equation for $M(r)$ can be now written

$$\left\{-\frac{r_0\gamma M}{r^2} - V_0\right\} - \lambda_1\left\{1 - \frac{1}{4\alpha_0}\left(\frac{r}{M}\frac{dM}{dr}\right)^2\right\} = 0.$$

Since λ_1 is simply a measure of an arbitrary constant speed, then this equation must be decomposable into

$$\left\{-\frac{r_0\gamma M}{r^2} - V_0\right\} = 0 \quad \text{and} \quad \left\{1 - \frac{1}{4\alpha_0}\left(\frac{r}{M}\frac{dM}{dr}\right)^2\right\} = 0. \qquad (7)$$

According to the first of these equations,

$$M(r) = -\frac{V_0 r_0}{\gamma}\left(\frac{r}{r_0}\right)^2,$$

which satisfies the second equation if $\alpha_0 = 1$. This solution is a special case of the more general admissable form (3) so that, finally, the mass-distribution function appropriate to an equilibrium model universe is

$$M(r) = m_0\left(\frac{r}{r_0}\right)^2 \qquad (8)$$

where, by comparing the two forms of $M(r)$, the value of the constant potential is found to be given by

$$V_0 = -\frac{\gamma m_0}{r_0}.$$

Since m_0, in (8), has dimensions of *mass*, it must be interpreted as the amount of mass contained in a sphere of arbitrarily chosen radius r_0. It is to be noted that the definitive constant value - lacking all arbitrariness - given to the constant potential in the present equilibrium case can only be interpreted to represent some kind of *absolute* ground state energy, or vacuum energy, associated with the system.

Finally, if (7) is compared with (6), it is can be seen how the second of (7) is equivalent to $B = 0$ so that, with (2), (5) and (8), the metric tensor for the equilibrium universe is given as

$$g_{ab} = \left(\frac{M}{\Phi}\right)\delta_{ab} = \left(\frac{2m_0}{r_0^2}\right)\delta_{ab}. \qquad (9)$$

4. The Model Fractal Universe

The equilibrium model universe is characterized by $\ddot{\mathbf{r}} = 0$, which means that *all* points in the space are dynamically equivalent. Consequently, there is no dynamical experiment in the space which can distinguish between any pair of points, and hence there is no way of determining the position of a global mass-centre. Since a unique origin for the mass distribution (8) cannot now be defined, then it must be considered true about arbitrarily chosen origins in the space, and this amounts to the statement that mass is distributed in a self-similar, or fractal, fashion with a fractal dimension of *two*.

A direct corollary of this argument is the fact that, if $M(r)$ has any form, other than (8), then potential gradients must exist, so that $\ddot{\mathbf{r}} \neq 0$ necessarily. As a consequence, it becomes possible to determine a unique global-mass centre and so the corresponding $M(r)$ cannot be describing a fractal distribution of mass, since such distributions are necessarily isotropic about *all* points in the space. So, in conclusion, the only possible *fractal* distribution of mass in the model universe is the one which has fractal dimension two.

5. A Fractal Universe, The Evidence

A basic assumption of the *Standard Model* is that, on some scale, the universe is homogeneous; however, in early responses to suspicions that the accruing data was more consistent with Charlier's conceptions of an hierarchical universe (Charlier, 1908, 1922, 1924) than with the requirements of the *Standard Model*, de Vaucouleurs (1970) showed that, within wide limits, the available data satisfied a mass distribution law $M(r) \approx r^{1.3}$, whilst Peebles (1980) found $M(r) \approx r^{1.23}$. The situation, from the point of view of the *Standard Model*, has continued to deteriorate with the growth of the data-base to the point that, (Baryshev et al (1995))

> ...the scale of the largest inhomogeneities (discovered to date) is comparable with the extent of the surveys, so that the largest known structures are limited by the boundaries of the survey in which they are detected.

For example, several recent redshift surveys, such as those performed by Huchra et al (1983), Giovanelli and Haynes (1985), De Lapparent et al (1988), Broadhurst et al (1990), Da Costa et al (1994) and Vettolani (1994) etc have discovered massive structures such as sheets, filaments, superclusters and voids, and show that large structures are common features of the observable universe; the most significant conclusion to be drawn from all of these surveys is that the scale of the largest inhomogeneities observed is comparable with the spatial extent of the surveys themselves. So, to date, evidence that the assumption of homogeneity in the universe is realistic does not exist. By contrast, evidence for the fractal nature of the matter distribution is becoming increasingly strong; for example, Coleman et al (1988) analysed the *CfA1* redshift survey of Huchra et al (1983), and found $M(r) \propto r^{1.4}$ for this sample; subsequently, the *CfA2* survey of Da Costa et al (1994), which is an extension of the *CfA1* survey out to about twice the depth, has been analysed by Pietronero and Sylos Labini (1995) to reveal $M(r) \propto r^{1.9}$. The pencil beam survey data accumulated in ESO Slice Project (Vettolani 1994), which reaches out to 800 Mpc, has been similarly analysed (Pietronero and Sylos Labini (preprint 1995)) to conclude that, within this data, the distribution of galaxies conforms to the fractal law $M(r) \propto r^2$ up to the sample limits and, according to Baryshev et al (1995), this same result of fractal distribution of dimension ≈ 2 has been found in the analysis of other deep redshift surveys such as those of Guzzo et al (1992) and Moore et al (1994).

To summarize, for more than twenty years evidence has been accummulating that material in the universe appears to be distributed in an hierarchical, or fractal way - in direct opposition to the requirements of the *Standard Model* - and the results of the most modern deep and wide angle surveys are consistent in suggesting the distribution law $M(r) \propto r^2$, valid about arbitrarily chosen centres. This empirical law, which describes a self-similar mass distribution of fractal dimension two, *is in direct conformity with the mass distribution law derived, for a universe in gravitational equilibrium, in this paper.*

6. Discrete Mass Implies Discretized Distance Scales

The model universe was defined, in the first instance, to consist of a *finite* amount of *discrete* material, and it was the finite quality which allowed the definition of the global mass-centre, and hence enabled the theory to be developed as it has been; in the following, the discrete quality of mass in the model universe is considered, and shown to imply the discretization of distance scales. At first sight, this seems to be a surprising conclusion but, when it is remembered that, according to the theory, concepts of space and time cannot be formulated in the absence of mass, then it appears reasonable to expect that a discrete matter distribution must imply discrete space.

We begin by considering the mass distribution function which, according to (3), is given by

$$M(r) = m_0 \left(\frac{r}{r_0}\right)^{\gamma}.$$

When γ is real then $M(r)$ varies continuously through real values with r, and so the discrete quality of the model universe cannot be made manifest in this case. However, the analysis which gave rise to the foregoing expression for $M(r)$ does not exclude the possibility of γ assuming complex values so that, with γ written as explicitly complex, the most general expression of $M(r)$ is

$$M(r) = m_0 \left(\frac{r}{r_0}\right)^{\gamma_0 + i\gamma_1}$$

for $i \equiv \sqrt{-1}$ and real γ_0 and γ_1. The function $M(r)$ now only takes real positive values at the set of discrete points

$$r_k = r_0 \exp(\frac{2k\pi}{\gamma_1}), \quad k = 0, \pm 1, \pm 2, ... \tag{10}$$

and so, from point to point, $M(r)$ varies discretely over real values, as required for the model universe. It follows that, for perfect rigour, the whole analysis to this point should be recast from a continuum form into a discrete form, where r is discretized according to (10). However, for the sake of brevity and convenience, the discrete analysis will only be applied from (7) onwards.

Defining the derivative in (7) in terms of differences, according to

$$\frac{dM}{dr} \equiv \frac{M_k - M_{k-1}}{r_k - r_{k-1}},$$

and using (10), the first of (7) is found to be satisfied by

$$M_k \equiv M(r_k) = -\frac{V_0 r_0}{\gamma} \left(\frac{r_k}{r_0}\right)^2, \quad r_k = r_0 \exp(\frac{2k\pi}{\gamma_1}), \quad k = 0, \pm 1, \pm 2, ...$$

whilst the second of (7) is found to be satisfied only when

$$4\alpha_0 = \left(1 + \exp(-\frac{2\pi}{\gamma_1})\right)^2.$$

Notice that, according to (10), there is no such thing as an origin $r = 0$; it then becomes natural to interpret r_0 as a form of 'reference surface' from which displacements are calculated. In this case, (10) gives, for the value of non-negative displacements,

$$\Delta_k = r_k - r_0 = r_0 \left\{ \exp\left(\frac{2k\pi}{\gamma_1}\right) - 1 \right\}, \quad k = 0, 1, 2, ...$$

If γ_1 is large compared to $2k\pi$, this gives

$$\Delta_k \approx \frac{2k\pi r_0}{\gamma_1}, \quad k = 0, 1, 2, ... \tag{11}$$

A crucially important point about the foregoing analysis is that is valid about *arbitrary* points in the space, because of the fractal nature of the equilibrium mass-distribution. So, although (11), taken as a statement about the nature of 'space' about any origin, appears paradoxical, this is only so when 'space' is imagined as something which has properties independently of its material content; but, when it is remembered that, here, 'space' is merely a *metaphor* for the relationships between material and that, in this case, the fractality of the matter distribution ensures it looks the same from all origins, then the idea of (11) being true about arbitrary points presents no difficulty of comprehension.

7. Quantized Redshifts

The considerations of the previous section, together with Hubble's Law, make the existence of the quantized redshift phenomenon axiomatic: specifically, Hubble's Law and (11) together give

$$cz = H\left(r_k - r_0\right) \approx H\frac{2k\pi r_0}{\gamma_1}, \quad k = 0, 1, 2, ...$$

Apart from the quantal aspect, one interesting thing about this redshift-distance relationship is that it inherently requires redshifts to have a non-trivial 'zero-surface' from which the Hubble Law is valid. Such a surface is, in fact, well known to be a feature of the real redshift phenomenon, and Sandage (1986) puts this surface at about $r_0 = 1.5$ Mpc.

8. Quantized Redshifts: The Evidence

The conclusions of §7, that redshifts necessarily occur as integer multiples of a basic unit, has been the substance of claims made for the past twenty years by Tifft (1976, 1980, 1990) and Tifft & Cocke (1984); these claims have generated considerable dissension, but not much reasonable discussion. However, in independent study by Guthrie & Napier (1996) have tested the *specific* hypothesis of Tifft & Cocke, *that quantization at 72km s^{-1} and 36km s^{-1} exists in the redshifts of low redshift spirals*, in a statistically rigorous manner using independent data sets, characterized by their high accuracy and totalling several hundred objects. They found that, after the redshifts were corrected for the solar motion about the galactic centre, then Tifft & Cocke's basic hypothesis is confirmed at the level of virtual certainty for the samples analysed.

A further consequence of the Guthrie & Napier analysis is that the detected quantization effect is so sharp, it puts a very narrow limit on the magnitude of the peculiar velocities of objects used for the analysis;

specifically, Monte-Carlo simulations indicate that the magnitude of peculiar velocities cannot exceed about 4km s^{-1}, otherwise Doppler effects completely mask the redshift quantization signal.

The claimed periodicity of 37.5km s^{-1}, of the Guthrie & Napier analysis allows an estimation of the parameters in the 'quantized redshift' formula given in §7, which is an approximation based on the assumption that γ_1 is large compared to $2k\pi$. Using $H \approx 50$km s^{-1} Mpc^{-1} and the Sandage value $r_0 \approx 1.5$ Mpc, leads to $\gamma_1 \approx 12.6$, which is *not* large compared to $2k\pi$. For the approximation to make sense within the context of its own logic requires $r_0 \approx 15$ Mpc, or greater.

9. An Heirarchy of Measurement Scales

Equation (10), which defines the sequence of possible radial shells definable from the origin, gives

$$r_k - r_{k-1} = r_0 \exp\left(\frac{2k\pi}{\gamma_1}\right)\left(1 - \exp\left(\frac{-2\pi}{\gamma_1}\right)\right),$$

which can be directly interpreted as the minimum possible distance interval definable at r_k. Since this interval increases with k, it follows that, from the perspective of an observer at the origin, there is an heirarchy of increasing local spatial scales at increasing distance and, by the comments at the end of §6, this heirarchy will be apparent from arbitrary origins.

To understand the behaviour of the temporal scales, it is necessary to refer to the defining relationship between time scales and distance scales given, for the general case, at (4) as

$$dt^2 = \left(\frac{\Phi^2}{r_0\gamma M^2}\right)g_{ij}dx^i dx^j.$$

Using the prescriptions of the mass-function and the metric tensor in the equilibrium universe given at (8) and (9) respectively, together with the equivalence $dx^i dx^j \delta_{ij} \equiv |d\mathbf{r}|^2$, this can be written

$$dt^2 = \left(\frac{r_0}{2\gamma m_0}\right)|d\mathbf{r}|^2,$$

where, as given in §3, m_0 denotes the amount of mass contained in a sphere of arbitrarily chosen radius r_0. This can be put in a form more useful for present purposes by noting, from (8), that $M(r)/r^2$ is a *global* constant of the system, and so $2\gamma m_0/r_0^2$ is also a global constant. Denoting this latter constant by α^2, the above equation can be written as

$$dt = \frac{1}{\alpha\sqrt{r_0}}|d\mathbf{r}|. \tag{12}$$

This expression is then the defining relationship between time scales and distance scales in the equilibrium universe. However, it is to be noted that, for given $d\mathbf{r}$, the elapsed time, dt, depends upon the arbitrary choice of the radius-parameter r_0; this can only mean that the choice of r_0 amounts to choosing the *clock* with which the passage of time is to be measured. If physical substance is to be assigned to the chosen 'clock', then a reasonable working hypothesis would be that it consists of the ensemble of material, mass m_0, contained within the radius r_0 sphere. Note that, according to this interpretation of (12), the more massive the clock, then the more slowly it records the passage of time.

To summarize, along with the heirarchical distribution of matter in the fractal universe, there are corresponding heirarchies of spatial and temporal measurement scales.

10. Kinematics

Whilst (12) defines the relationship between time and distance scales in the equilibrium universe, it also necessarily defines the equation of motion for particles in this universe, given by

$$|\dot{\mathbf{r}}| = \alpha\sqrt{r_0} \equiv \sqrt{-2V_0}, \tag{13}$$

where α is the universal constant defined in §9 and V_0 is the ground state energy of the system, identified in §3; it is clear from (13) that the choice of r_0 amounts to choosing the clock used to define the velocity $\dot{\mathbf{r}}$. There are three fundamental peculiarities arising from this equation, considered in turn below.

Firstly, (13) says that, for a given clock, *all* particles in the model universe have velocities of the same magnitude and this velocity corresponds to the ground state energy of the equilibrium system, identified in §3; the absence of any other constraint implies that the directions of these velocities must be uniformly random. Thus, according to the equilibrium model, the distribution of material particles in the model universe has kinematic properties which exactly mirror those existing in an isotropic distribution of photons. With the exception of any statements about the distribution of mass in these material particles, what emerges is a material analogue of the cosmic background radiation. This seems very odd when set against conventional experience, but it must not forgotten how this experience relates exclusively to a world of electromagnetic and non-equilibrium gravitational forces, neither of which is part of the equilibrium model, and both of which occur on a scale far below that presumed for this model.

Secondly, by definition, (13) assumes the existence of some absolute rest-frame which, at the beginning of this development, was assumed given by the global mass-centre. However, in §4, it was indicated that, in an equilibrium universe, it becomes impossible to locate the global mass-centre and therefore impossible to define an absolute state-of-rest in terms of it. Since (13) is an equation which arises from the assumption of universal equilibrium, it follows that, implicit to the whole development, there must be another means of determining the absolute rest-frame and the answer lies in the considerations of the previous paragraph: Specifically, the predicted kinematic structure will lead to a Doppler redshift field that will appear (statistically) isotropic when viewed from the absolute rest-frame, but will be subject to a dipole displacement when viewed from any other frame. So, the absolute rest-frame is that in which no dipole effects exist in the observed redshift field.

Thirdly, in (13) α is a universal constant, and r_0 is the radius of the sphere which contains mass m_0; however, the model universe is defined to be finite, and so it follows that there is a limiting value of r_0 defining the smallest sphere which contains the total of the universal mass. Denoting this value as r_* then, by (13),

$$|\dot{\mathbf{r}}| \leq \alpha r_*^{1/2} \equiv \sqrt{2\gamma m_*/r_*}. \tag{14}$$

That is, although the choice of clock in the model universe is arbitrary, there is one *fundamental* clock, which consists of the whole ensemble of mass in the model universe, and according to which, velocities attain their maximum value.

11. The Nature of Light

To the extent that the presented theory possesses the concepts of a universal time together with a separate three-dimensional physical space, then it is a 'classical theory'. However, consideration of (10), according to which a radial displacement from a given origin can only have certain admissable values, shows immediately that a photon can no longer be considered as something with a continuous trajectory, but must be considered as a sequence of *resonances* at discrete locations. The theory tells us nothing about the rate at which these resonances propagate, and so further progress can only be had by introducing an ad-hoc propagation model. Since it is necessary for any such model to be consistent with the kinematic structure of the theory, then it must have the general form of (13), but with α replaced by another value appropriate to light-propagation. So, consider the form $c_0 = \beta\sqrt{r_0}$ where c_0 is the speed of light measured by the r_0-clock and β is a universal constant. Since c_0 can never be zero, then this equation implies that r_0 must have a minimum non-zero value which represents the minimum dimension of a physical clock. Consequently, we can write

$$c_0 = c_{min}\sqrt{\frac{r_0}{r_{min}}} \equiv \beta\sqrt{r_0} \qquad (15)$$

for a simple light-propagation model which is consistent with the kinematic structure defined by (13).

12. Peculiar Velocities

In the context of the *Standard Model*, the phrase 'peculiar velocities' refers to real motions that galaxies might possess, defined with respect to some fundamental rest-frame, and generated by local gravitational gradients. In the present context, the basic assumption is that the peculiar motions arise wholly out of the kinematic structure of the equilibrium universe and, in the following, the extent to which the observations support this assumption is considered.

In effect, any practical determination of the peculiar velocity of a distant object involves an estimation of the object's distance made on the basis of magnitude information; this is then used to estimate the corresponding *cosmological redshift* which should be associated with the object. This estimated cosmological redshift is compared with the actually measured redshift of the object, and the difference between the two redshifts is assumed to give that component of the measured redshift which arises in consequence of a radial Doppler effect; the radial component of the object's peculiar velocity is then inferred from that. More specifically, in the conventional way of doing things, the 'Doppler shift' of an object at distance r_0 is estimated as a wavelength-shift defined relative to the observer's local measurement standards, and the peculiar velocity calulated from that.

However, by the considerations of §9, it is known that, from the point of view of any observer, measurement standards vary with radial location. So, suppose the 'Doppler shift' of the r_0-object is estimated as a wavelength-shift defined relative to the measurement standards at r_0, rather than to the observer's measurement standards, and suppose this estimate is labelled $z_0^{(D)}$; in this case, a Doppler shift is being estimated purely in terms of r_0-scales, and so any expression which relates $z_0^{(D)}$ to velocities, $z_0^{(D)} = f(v/c)$ say, must define v and c in terms of the r_0-clock. Therefore, using (13) and (15), the Doppler shift of an object estimated in terms of the scales at the object can be expressed as

$$z_0^{(D)} = f(|\dot{\mathbf{r}}|/c_0) = f(\alpha/\beta),$$

which is a *global* constant, since α and β are global constants. If the estimated cosmological component of an object's measured redshift is taken as an indicator of the measurement scales at the object, then this latter result simply means that the Doppler component of the wavelength-shift *expressed in terms of our local scales* must increases in direct proportion to the cosmological component of the wavelength shift. Consequently, the corresponding estimates of peculiar velocity magnitudes will appear to increase linearly with distance. So, they will be small for small r_0 and large for large r_0.

13. Peculiar Velocities: The Evidence

Available evidence falls into three categories: (a) objects with redshifts $(0,500)$km s^{-1}, (b) objects with redshifts $(800,2000)$km s^{-1} and (c) objects with redshifts $(2000,15000)$km s^{-1}. All the evidence is indirect, and differs in type between the cases.

For the first category, Karachentsev & Makarov (1996) analyse the local velocity field using a sample of 103 galaxies with maximum redshifts of 500km s^{-1}. There are two surprising results arising from their analysis, only one of which they note: the noted result is that the dispersion of the radial components of the peculiar velocities is 72 ± 2km s^{-1} throughout their sample volume and this value is *the same for dwarf and giant galaxies*. This is contrary to the standard expectation which, by the equipartition of kinetic energies in a random 'gas', would have the small objects moving more rapidly that the large objects; according to the presented view, all objects have peculiar velocities of identical magnitudes, independently of their size, and so this view is consistent with the Karachentsev & Makarov result. The second, un-noted, point arises from the fact that the quoted values for the dispersion of peculiar velocities, given for increasing sample volume, have a *remarkably* stable value; this value varies by no more than about 3% of the mean value, 72 ± 2km s^{-1}, when the sample size gets above 12 objects and persists up to the full sample of 103 objects. The significance of this is profound, since it indicates very strongly that the peculiar velocities have *non-Gaussian* statistics - a conclusion which is also directly contrary to standard expectations, but which is consistent with the presented view, since the only randomness arises from the directions of peculiar velocities.

For the second category, Guthrie & Napier (1996) analysed a large sample of galaxies in the range $(800,2000)$km s^{-1}, primarily to test the Tifft hypothesis that redshift quantization is a real phenomenon. In the course of their analysis, they peformed Monte-Carlo simulations in which the real redshift data was perturbed by normal random error; they found that when the mean of this error exceeded about 4km s^{-1}, then the signal indicating the presence of redshift-periodicity disappeared.

From this, they conclude that the peculiar velocity magnitudes of the objects in their data base have an upper bound of about 4km s^{-1}- at face value, this latter figure conflicts in an obvious way with the Karachentsev & Makarov values for peculiar velocities. However, the alternative view, expressed in §12, is that peculiar velocities will be increasingly recognized to have very strange properties that will allow a consistent resolution of such apparent conflicts.

For the third category, the most extensive *specific* peculiar velocity survey completed to date is that by Lauer & Postman (1994, 1995) which had the specific aim of measuring the velocity of the local group with respect to an inertial frame defined by the 119 Abell clusters within 15000km s^{-1}. Since this inertial frame was to be defined with respect to a very large amount of matter distributed over the whole sky, it was expected to be approximately stationary in the CBR frame, with the effect that the calculated local-group velocity should approximate the COBE vector. However, the analysis of the radial component of the Abell peculiar velocities appears to indicate that these 119 clusters are participating in a bulk flow of approximately 689km s^{-1} with respect to the CBR - a result which Lauer & Postman say surprised them; they conclude that, if the CBR can be considered as a valid frame of rest, then the calculated bulk flow must arise from the gravitational action of large material concentrations beyond 100 Mpc (cf the discussions of Baryshev et al (1995), §5). Furthermore, as Strauss, Cen, Ostriker, Lauer & Postman (1995) observe, this result is extremely difficult to understand on the basis of the *Standard Model*, or any of the popular variants.

By contrast, the results of these analyses can be readily understood from from the perspective of the presented work: Specifically, from the considerations of §3, in which the impossibility of giving a dynamical meaning to the notion of a mass-centre in an equilibrium universe was indicated, then the Lauer & Postman concept of an Abell clusters inertial frame is dynamically meaningless - if the observed 'fractal two' nature of the real Universe is taken to indicate a condition of global equilibrium. It follows that the figure of 689km s^{-1} quoted for the supposed bulk flow of the 119 Abell clusters is simply an estimation of a weighted mean of the radial components of the estimated peculiar velocities, measured in the CBR frame, and has no dynamical significance whatsoever. From this viewpoint, the bulk flow is not a bulk flow at all - therefore not requiring any mass-concentrations whatsoever to explain it - and the figure of 689km s^{-1} calulated for the weighted mean velocity of the Abell clusters can be understood in terms of the individual clusters having having large measurable radial velocities (cf (13), and the associated clock-assumption) with a large dispersion arising from inhomogeneities in their distribution over the sky.

To summarize, the Karachentsev & Makarov analysis, involving objects with redshifts in the range $(0,500)$km s^{-1}, leads to the general inference that the statistics of peculiar velocities in this range are independent of object masses, contrary to standard expectations, and exhibit a strong uniformity independently of sample size which is contrary to the behaviour expected on the basis of the standard assumption that peculiar velocity magnitudes are random normal variables. The Guthrie & Napier analysis, involving objects with redshifts in the range $(800,2000)$km s^{-1}, leads to the inference that the magnitudes of peculiar velocities (assumed to have Gaussian statistics) of the objects concerned have an upper bound of ≈ 4km s^{-1}, whilst the Lauer & Postman analysis, involving objects with redshifts in the range $(2000,15000)$km s^{-1}, leads to the inference that the peculiar velocities of the objects concerned are of the order 700km s^{-1}. It is therefore possible to conclude that, overall, the observations provide *qualitative* support for the kinematic structure described by (13). An extremely interesting question is whether future observations will provide support for the *quantitive* kinematic structure described by (13); the resolution of this question will require considerably more data than is currently available. However, Lauer & Postman are planning a survey out to 24000km s^{-1}, and the results of this might begin to provide an answer to the question.

14. Summary

A cosmology is derived by imposing the most simple possible *Cosmological Principle* that the model universe is in a state of gravitational equilibrium. The resultant cosmology gives a *unique* specification of the mass distribution function and, according to this function, material in the model universe is distributed in a fractal fashion, having fractal dimension *two*. This prediction is in exact accordance with the most recent analyses of modern wide-angle and deep pencil-beam surveys, (Baryshev et al 1995).

Fractality implies structure, and structure implies discreteness and this was one of the assumed properties of material in the model universe. Analysis then showed that this material discreteness necessarily implies a discretized distance scale which, together with Hubble's Law, makes the existence of a quantized redshifts in the model universe axiomatic; this is consistent with claims made by Tifft (1976, 1980, 1990) and Tifft & Cocke (1984) over the past 20 years, and with the results of a recent rigorous and independent analysis performed by Guthrie & Napier (1996).

The theory then predicts very strange behaviour of the peculiar velocity field at all distance scales; specifically, assuming a specific model of light propagation and the idea that cosmological redshift is an indicator of local measurement scales, it states that the estimated magnitudes of the peculiar velocities of objects should increase linearly with distance, r. Whilst there is insufficient evidence available at present to test this prediction quantitatively, there is evidence arising from the Guthrie & Napier analyses (1990,1991,1996) which implies the peculiar velocities are unexpectedly small at small distances, and evidence arising from recent peculiar velocity surveys (Lauer & Postman 1994) to suggest the peculiar velocities are unexpectedly large at large distances. Additionally, there is evidence from the Karachentsev & Makarov analysis that the statistics of peculiar velocity magnitudes, out to small distances, do not conform to any of the standard models of the peculiar velocity but are broadly consistent with the those expected of the presented model. These results, taken together, are consistent in a qualitative sense with the predicted behaviour.

References

Alpher, R. A., Herman, R.: 1949, *Phys. Rev.*, **75**, 1089

Baryshev, Yu V., Sylos Labini, F., Montuori, M., Pietronero, L.: 1995, *Vistas in Astron.*, **38**, 419

Broadhurst, T. J., Ellis, R. S., Glazebrook, K.: 1990, *Nature*, **355**, 55

Broadhurst, T. J., Ellis, R. S., Koo, D. C., Szalay, A. S.: 1990, *Nature*, **343**, 726

Charlier, C. V. L.: 1908, *Astronomi och Fysik*, **4**, 1

Charlier, C. V. L.: 1922, *Arkiv. for Mat. Astron. Physik*, **16**, 1

Charlier, C. V. L.: 1924, *P.A.S.P.*, **37**, 177

Coleman, P. H., Pietronero, L., Sanders, R. H.: 1988, *A&A.*, **200**, L32

Da Costa, L.N., Geller, M.J., Pellegrini, P.S., Latham, D.W., Fairall, A.P., Marzke, R.O., Willmer, C.N.A., Huchra, J.P., Calderon, J.H., Ramella, M., Kurtz, M.J.: 1994, *ApJ.*, **424**, L1

De Lapparent, V., Geller,M. J., Huchra, J. P.: 1988, *ApJ.*, **332**, 44

De Vaucouleurs, G.: 1970, *Science*, **167**, 1203

Eddington, A. S.: 1988, CUP, p371, reprint of 1926 ed

Gamow, G.: 1953, *KongDansk. Ved. Sels*, **27**, 10

Giovanelli, R., Haynes, M. P.: 1985, *AJ.*, **90**, 2445 *ApJ.*, **300**, 77

Guillaume, C-E.: 1896, *La Nature*, **24** series 2, 234

Guthrie, B., Napier, W. M.: 1996, *A&A.*, **310**, 353

Guzzo, L., Iovino, A., Chincarini, G., Giovanelli, R., Haynes, M. P.: 1992, *ApJ. Lett.*, **382**, L5

Huchra, J., Davis, M., Latham, D.,Tonry, J.: 1983, *ApJ. Sup.*, **52**, 89

Karachentsev, I. D., Makarov, D. A.: 1996, *AJ.*, **111**, 794

Lauer, T. R., Postman, M.: 1994, *ApJ.*, **425**, 2, 418

Lauer, T. R., Postman, M.: 1995, *ApJ.*, **440**, 1, 28

Moore, B., Frenk, C. S., Efstathiou, G., Saunders, W.: 1994, *M.N.R.A.S.*, **269**, 742

Peebles, P. J. E.: 1980, *The Large Scale Structure of the Universe* Princeton University Press, Princeton, NJ

Penzias, A. A., Wilson, R. W.: 1965, *ApJ.*, **142**, 419

Pietronero, L., Sylos Labini, F.: 1995, *Birth of the Universe and Fundamental Physics*, ed Occhionero, F. Springer Verlag, in press

Pietronero, L., Sylos Labini, F.: 1995, *preprint*

Regener, E.: 1933, *Zeitschrift für Physik*, **80**, 666

Regener, E.: 1995, *Apeiron*, **2**, 85 (English translation by Gabriella Moesle of 1933 article)

Roscoe, D. F.: 1995, *Phys. Essays*, **8**, 79

Sandage, A.: 1986, *ApJ*, **86**, 307

Sciama, D. W.: 1971, *Modern Cosmology*, CUP
Strauss, M. A., Cen, R. Y., Ostriker, J. P., Lauer, T.R., Postman, M.: 1995, *ApJ*, **444**, 2, 507
Tifft, W. G.: 1976, *ApJ.*, **206**, 38
Tifft, W. G.: 1980, *ApJ.*, **236**, 70
Tifft, W. G.: 1990, *ApJ. Sup.*, **73**, 603
Tifft, W. G. & Cocke, W. J.: 1984, *ApJ.*, **287**, 492
Vettolani, G.: 1994, *Studying the Universe With Cluster of Galaxies*, Proc. of Schloss Rindberg Workshop

POSSIBLE NEW PROPERTIES OF GRAVITY

TOM VAN FLANDERN
Meta Research
6327 Western Ave. NW
Washington, DC 20015

Abstract. If static gravity or spacetime curvature information is carried by classical propagating particles or waves, a modern Laplace experiment places a lower limit on their speed of $10^{10}c$. The so-called Lorentzian modification of special relativity permits such speeds without need of tachyons. But there are other consequences. If ordinary gravity is carried by particles with finite collision cross-section, such collisions would progressively diminish its inverse square character, converting to inverse linear behavior on the largest scales. At scales greater than several kiloparsecs gravity can apparently be modeled, without need for dark matter, by an inverse linear law. The orbital motions of Mercury and Earth may also show traces of this effect. If gravity were carried by particles, a mass between two bodies could partially shield each of them from the gravity of the other. Anomalies are seen in the motions of certain artificial Earth satellites during eclipse seasons that behave like shielding of the Sun's gravity. Certain types of radiation pressure might cause a similar behavior but require many free parameters. Particle-gravity models would change our understanding of gravitation and our views of the nature of time in relativity theory.

1. Properties of Gravity

Gravity has some curious properties. One is that its effect on a body is apparently independent of the mass of the affected body. Heavy and light bodies fall in a gravitational field with equal acceleration. Another is the infinite range of gravitational force. Truly infinite range is not possible for forces conveyed by carriers of finite size and speed.

Another curious property of gravity is its apparently instantaneous action. By way of contrast, light from the Sun requires about 500 seconds

Astrophysics and Space Science **244**:249-261,1996.

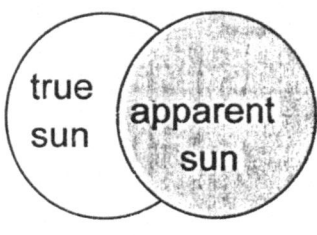

Figure 1.

to travel to the Earth. So when it arrives, we see the Sun in the sky in the position it actually occupied 500 seconds ago rather than in its present position. (Figure 1.) This difference amounts to about 20 seconds of arc.

From our perspective, the Earth is standing still and the Sun is moving. So it seems natural that we see the Sun where it was 500 seconds ago, when it emitted the light now arriving. From the Sun's perspective, the Earth is moving. It's orbital speed is about $10^{-4}c$, where c is the speed of light. Light from the Sun strikes the Earth from a slightly forward angle of 10^{-4} radians (the ratio of Earth's speed to light speed), which is 20 arc seconds. This displacement angle is called *aberration*, a classical non-relativistic effect. Length contraction and time dilation effects are four orders of magnitude smaller, since they are proportional to the square of the ratio of speeds.

We might logically expect that gravity should behave like light. Viewing gravity as a force that propagates from Sun to Earth, the Sun's gravity should appear to emanate from the position the Sun occupied when the gravity now arriving left the Sun. From the Sun's perspective, the Earth should "run into" the gravitational force, making it appear to come from a slightly forward angle equal to the ratio of the Earth's orbital speed to the speed of gravity propagation.

This slightly forward angle would tend to accelerate the Earth, since it is an attractive force that does not depend on the mass of the affected body. Such an effect is observed in the case of the pressure of sunlight, which of course does depend on the mass of the affected body. The slightly forward angle for the arrival of light produces a deceleration of the bodies it impacts, since light pressure is a repulsive force. Bodies small enough to notice, such as dust particles, tend to spiral into the Sun as a consequence of this deceleration. This process is the Poynting-Robertson effect.

Observations indicate that none of this happens in the case of gravity. There is no detectable delay for the propagation of gravity from Sun to Earth. The direction of the Sun's gravitational force is toward its true, instantaneous position, not toward a retarded position, to the full accuracy of observations. No perceptible change in the Earth's mean orbital speed has yet been detected, even though the effect of a finite speed of gravity is cumulative over time. Gravity has no perceptible aberration, and no Poynting-Robertson effect – the primary indicators of its propagation speed. Indeed, Newtonian gravity explicitly assumes that gravity propagates with infinite speed.

2. The Speed of Gravity

The absence of detectable aberration implies that, to the extent that gravity is a propagating force, its speed of propagation must be very high compared to that of light. In the early 19th century, Laplace (1966) used the possible error in the determination of the absence of an acceleration of the Earth's orbital speed to set a lower limit to the speed of gravity of about $10^7 c$. Using the same technique with modern observations, Van Flandern (1993) improved that lower limit to $10^{10} c$.

General relativity (GR) explains this result without involving faster-than-light propagation. GR suggests that gravity is not a force that propagates. Instead, the Sun curves spacetime around it; the Earth simply follows the nearest equivalent of a straight line available to it through this curved spacetime. It has been known since the time of Sir Arthur Eddington that the curved spacetime explanation is not required by general relativity (see Van Flandern, 1994), or certain other variants that preserve agreement with the classical observational tests of the theory. Other authors have proposed minor modifications of the field equations to replace spacetime curvature tensors with gravitational energy-momentum density tensors (Rosen 1940). There is even some direct experimental evidence against the curved space-time explanation that is provided by neutron interferometers (Greenberger and Overhauser 1980). The results are incompatible with the geometric weak equivalence principle because the interference depends on mass.

How does spacetime far from a large mass get its curvature updated without detectable delay so that orbiting bodies accelerate through space toward the true, instantaneous position of the source of gravity? Computer experiments based on arguments posed by Eddington (1987) show that binary pulsars are especially sensitive to this test. To satisfy observations, it is not sufficient that each massive companion of a binary pulsar acts from its retarded position; nor from its linearly extrapolated position over one light-time, as electromagnetic forces do. It is not even sufficient for each companion to accelerate via the full curvature that spacetime would have had one light-time ago. Some information is being propagated between source and affected body faster than light.

We can illustrate this dilemma for GR with examples involving black holes. A black hole emits no light because escape velocity is greater than c. Yet it still has gravity. This is explained as due to the presence of a

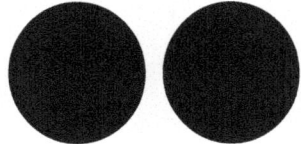

Figure 2.

"fossilized" field, a curvature of spacetime outside the black hole's event horizon that remains after the star that created the hole collapsed. But the black hole may well be an orbiting companion of a normal star. How does the "fossilized" field know about accelerations of the center of mass behind the event horizon caused by the normal star, so that it can accurately keep pace?

There are two problems here. 1) The curvature of spacetime created by the normal star is sufficiently different at points inside the event horizon of the black hole from what it is for points outside that nothing outside the event horizon could remain in proximity to something inside for very long without some sort of linkage across the horizon. 2) The curved spacetime generated by the normal star should require an infinite time to reach the center of mass of the black hole, leaving the singularity in the black hole unaware of the current state of curvature of spacetime that it must respond to without detectable delay.

A second example consists of two identical black holes that make a close approach, and then recede again to infinity. (Figure 2.) Despite the complex interactions between black holes when they draw close, an observer riding the balance point between the two could remain there indefinitely, and recede again to infinity, without experiencing strong gravitational forces or being drawn toward either hole, because of the balance and symmetry of the example. This would be true even if the event horizons of the two holes came to overlap, allowing the observer to peer into the spacetime formerly hidden behind both event horizons!

Such paradoxes could not be constructed if GR were not trying to insist that gravitational information must not propagate faster than light. But abandoning the light speed limit does not mean abandoning GR. The main properties of the theory, including its satisfaction of the four classical observational tests, can be retained in flat spacetime versions of the theory as in the papers already cited.

If we assume a propagation speed greater than the speed of light, we contradict a corollary of special relativity (SR), wherein no communication faster than light speed in forward time is possible. SR is a well-tested and confirmed theory. But the emphasis has been on the experimental tests that SR has already passed, demonstrating the reality of time dilation, length contraction (indirectly), and the increase in inertial mass with speed, as well as the independence of measured light speed on the motion of its source, etc.

However, there are two postulates underlying SR. The first, called the "covariance" postulate, requires that no inertial frame be "special" since all are equivalent for formulating the laws of physics. Since almost all SR experiments of the past have been done in the "laboratory", it has not been possible to confirm this frame-independence postulate experimentally. Two historical experiments have made the attempt: the Sagnac experiment in 1913, and the Michelson-Gale experiment in 1925. Both utilized rotating reference frames, and both obtained non-zero fringe shifts in Michelson-Morley-type experiments performed on rotating platforms. Both published results claiming experimental contradictions of SR. SR has developed an "explanation" and incorporated the Sagnac effect for rotating frames as a standard part of the model.

Both Sagnac and Michelson favored an alternate formulation of SR that allows a "universal time", as originally advanced by Lorentz (1931). The modern formulation of this idea is referred to as the "Mansouri-Sexl" (1977) transformation. The respective equations for Einstein SR and the Lorentzian (Mansouri-Sexl) alternative are:

Einstein SR equations:

$$t = \gamma(T - \frac{vX}{c^2}) = \frac{T}{\gamma} - \frac{vx}{c^2} \quad x = \gamma(X - vT)$$

Lorentzian SR equations:

$$t = \frac{T}{\gamma} \quad x = \gamma(X - vT)$$

Both of these transformations relate coordinate X and time T in one inertial frame ("the laboratory") to x and t in a frame moving relative to the laboratory in the X direction with speed v. The dilation-contraction factor, $\gamma = 1/\sqrt{1 - v^2/c^2}$, is always ≥ 1.

These two sets of equations differ only by the term $-vx/c^2$. The reality of this term has not been tested by past experiments because the term is zero (or constant) when only a single clock represents time in the "moving" frame. This is because the single clock is usually placed at $x = 0$ by definition of the origin in the frame. This term plays a crucial role in the formula for the addition of velocities in SR, which in turn plays a central role in the proof that nothing can propagate faster than light speed in forward time.

Can we now perform a test to distinguish between these forms of SR? The Global Positioning System (GPS) is a network of 24 satellites carrying atomic clocks, now in various orbits around the Earth. The satellite orbits are about 80 light milliseconds in radius, and v/c for the satellites is about 10^{-5}; predicted discrepancies can be on the order of 800 nanoseconds, easily detectable.

For the actual GPS satellite network, the rate of each orbiting clock has been pre-adjusted while still on the ground so that the average length of the second will be the same for orbiting clocks as for ground clocks. This is equivalent to setting $\gamma = 1$ in the preceding transformation equations. That should not affect the distinguishing term in question here, since no γ factor appears in that term. Nonetheless, simultaneous and continuing synchronization between all satellites and all ground clocks to a precision of a few nanoseconds has been achieved.

Whether or not GPS clocks can still be related to each other with the Lorentz SR time transformation is still under debate. But Lorentz transformations are just one of a family of transformations in which the speed of light is constant (Robertson and Noonan, 1968). The dependence of the speed of light on the speed of the observer depends on the method of synchronization of clocks, since speed measurement involves more than one point in space and instant of time.

The GPS system has shown that, in the classical "Twins Paradox", the traveling twin could have carried along a second clock preset in epoch and rate such that it reads the correct time back on Earth throughout the traveler's journey. That is what GPS satellites clocks are – clocks in relatively moving frames that maintain their synchronization with ground clocks even as they travel at high relative speeds and change frames relative to the ground clocks.

If there is no $-vx/c^2$ term in the transformations the proof that nothing can propagate faster than light fails, and gravity itself might then have no aberration. Thus it could be that gravity propagates much faster than light. In like manner, non-locality in quantum physics may be understood.

3. Particle Models of Gravity

One simple particle model of gravity assumes that space is filled with a flux of rapid randomly moving particles. Ordinary matter is almost transparent, much as it is to neutrinos. A downward force is felt on Earth because of a differential caused by absorption of upward flux by the Earth. Bodies will shield one another to some degree (Figure 3), resulting in an acceleration of each body toward the other that depends on the mass within the bodies and inversely on the square of the distance between them.

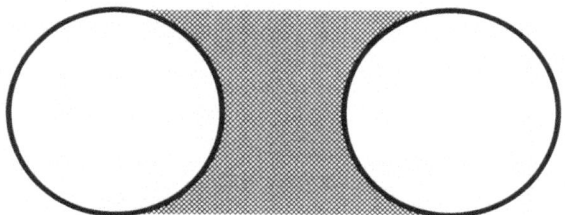

Figure 3.

The 18th century physicist LeSage is usually credited with the first particle model of gravity, although LeSage indicates (1784) that he was inspired by earlier writers. A flux of tiny, rapidly moving particles in space is one way to explain the gravitational force, including relativistic effects (Van Flandern, 1994).

Particle gravity models imply that bodies will experience resistance as they move through the flux. Resistence is minimized if the particles momentum arises from very high speed coupled to very low mass. The ratio of the mass of a particle to the mass of a single absorber within a body is constrained by the absence of observable resistance for large orbiting bodies. Actual detection of resistance could support a particle model but would be difficult to distinguish from effects of tidal friction or other causes of orbital acceleration. The lack of acceleration of the Earth's mean orbital speed around the Sun sets an upper limit to the size of any resistance-induced acceleration.

Absorption of the particle flux must heat the bodies. This heat must be fully re-radiated to maintain thermodynamic equilibrium. The largest planets do radiate more heat back into space than they receive from the Sun. This radiation, traditionally ascribed to such sources as radioactivity in the planetary cores, could mask a component associated with a particle flux.

4. The Range of Gravity

In a particle model gravity has a finite range. The particles must have finite dimensions and speeds. There must exist some characteristic distance, r_G, that a particle can travel before it will scatter off anther particle. If two large bodies are separated by much more than the distance r_G, any effect they have on one another will be diluted and eventually canceled by scattering.

The following formula represents a possible modification of Newtonian gravity to account for the range limitation. It assumes that back-scattering into the shadow between bodies occurs uniformly with distance, and at a

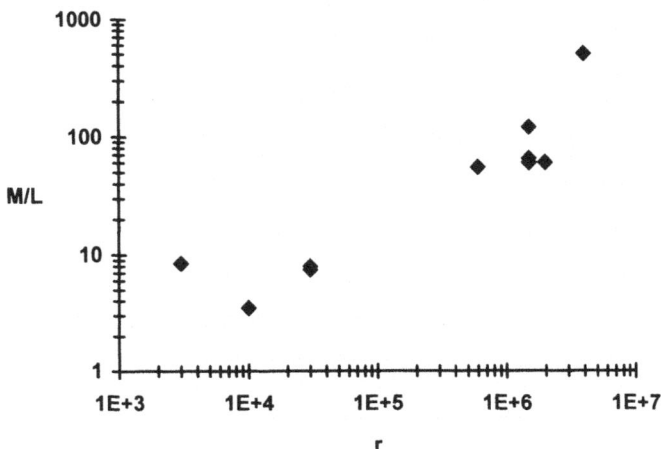

Figure 4. M/L (mass-to-light ratio) versus r (linear scale-size in light-years).

rate proportional to the particle deficit:

$$\ddot{\vec{r}} = -\frac{GM}{r^3}\vec{r}e^{-(r/r_G)}$$

G is the gravitational constant, M the mass of the attracting body, r the distance from the attracting body and r_G the characteristic range of gravity (the mean free path). This reduces to Newtonian gravity as r_G approaches infinity. Some arguments suggest that r_G might be only 1-2 kiloparsecs.

Most astronomers assume that the Newtonian law of gravity applies and invoke invisible "dark matter" in amounts that increase radially with r in galaxies. This cancels one power of r in the inverse square attraction. Astronomers utilize the M/L ratio of galaxies, where M is mass and L is luminosity or light. This ratio is unity for ordinary stars, but is observed to be much larger in large scale systems, from which the presence of dark matter is inferred.

Some astronomers note that the universe does seem to obey an inverse linear law at large scales. Figure 4., which uses data from Wright, Disney, and Thompson (1990), illustrates inferred M/L ratios over a variety of scales. The general trend is linear over three orders of magnitude in scale. The same authors also discuss computer experiments showing that an inverse linear law of gravity is more effective in predicting observed shapes of interacting galaxies than is an inverse square law.

For small values of r, the non-Newtonian exponential factor in the gravitational acceleration formula simplifies to $(1 - r/r_G)$. For the Earth, this factor differs from unity by 4.85×10^{-9} kpc / r_G . For Mercury, this difference would be 1.9×10^{-9} kpc / r_G, since it varies linear with orbit size. For any given distance from the Sun, the factor is constant, and therefore behaves as if the gravitational constant G were slightly modified and slightly variable with distance.

Observationally, orbit determinations using radar ranging data are dominated by Mercury observations for determining the effective value of G because of Mercury's large eccentricity. In Kepler's third law, $n^2 a^3 = GM_\odot$ (where n = mean motion, a = semi-major axis, $M_\odot$ = mass of Sun), radar observations of Mercury's mean motion n_1 and semi-major axis a_1 are used to determine $GM_\odot$. This value is then used for the Earth's orbit, for which a_3 (semi-major axis of third planet) is much better determined by ranging data than n_3 because n_3 becomes indeterminate from radar data for a circular orbit. So n_3 is effectively measured with respect to n_1 rather than independently determined. When the radar-determined orbits are compared with optical data over the past century or more, the optical data being very sensitive to the true value of n_3 for Earth, the error in n_3 determined from radar through Kepler's law and n_3 determined optically would be a function of the difference between the effective value of G for Mercury and that for Earth: $(n_{\text{radar}} - n_{\text{optical}})_3 = n_3(a_3 - a_1)/(2r_G) = 0.19/r_G$ arc seconds per century ($''$/cy). In the latter form, r_G must be measured in kpc.

At the same time, the difference in effective gravitational constant between a planet's perihelion and its aphelion causes the longitude of perihelion to rotate by a comparable amount. For Mercury, this rotation rate is: $n_1 a_1/[2r_G(1 - e_1^2)^{1/2}] = 0.52/r_G$ $''$/cy. Since Mercury's perihelion direction dominates the determination of a fixed direction in inertial space for the radar data, this motion will cause a rotation of the radar inertial frame at the rate just specified, which is not negligible.

The combination of the two effects just described, one for the Earth's mean motion and the other for the direction of the origin, will cause the radar mean motion of the Earth to exceed the optical mean motion by $0.71/r_G$ $''$/cy. Such a discrepancy is actually observed, has a magnitude corresponding to $r_G \approx 1$, and has remained an unexplained puzzle over the past 5-10 years. This a priori derivation of the effect supports the basic idea of particle models for gravity and may limit values of r_G.

The model predictions could now be tested against observations of other planets to determine if it is consistent with other existing solar system data.

5. Gravitational Shielding

Particle gravity models differ from Newtonian gravity in the ability of matter to shield other matter from the effects of gravity. Ordinary matter must be extremely transparent to the flux of particles, and is indeed mostly

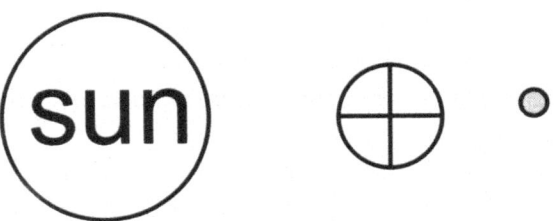

Figure 5.

empty space. There must exist, however, some density of matter through which no flux of particles can penetrate. Matter behind such a wall could not contribute to the gravitational field of a body.

In the 19th century, J.C. Maxwell used the analogy of a swarm of bees blocking sunlight. If two equal swarms of bees are superimposed, twice as much light will be blocked - unless the swarms are so dense that some bees overlap bees in the other swarm, in which case less than twice as much light is blocked. If one swarm is so dense that it blocks all the light, then the second swarm adds nothing to the light loss.

If only the outer layers contribute to an external gravitational field the ratio of gravitational to inertial mass will depart from unity – a situation which can be reconciled with the results of the most sensitive Eötvös type of experiments (Van Flandern, 1995).

The effect is usually referred to as "gravitational shielding", since a portion of the gravitational field that would exist in Newtonian gravity is shielded. At a point in space, the gravitational acceleration induced by a body of mass M at a distance r when another body intervenes is:

$$\ddot{\vec{r}} = -\frac{GM}{r^3}\vec{r}e^{(-K_G \int \rho dr)},$$

where ρ is the density of the intervening body over the short distance dr. The integral is taken through the intervening body along the vector joining the point in space and body M, and K_G is the shielding efficiency factor in units of cross-sectional area over mass.

To test for such an effect in nature, one needs to examine a test body orbiting near a relatively dense intermediate body, where the intermediate mass occasionally intervenes in front of a more distant large body (Figure 5). We then seek evidence that the distant mass exerts less than its full effect on the test body at times when the intermediate mass is aligned between them. There is no way to determine in advance how big this effect might be since K_G is unknown.

Probably the most suitable test for this effect in the solar system arises from the two Lageos artificial satellites. The Earth's core provides the dense

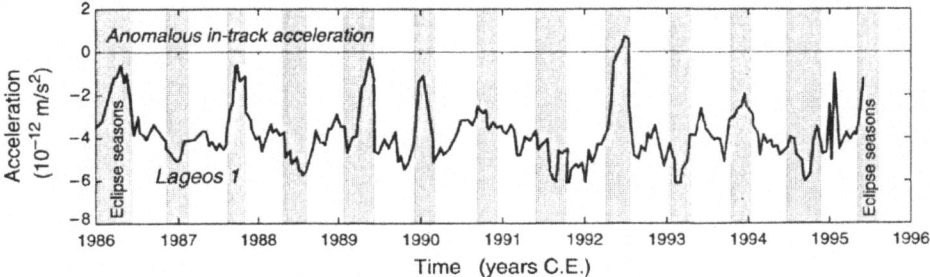

Figure 6. Lageos 1 anomalous in-track acceleration. Shading denotes eclipse seasons.

intermediate mass, and the Sun is the distant body. Both satellites are in orbits high enough, and the 400-kg satellites are massive enough, to be very little affected by most non-gravitational forces such as atmospheric drag or solar radiation pressure. Both satellites are covered with retro-reflectors that reflect light back along the incoming direction. This enables these satellites to have their positions measured by laser ranging from ground stations. The orbits can be determined with a precision on the order of a centimeter or better.

Lageos 1 has been in orbit for 20 years, and Lageos 2 for about 4 years. Both are in nearly circular orbits roughly an Earth radius high, and circle the globe roughly once every four hours. Lageos 1 revolves retrograde with an inclination of 110°, which causes its orbit plane to precess forward. Lageos 2 is in a direct orbit with an inclination of 53°, precessing backward. As a consequence, Lageos 2 has "eclipse seasons" – periods of time when the satellite enters the Earth's shadow on every orbit for up to 40 minutes – that are more frequent and more variable in length than for Lageos 1. As precession changes orbit orientation, each satellite may go many months continuously in sunlight, without eclipses. For Lageos 2, it is possible for two consecutive eclipse seasons to merge into one long season, as happened in late 1994 through early 1995.

During eclipses any gravitational shielding effect that may exist will be operative. Several other types of non-gravitational forces also operate only during eclipses. Solar radiation pressure shuts off during eclipses, as does some of the thermal radiation from the Earth. Light, temperature, and charged particles are all affected, and at the one centimeter level must all be considered.

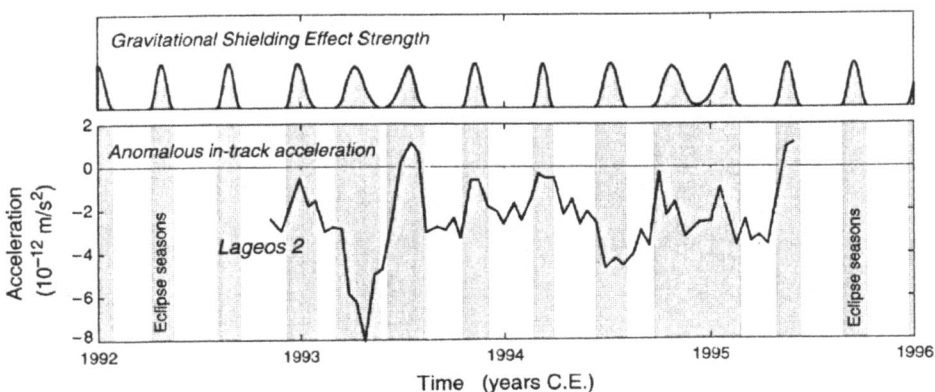

Figure 7.   Lageos 2 anomalous in-track acceleration. Shading denotes eclipse seasons. Theoretical gravitational shielding effects appear avove the observed anomalous acceleration for comparison.

Both Lageos satellites exhibit unexpected anomalous in-track accelerations. Figures 6 and 7 show the effect for each satellite. The anomalous in-track acceleration (negative because it operates as a drag force) in units of 10^{-12} m s^{-2} is plotted against time, shown as a 2-digit year. Eclipse seasons are indicated with shading. An average negative acceleration throughout the data can be explained as a combination of radiation, thermal, and charge drag forces. But the data show substantial deviations from this average drag, especially during eclipse seasons, and these are not easily explained (Rubincam, 1990).

Other authors, most recently Slabinski (1996), have succeeded in modeling the bulk of the anomalous acceleration, including the eclipse season variations, for Lageos 1. This was accomplished by using about a dozen empirical corrections, and the assumption that albedo variations over the satellite surface combined with spin orientation and precession to produce these variations. The surface of Lageos 1 was supposed to be very uniform and highly reflective. For these models to be viable, it must be assumed that some factor, perhaps rocket exhaust at the time of injection into orbit, dirtied the surface and produced the albedo variations. Lageos 2 was launched to avoid such problems. Yet the preliminary data available so far suggest that the anomalous acceleration is still present.

The upper panel of Figure 7 shows the theoretical gravitational shielding effect for Lageos 2, calculated with $K_G = 2 \times 10^{-18}$ cm^2 g^{-1}. The amplitude of the effect would be essentially the same for Lageos 1 and Lageos 2. Lageos 1 is affected by radiation forces and/or other effects that may reinforce or oppose any shielding effect. The data allow (though they do not require) a shielding effect.

It has been proposed to launch a satellite inside a large, spherical shell. The shell would protect the satellite from the non-gravitational forces. The shell would have sensors that would allow adjustments to keep the interior satellite near its center. The interior satellite would move under the influence of gravitational forces alone, protected from external radiative, thermal, and charge influences. Such a configuration would allow the unambiguous detection of a gravitational shielding effect, if one does exist.

The author is indebted to Erricos Pavlis at NASA Goddard Space Flight Center for supplying the Lageos data, and he thanks Boris Starosta for preparing the figures.

References

Eddington, A.: 1987, *Space, Time, and Gravitation*, Harper & Bros., New York, pg. 94

Greenberger, D. M. and Overhauser, A. W.: 1980, *Sci. Amer.*, **242**, 66

Laplace:, 1966 *Mechanique Celeste*; English translation reprinted by Chelsea Publ., New York

LeSage, G. L.: 1784, *Berlin Mem.*, 404

Lorentz, H. A.: 1931, *Lectures on Theoretical Physics*, vol. 3, Macmillan & Co., London, pp. 208-211

Mansouri, R. and Sexl, R. U.: 1977, *Gen. Rel. & Grav.*, **8**, 497

Robertson, H. P. and Noonan, T. W.: 1968, *Relativity and Cosmology*, W.B. Saunders Co., Philadelphia, cf. pp. 46-50

Rosen, N.: 1940, *Phys. Rev.*, **57**, 147

Rubincam, D. P.: 1990, *J. Geo. Res.*, **95**, 4881

Slabinski, V. J.: 1996, *Celestial Mechanics*, in press.

Van Flandern, T.: 1993 *Dark Matter, Missing Planets and New Comets*, North Atlantic Books

Van Flandern, T.: 1994, *Meta Res. Bull.*, **3**, 9

Van Flandern, T.: 1995, *Meta Res. Bull.*, **4**, 1

Wright, A. E., Disney, M.J. and Thompson, R. C.: 1990, *Proc. Astr. Soc. Australia*, **8**, 334

ON THE COSMIC LIMITS OF PHYSICAL LAWS

LEOPOLD HALPERN
Department of Physics
Florida State University, Tallahassee, FL 32306

Abstract. Views on the completeness of the general theory of relativity are reviewed. An approach to a generalization of the theory based on a modification of the principle of inertia is outlined.

The Euclidean geometry of three dimensions may be considered the first successful physical theory, as it allowed predictions of many of the properties of uniform matter. These predictions are logically derivable from axioms — basic assumptions obtained by idealization from the shapes of simple real objects. Such idealizations (points, lines, surfaces, etc.) facilitate the logical conclusions, but they are doomed to conflict in the description of nature when the limits of their identifiability with empirical objects play a role. This difficulty helped to ferment the philosophical idea of atomism of matter, in which, however, idealized features of Euclidean space remained unmodified.

In his differential calculus, Newton made use of the idealized properties ascribed to space, assuming that space remains unmodified by physical bodies. Only in the last century was it realized that Euclidean geometry is not the only possibility, and Riemann recognized that the point-dependent geometry which he had constructed was not independent of the physical laws but rather complementary to them — spatial curvature can be equivalent to universal mechanical forces. This led to the geometrization of such forces by H. Hertz. Riemann also recognized the ambiguity of the continuum for physics and created a model of a discrete space to avoid it. Einstein, after he had discovered the principle of relativity, succeeded in geometrizing gravitation in the four-dimensional space-time continuum of (pseudo)-Riemannian geometry. General relativity, like Maxwell's electrodynamics, is a macroscopic theory and describes gravitational physics satisfactorily far beyond the conditions of the solar system. Vacuum solutions of the Einstein-Hilbert field equations show however the appearance of closed sharp surfaces called

Astrophysics and Space Science **244**:263-268,1996.

horizons, beyond which the relations with matter outside should manifest modifications that go far beyond our experience (one-way membranes), and the matter inside is furthermore doomed to collapse to a point within a period that remains finite for comoving observers.

Here we have a prediction of the best macroscopic theory available that unavoidably involves critically sharp surfaces and even points — features that were just characterized as manmade idealizations unlikely to be precisely identifiable with anything in nature.

Einstein himself did not believe that the left hand side of the field equations will have solutions that are everywhere physical as long as a right-hand matter term that remains alien to the geometric description of the purely gravitational left side can exist. He believed, however, that the left side with its second order derivatives would not need significant modifications; to incorporate the right-hand matter side into the geometric formalism he considered modifications of the topology as well as of the geometry, searching for unified singularity free solutions. Two of his last letters still stress that he expects ambiguities in physical theories to occur unless the above problem of unification is solved [1].

He also classified as singularities horizons which do not show locally singular behaviour because of the aforementioned global features to which they give rise. Einstein's work with N. Rosen [2] emerged from the attempt to eliminate the domain at and inside the Schwarzschild horizon in favour of geometrical properties related to local matter.[1]

Einstein also stressed in his letters that the mainstream of physicists in the fifties does not share his views. Indeed even today unmodified general relativity is seen as the only physics describing macroscopic situations and thus the resulting description inside the horizon is also accepted as physical. The confidence is vaguely expressed that quantum effects will bail them out when the physics of the collapse becomes too absurd near the point singularity.

Schrödinger, who had worked in friendly competition with Einstein on a (geometrically) unified theory of gravitation and matter, had a similar distrust of the prevailing interpretation of the physics of the Schwarzschild horizon; however, he considered horizons for models of the universe and coined in their characterization an analogy with Xeno's paradox of the race of Achilles and the turtle. Earlier, Schrödinger had also discovered the first elementary particle reaction in an (external) gravitational field — the pair creation of fermions in the metric of an expanding universe. Such processes and even more their virtual counterpart — the analog of the Lamb shift in

---

[1] I thank the late Prof. N. Rosen for directing my attention to this paper.

classical gravitational fields, are expected to modify the gravitational law [3].

The related Casimir effect in gravitational fields should play a role even in macroscopic situations [4]. An estimate of the magnitude of such effects can however not be hoped for because we lack knowledge of elementary particle and quantum physics in connection with the universality of the gravitational interaction. Quantum theory gives a rough indication of the nonlinear curvature terms in the gravitational Lagrangian [5]:

$$L_g = \sqrt{g} \left(R + \alpha R^2 + \beta R_{ik} R^{ik} \right)$$

The resulting field equations may have other vacuum solutions besides those of general relativity, but they also give rise to difficulties in the quantum theory.

The author has suggested a different approach to these problems [6, 10]. Gravitational collapse occurs in macroscopic situations and can thus be described by classical equations. The Einstein-Hilbert equations, which dominate classical general relativity, clearly do not suffice to describe phenomena such as the Schrödinger pair creation and the associated virtual effects which should appear before a collapse through a horizon takes place [7].

We try here to obtain the necessary modifications of the classical field equations — not in fragments deduced from an incomplete quantum-perturbative procedure — rather from very different general considerations which may finally even modify the quantum formalism.

Galilei's principle of inertia determines the orbits of the Galilei group (solutions of Newton's homogenous differential equations) as the paths of idealized motion in Euclidean space, if all other matter is remote from interfering in a noticeable way with the moving particle.

We have suggested a generalized relativistic version of this principle [10]: unperturbed particle motion follows the projection π of timelike orbits of the anti De Sitter group $G = SO(3, 2)$ onto the anti De Sitter universe $B = G/H$, where $H = SO(3, 1)$, the Lorentz subgroup of G, B (the space of right cosets of G with respect to H) is the anti De Sitter universe, our habitat, and π is the natural projection $G \to B$. The groups G, H are simple and the Cartan-Killing metric γ on G, projected on B by π', provides us with the anti De Sitter metric $g = \pi'\gamma$ on B with a unit of length which is the radius of this universe. We need to assume the dimension of a physical length only on B.

$P(G, H, B, \pi)$ forms a principal fibre bundle, and the metric γ determines a connection which makes all vectors perpendicular to the fibres horizontal. Timelike horizontal geodesics on G project on geodesics on B, which are candidates for ordinary particle orbits. Geodesics with a non-horizontal "spin components" on G project onto the orbits of spinning test particles on B (which for the special case of our present metric γ and g are however still geodesics).

A remarkable feature of this modified principle of inertia is that it includes not only the motion of test particles but even the Einstein-Hilbert equations: The Cartan-Killing metric γ fulfills the relation $R_{uv} = \frac{1}{4}\gamma_{uv}$:

$$R_{uv} - \frac{1}{2}\gamma_{uv}R + \gamma_{uv} = 0 \tag{1}$$

Projection by π' on B results there in the Einstein-Hilbert equations with a cosmological term, expressed in units of the radius of the universe.

The modified principle of inertia, which like that of Galilei is based on the orbits of groups, yields also the gravitational field equations and even the orbits of spinning test particles. The dimension of a length with the above unit can be associated with geodesic coordinates on B.

Our construction leads from the uninteresting case of the homogenous universe straightforwardly to a generalization which is wider than that of general relativity: Consider other metric solutions γ of eq. (1) on P. We have still a set of six vertical killing vector fields A_M tangent to the fibres with the commutation relations of the Lie algebra of H. Horizontal vectors are perpendicular to the A_M with respect to γ. The projection of eq. (1) on B has now the general form

$$B_{ik} - \frac{1}{2}g_{ik}(B+1) = \frac{3}{2}\kappa\left(H_{ijhl}H_k^{jhl} - \frac{1}{4}g_{ik}H_{abcd}H^{abcd}\right) \tag{2a}$$

$$H_{jhl}^k \, ; \, _k = 0 \tag{2b}$$

Here B_{ik} is the Ricci tensor and B the curvature invariant of the metric g on the base B, whereas H_{ijhl} are the components of the curvature tensor resulting from the general connection ω on P. The geometry on B resulting from the projection of γ is of the Einstein-Cartan type with a connection which is the sum of a Riemannian connection and a contortion term which does in general not vanish. B_{ijkl} is formed with the Riemannian term alone. Our unit, as seen from the cosmological member, is still of the order of the "radius" of the universe. The constant κ has the dimension of a length squared; one might expect it to be of magnitude unity, but this would give wrong physical results in the solar system. The constant must be very small, and it is suggestive to choose the Planck length as the unit.

The semicolon in eq. (2b) denotes a covariant derivative formed with the Riemannian part of the connection alone.

Our construction has the remarkable property that right translations on P by elements of H (e.g., elements of $SO(3)$), transform each fibre into itself, whereas left translations act also horizontally and act thus via the projection on the points of B. This serves as a model of the action of the rotation group on the internal spin space and on space-time.

The term bilinear in H of eq. (2a) vanishes identically for all solutions of the Einstein-Hilbert equations on B, so that such solutions are also among the solutions of eq. (2).[2]

The contortion terms of the connection can appear in solutions of eq. (2) even when it was zero initially, because it is not conserved by itself. This suggests associating torsion with matter. Schrödinger's "worrying phenomenon" — the creation of matter pairs — is thus taken into account [7]; it should appear in averaged form in any realistic macroscopic theory of gravitation and matter. The present case limits it however to matter of vanishing rest mass. The right hand term of eq. (2a) modifies the collapse so that collapsing matter does not move any more along geodesics. The term grows faster than the density of matter with diminishing distance from the center. Eventually the collapse should come to a stop (roughly speaking, because Yang-Mills fields are vector fields and thus counteract the gravitational attraction). If κ is the gravitational constant, a rough estimate indicates that the right hand term for a collapse of one solar mass should acquire significance only near a radial distance of 10^{-13} cm, unless the modified field equations surprise us with new alternative physical solutions.

The generalized principle introduced thus leads the way to a generalized macroscopic theory of gravitation which tends to relax some of the problematic results of classical general relativity. The theory remains incomplete at this stage, still requiring alien sources for matter with rest mass, but it appears to be a path in the right direction. The gauge structure of the theory suggests (if not prescribes) also a generalization of quantum mechanics from the field of complex numbers beyond the generalizations of quaternions to a more general algebra. A structure of this kind was considered early on by the school of Sommerfeld [8]. Here it appears however with a much wider outlook for physical interpretation.

Finally, the present stage of the theory already more than promises the transition to a discrete formulation which opens the door to hope for relief from "the curse of the thirteenth fairy" [9] — the ambiguities of the continuum. The technical details are far too involved to be outlined in this short communication. It should however be stressed that, at the present stage, all these promising outlooks do not show any distinct hope to derive phenomena like the quantization of redshifts which is one of the main subjects of this conference.

---

[2] I thank Dr. J. Åman of the Institute of Theoretical Physics of the University of Stockholm for pointing this out to me.

References

1. Einstein, A.: 1953, letters to J. Moffat, May 29 and August 24, 1953, courte John Moffat.
2. Einstein, A., Rosen, N.: 1935, *Phys. Rev.*, **48**, 73
3. Halpern, L.: 1962, *Ark. f. Fysik*, **39**, 539
4. Gables, R.: 1970, Thesis, Univ. of Windsor
5. DeWitt, B.: 1952, Thesis, Harvard Univ.; DeWitt, B., Utiyama, R.: 1962, *J. Math. Phys.*, **3**, 609
6. Halpern, L.: 1993 *Spinors, Twistors and Clifford Algebras*, Z. Oziewicz et al., eds., Kluwer, Dordrecht; Halpern, L.: 1994, *Int. J. Theor. Phys.*, **33**, 401; Halpern, L.: 1994, *Found. of Physics*, **24**, 1697
7. Schrödinger, E.: 1939 *Physica*, **6**, 889; Schrödinger, E.: 1940 *Proc. Irish Acad. A*, **46**,24
8. Sommerfeld, A.: 1944, *Atomic Structure and Spectral Lines*, Vol. 2, Ch. 4, p. 238 ff., Vieweg Braunschweig
9. Schrödinger, E.: 1997, *Science and Humanism*, Cambridge
10. Halpern, L.: 1996, ref. 6 and Research Project, 1996, unpublished

CHANGES IN CONCEPTS OF TIME
FROM ARISTOTLE TO EINSTEIN

MENDEL SACHS
Department of Physics
State University of New York at Buffalo

Abstract. The meaning of time and motion is discussed, at first tracing conceptual changes from Aristotle to Galileo/Newton to Einstein. Different views of 'time' in 20th century physics are then examined, with primary focus on the revolutionary changes that came with the theory of general relativity. Implications of its new view in all domains of physics are discussed – from elementary particles to cosmology. The special role of Hamilton's quaternion calculus in equations of motion in general relativity is shown.

1. Introduction

The problem of 'time' has been around since the ancient times, in Greece and Asia. There have been common threads of thought on this subject throughout the millennia. In the contemporary period of physics and philosophy, different concepts of time may be traced to the earliest views and controversies on this subject. There are presently different concepts of time, applied to different sorts of physical situations, though they are all called by the same name, "time". In this talk, I wish to survey some of the older ideas and then follow them through to ideas of the twentieth century, finally highlighting the view of time in Einstein's theory of general relativity, as well as other modern views, as in the quantum theory of measurement and in irreversible thermodynamics, regarding the question of the 'arrow of time'.

2. Early Concepts

In ancient Greece, Aristotle defined time as the potentiality for the motion of matter. He said that the reason that a thing moves is that it absorbed its

Astrophysics and Space Science **244**:269-281,1996.

motion from an earlier motion, and the earlier motion was preceded by a still earlier motion, and so on, ad infinitum. Thus he concluded that time, as the potentiality of the motion of matter, extends to the infinite past and it will continue into the indefinite future. In Aristotle, 'time' relates to the motion of matter in this way, but it is still considered as a 'thing in itself' – like a ladder that a person has available to climb or not to climb [1].

In the fourth century, Augustine wrote that he could not accept Aristotle's conclusion about the infinity of time since the universe was created at a definite 'beginning' of time, in accordance with the Biblical Scriptures – when the creation of the universe happened, *ex nihilo*. What he said was that when God created the universe at this initial time, along with the laws of nature, he also created time, simultaneously. Thus, there is no 'time' to talk about, 'before' the creation of the universe [2]. From my reading of the twelfth century scholar, Moses Maimonides, he proposed a variation of Augustine's 'time', wherein the time that was created with the matter of the universe and its laws was to be a manifestation of matter, rather than a 'thing-in-itself'. Indeed, the latter view is closer to Einstein's interpretation of time in his twentieth century theory of general relativity, as we will discuss later [3].

3. Development of the Modern Viewpoint

In the seventeenth/eighteenth centuries, we come to Galileo and Newton – the fathers of the modern era of physics. Galileo focused on the descriptive aspect of motion – 'kinematics' – but maintained the existence of underlying objective causes for this motion, though the latter was not quantified until Newton, in his formulation of the laws of motion and his invention of calculus to facilitate it. Galileo also proposed the idea that 'motion' per se, is strictly a subjective concept. Thus he ruled out the idea of an absolute space or time measure, in principle, though he did not discuss the time measure in this way. His assertion of the relativity of motion in the description of the physical laws was indeed a very important precursor for Einstein's theory of relativity, which was to underlie the twentieth century theory of relativity. Galileo then superseded Copernicus, who concluded that all of the planets must orbit about the sun, which is at an absolute center of the universe. On the other hand, Galileo reflected (in his discovery of 'Galileo's principle of relativity'), that the laws of motion should be unchanged if we describe the earth orbiting about the sun or if we should describe the sun orbiting about the earth! That is, he saw that 'motion', per se, is strictly a subjective concept in the laws of motion [4].

Newton did claim that there is an objective space and time that are not relative entities, and do have absolute origins. He did, however, concede that the laws of motion entail only relative spatial measures. In Newtonian mechanics, one may define the 'time' as a parametric representation of the spatial trajectory of a material object. That is, the continuous change of the time parameter from one point of a trajectory to another, corresponds to the change in the spatial location of the object. The vector $\vec{r}(t)$ (in modern parlance) is the basic variable of the 'thing' of matter, that solves the laws of that matter. Thus, we put the discrete things of matter into space and time, and the laws predict how they move from the spatial point $\vec{r}_1(t_1)$ to the point $\vec{r}_2(t_2)$, where the vectors $\vec{r}$ are the 'dependent variables', that solve the laws of nature, and 't' is the continuously changing 'independent variable'. This trajectory could be the path of a falling apple or that of the orbit of a planet. We see, here, then, that the 'time' parameterizes the body's spatial motion in terms of the trajectory in space [5].

In the nineteenth century, the laws of thermodynamics were discovered. The 'first law of thermodynamics' is a consequence of the law of 'conservation of energy'. The term, "conservation" relates to the time measure in a way we will discuss later. The 'second law of thermodynamics' deals directly with the entropy of a system – a measure of its intrinsic disorder – and its change as time changes. This law asserts that if a system of matter, at some initial time, is in a non-equilibrium condition, with maximum order (minimum entropy), and if it is then left on its own, it will naturally proceed toward the equilibrium state and maximum entropy (minimum order), and it will remain in that state, if left on its own. The main point here is that this is an 'irreversible' process in time, from non-equilibrium to equilibrium, for a complex system of matter. The quantitative change of the entropy of this system is parameterized in 'time' [6].

It is important to note that this 'entropy-time' is not the same concept as Newton's parameterization of a trajectory of a single bit of matter. Under particular physical circumstances, it is possible to correlate these two types of time measure, but it is important to note that they are conceptually distinct from each other. For example, the 'entropy-time' is uni-directional while the Newtonian time is reversible. That is, in Newtonian mechanics, $t \rightarrow -t$ reverses the body's motion (mathematically), so as to re-trace its original path. An example of 'entropy-time', on the other hand, is the irreversible decay of an unstable particle. The lifetime of such a particle is defined in terms of the average number of such particles left after a particular time has passed. It is an irreversible process. For example, the lifetime of a mu meson is the order of 10^{-6} seconds. This means that the average mu meson would decay to an electron and two neutrinos in this amount of time, out of a very large number of mu mesons – some living a much longer time and others a much shorter time. Once decayed, the mu meson is never restored – the process is irreversible. Another example is the irreversible cellular decay of a human body, to affect its natural aging.

In the twentieth century, Emma Noether discovered that a necessary and sufficient condition for the incorporation of laws of conservation with the other laws of nature is that all of the laws of nature must be covariant (i.e. unchanged in form) with respect to continuous, analytic changes of the space and time coordinates. The law of energy conservation follows when this invariance is with respect to the changes of the time reference [7]. Thus we see that the unchangeability of the energy (for any observer, in terms of his own time measure) may define the 'time' as a parametric representation of the constancy of energy of a material system. This 'time' concept corresponds with Newton's 'trajectory-time' but not with the 'entropy-time' entailed in irreversible thermodynamics or the aging of a biological system.

Another important definition of 'time' that also relates to an irreversible process is its role in quantum mechanics, as a theory of measurement. The view here is that a large (macro-) apparatus measures the physical properties of a small (micro-) matter system – a particle or a molecule – *irreversibly*. One of the two (physically equivalent) equations that underlies the measurement process in quantum mechanics is Heisenberg's equation of motion (the other is Schrödinger's wave equation). On the left side of Heisenberg's equation we have the commutator $[H, P]$ of the Hamiltonian operator H, representing the act of measuring the energy of some micro-matter , and the operator P, representing the act of measuring some other property of this same matter. This commutator, in acting on the states of a system then is equal to the time-rate of change of the operator P in acting on the states of the system (multiplied by $i\hbar$). That is,

$$(HP - PH)\psi \equiv [H, P]\psi = i\hbar\frac{\partial P}{\partial t}\psi. \tag{1}$$

This equation then tells us how the simultaneous measurements of the energy and the property P of the microsystem changes the latter operator in time – because, it is said, the measurement of energy interferes with the knowledgeable information about P simultaneously. The way that this appears in the formal quantum mechanical expression is in the nonvanishing of the transitions (in 'time') that are induced by P, from one state function ψ_m to another ψ_n, when the measurement of P is carried out. Only in special cases are the properties P_o simultaneously measurable with the energy of the system (in which cases the commutators $[H, P_o] = 0$). The theory relates the actual observables, then, to the weighted values in a statistical calculation, wherein the weighting functions used in the averaging process are the *spreads* of wave functions, $\psi = \sum a_n\psi_n$. It is the latter predictions (not following from the equations of quantum mechanical operators themselves) that relate to the 'time-irreversible' measurements.

In this case, then, of the quantum theory of measurement, the 'time' relates to a parameterization of the operators involved in the measurement process that is irreversible. This is somewhat akin to the irreversibility of the entropy change according to the second law of thermodynamics. But this is not the same time concept as that in Aristotle, Maimonides, Newton or Einstein.

4. Time in the Theory of Relativity

This brings us to the view of 'time' according to the theory of relativity, of twentieth century physics. Einstein discovered, in his initial studies of special relativity, that there are no solutions for Maxwell's equations describing the propagation of light, in any frame of reference defined as relative to any other frame of reference (say that of an observer) that does not predict that the speed of light in a vacuum is anything other than the speed c. He found this to be a consequence of his tacit assumption that the form of Maxwell's equations remains in one-to-one correspondence in all relatively moving inertial frames of reference. He then generalized this conclusion to apply to all of the laws of nature, in addition to electromagnetism; later he generalized it to include arbitrary types of relative motion (his theory of general relativity). The invariance of the forms of the laws of nature – 'covariance' – then implied that there must be a universal set of transformations of the space and time measures that would leave the laws unchanged in all frames. In the case of special relativity, these were found to be the 'Lorentz transformations'. Thus the assertion follows that all of the laws of nature, when compared in all possible inertial frames of reference (by means of applications of the Lorentz transformations) must remain totally objective, i.e. in one-to-one correspondence. This is the assertion that underlies the theory of special relativity [8].

Some comments are now in order that contrast the meaning of 'time' in relativity theory and in the classical views. First, in contrast with Newton's action-at-a-distance, wherein the interaction between entities, spatially separated, is simultaneous, the relativity theory predicts that the interaction between material entities propagates at a finite speed. Thus, a cause (emission of a signal) here is correlated with an effect (absorption of the signal) there, at a *later time*. Thus the cause-effect relation correlates with a progression of 'time'. That is, the time measure is an abstract parametric change from 'cause' to 'effect' [9].

In relativity theory, then (both in the special and general forms), the space and time measures are not physical 'things-in-themselves', as they are in Newton. Rather, they are the language elements that are there to facilitate an expression of a law of matter – just as the 'words' of a verbal language and the syntax of a language system are there to express its meanings in its sentences. The transformations of the space and time measures, like the translations of languages, then are there to preserve the meanings expressed, verbally or mathematically.

Thus we correlate the space and time 'words' of the laws of nature with spatial and temporal measures, but these 'words' in themselves are not physical entities. Thus, 'space' cannot do anything physically, such as expand, nor can 'time' do anything physically, such as shrink! The Lorentz transformations serve the role of preserving the forms of the laws of nature in compared inertial frames of reference. To do this, it is found that the spatial and temporal scales must contract or expand in moving frames of reference, compared with those of the observer, in order to preserve the form of the law of nature. But this change of scale of spatial and temporal measures does not mean that anything physical is happening to matter in these moving frames of reference, by virtue of their motion; a stretching of a spatial measure or the contraction of a time measure does not mean that material sticks get longer nor that the duration of a moving object increases compared with such entities in the observer's frame of reference. It only signifies, e.g. that when expressing a time measure in the moving frame, one may put six numbers on the face of the clock rather than twelve, but it does not signify that anything physical is happening to the workings of the clock behind its face! Nor does it signify that the irreversible cell decay of a human body in the moving reference frame is slower than that of the observer. If it did so, as many contemporary physicists believe, then we would arrive at a genuine logical paradox. For motion, per se, is strictly a subjective aspect of the description of matter – as Galileo discovered – it does not have any absolute, objective consequences. That is to say, if A ages more slowly than B because A is moving away from B, then it is equally true to say, from the view of relativity theory, that B ages more slowly than A because from A's reference frame B moves away from A – whether the relative motion is uniform or not. Thus we would have to conclude that A becomes both older and younger than B, physically, after they would meet each other again. But there is no actual paradox when one interprets the time contraction as it was meant in the original formulation of the theory. That is, that the time contraction in the Lorentz transformation from one reference frame to the other that moves relative to it is only a scale change for the measure of time, that the observer applies to the moving objects' reference frame, in order to preserve the forms of the laws of nature in that frame – whether the observer is A or B. Thus there is no logical paradox here [10]

One other paradox that is encountered with the faulty interpretation of the time parameter in relativity theory and its transformations, in terms of physical (rather than scale) changes, is in the claim of the possibility of time travel. For if the union of the space and time into spacetime, in relativity theory, means that one may travel in time as one does in space, as many today claim to be the case, then it should be possible for one to travel to the past, physically, meet one's father *before he met his mother*, and to kill him. Then, the murderer could not exist at the moment of the crime. But he does exist at that moment!

Of course, this is nonsensical because the assertion of the traveler's 'trip' into the past and meeting and murdering his youthful father, would be a sequence of (irreversible) physical experiences. On the other hand, the 'time' (backward or forward) in relativity theory (as in Newton's theory) is simply a parametric measure, expressed with a particular scale and applied to a language to facilitate a description of physical processes; it is not the physical processes themselves!

One other paradox encountered because of this faulty interpretation has to do with Goedel's cosmological solution in general relativity [11]. With the assumption of a constant mass density of the universe in the energy-momentum tensor on the right-hand side of Einstein's field equations, Goedel found a solution of these field equations that entails a rotational motion of the universe as a whole and a time axis configuration that, instead of being unidirectional, is cyclic.

With the faulty interpretation of 't' as a physical process, this might imply, from the geodesic associated with his solution, that a person P, when he reaches a certain (older) age could meet himself P', when he was younger. The younger version of this person, P', would not believe the older one, P, when P tells him that he is many-valued, that he is P' as well a himself! P', thinking that P is an insane person, would then continue to live his life until he would be P, once again meeting P' and telling him the same story, and so on, ad infinitum.

But this conclusion is just as nonsensical as the former paradox - because the parameter t, in the geodesic path, is not a physically evolving process in itself! Goedel himself saw the error in his interpretation about time travel from his cosmological solution when he commented that 'this (cosmological solution) leads one to seek deeper understanding of the content of general relativity'.

On the experimental side, if it is claimed that any sort of matter (the hands or the digital reading of a clock, the aging of a human body, etc.) have 'slowed down' by virtue of its motion relative to a 'fixed clock', then it is incumbent on the physicists to find the cause-effect relation – the physical force – that is responsible for this physical change of one such clock compared to the other. My point is that the Lorentz transformation (or the corresponding coordinate transformations of general relativity) do not predict such (objective) physical changes, in themselves. This is for the same reason that the verbal language translation from English to Spanish applied to a sentence expressing an idea about a physical body, does not, in itself, predict any physical change of that body!

5. Time in General Relativity

Next we come to the interpretation of 'time' in the theory of general relativity. The theory of special relativity and the theory of general relativity are based on the same underlying 'principle of covariance'. As in special relativity, the time parameter in general relativity is not more than an abstract language element whose measure must be adjusted to the reference frame in which the laws of nature are expressed, in compared reference frames. But a new cornerstone of Einstein's theory of general relativity is its 'principle of equivalence'. I believe that the most general way to express this principle is to assert that a test body will move naturally along a geodesic. Indeed, it would take external energy to move the test body off of its geodesic path.

This is akin to Galileo's 'principle of inertia', stating that an unobstructed body must move naturally in a straight line at a constant speed - the path that is the geodesic natural to Euclidean geometry. In the thought experiment that led Galileo to his conclusion, the assumption was made that an external force, such as the force of gravity, oriented toward the center of the earth, does not act on the body when it moves on a horizontal surface, rather than on an inclined plane.

Einstein discovered that Euclidean geometry is an inadequate logic for the spacetime language in the expression of the laws of matter. He was then led to the non-Euclidean, differential geometry of Riemann, wherein the geodesics are curves rather than straight lines. Thus we take the geodesic of the spacetime to be the natural trajectory of the test body. The equation that determines this geodesic (the 'geodesic equation') is then taken as the equation of motion of the test body. If the curve traced by the test body's motion matches the observed path assumed previously as due to an external force field, then we say that there is an equivalence between the geometrical field that yields the correct curved path trajectory and the action of an external force field. This is the content of the 'principle of equivalence' of general relativity theory [12].

The geodesic equation, that is to serve as the equation of motion of the test body in a curved spacetime, is derived as follows. Start with the standard form of the Riemannian metric

$$ds^2 = g_{\mu\nu}dx^\mu dx^\nu. \tag{2}$$

We then minimize its square root with respect to the space and time coordinates, in terms of the variation

$$\frac{\delta}{\delta x^\mu} \int ds = 0, \tag{3}$$

yielding the geodesic equation

$$\frac{d^2 x^\mu}{ds^2} + \Gamma^\mu_{\nu\kappa} \frac{dx^\nu}{ds} \frac{dx^\kappa}{ds} = 0 \tag{4}$$

where $\Gamma^\mu_{\nu\kappa}$ are the 'affine connection' coefficients. They depend on the derivatives of the metric tensor $g_{\mu\nu}$ - which, in turn, are the solutions of Einstein's tensor field equations. With $\mu = j = 1, 2, 3$, we have the equation of motion of a spatial trajectory, with respect to changes in the differential metric ds, representing the parametric 'time change' along the object's trajectory in space. The latter is the 'time measure' represented as a one- parameter, real number-valued measure, as in the time definition in Newton.

It is not a trivial question to ask: How does one take the square root of the Riemannian (squared) metric (2), to generate the geodesic equation (4)? What is usually done is to assert that the square root is just $\pm(g_{\mu\nu} dx^\mu dx^\nu)^{1/2}$ and then to simply discard the minus sign. One cannot do this arbitrarily since it is a double-valued function at each spacetime point in this calculation, but in actual fact ds is single-valued!

What I have done to take the square root of the squared metric ds^2 was first to recognize that the irreducible representations of the Einstein group (the group associated with covariance in general relativity) obey the algebra of quaternions, then leading to the factorization

$$ds^2 = ds\widetilde{ds}, \quad \text{where} \quad ds = q_\mu dx^\mu$$

and $\widetilde{ds}$ is the quaternion conjugate (analogous to the conjugate of a complex function), corresponding to the 'time reversal' of ds.

With this (quaternion-valued) metric ds, one arrives at the same form for the geodesic equation (4), except that it now corresponds to four (rather than one) independent 4-vector equations at each spacetime point. That is,

$$\left[\frac{d^2 x^\mu}{ds^2} + \Gamma^\mu_{\nu\kappa} \frac{dx^\nu}{ds} \frac{dx^\kappa}{ds}\right]_{\alpha\beta} = 0 \tag{5}$$

where $\alpha, \beta = 1, 2$. Because of the extra degrees of freedom in this equation of motion, the predictions for the motion of a test body in general relativity, along its spatial trajectories, is different than in the conventional formalism

[13]. The relation of the quaternion algebra to the abstract time measure ds was a profound insight that William Hamilton had in the last century, when he discovered the algebra of quaternions. He said the following in his writings [14].

> It appeared to me ... to regard ALGEBRA as ... the Science of Order in Progression ... continuous and unidimensional: extending indefinitely forward and backward .... Although the successive states of such progression might (no doubt) be represented by points upon a line, yet I thought that their simple successiveness was better conceived by comparing them with the moments of time, divested however of all reference to cause and effect, so that the 'time' here considered might be said to be abstract, ideal or pure, ... In this manner, I was led to regard ALGEBRA as the Science of Pure Time ... and preparatory to the study of quaternions ....

Thus we see that Hamilton anticipated that the most basic representation of the measure of time must be in terms of variables that obey the algebra of quaternions. Here, the quaternion time measure is a four-parameter set, rather than the one-parameter set in the conventional geodesic equation, to parameterize the spatial trajectory of the test body that is subject to the influence of all other matter of a material system. It is an important feature of the quaternion algebra, and its representation of time, that quaternion variables are not commutative under multiplication.

The generalized quaternion-geodesic equation of motion (4) led to a few new predictions in my research program in general relativity, thus far:

1) The geodesic solutions predict that the natural motion of a test body entails a rotation (as well as translational motion), relative to any axis of rotation, depending on the observer's frame of reference. This result predicts, for example in astrophysics, that galaxies must naturally rotate, aside from any extra contribution to their rotation from other matter coupled to them (e.g. other galaxies and 'dark matter') [15].

2) The general motion of a test body, subject to its material environment, such as a constituent galaxy of the entire universe, or a constituent star of a galaxy, has an oscillatory, spiral configuration. This might explain (at least in part) the spiral configurations of galaxies. It also predicts that the 'big bang', at the beginning of each oscillatory cycle, is not singular and it is not characterized by an isotropic and homogeneous matter distribution background. Rather, the oscillatory motion is characterized by a spiral expansion and contraction, with two inflection points at the times of change from expansion to contraction and vice versa – analogous to the oscillatory motion of a simple pendulum. The reason for this is that the terms in general relativity that play the role of 'force' are not positive-definite (these are the affine connection coefficients). Thus, under some conditions of extreme

density and relative speeds, the force on a test body is predominantly repulsive (causing the expansion) and when the density is sufficiently rarefied and speeds slow enough, the force is predominantly attractive (causing the contraction) [16].

3) In a first approximation, the spiral configuration in the expansion phase, over a short part of its path, predicts the motion of a galaxy in accordance with the Hubble law (a distance- speed linear relation) – in agreement with the empirical facts [17]. The oscillatory universe cosmology also answers the question: How did the matter of the universe get into the state of maximum density and instability (when the 'big bang' happened) in the first place? – at the 'alleged' singular, absolute beginning of the universe, according to the present-day single big bang model cosmology. Answer: It came to the 'big bang' state at the beginning of the present cycle of an oscillating universe from a previous contraction phase of an ever-oscillating universe – between expansion and contraction.

Indeed, modern day cosmologists who adhere to the single big bang model, reject the idea of the subjectivity of time measure according to Einstein's theory of relativity. To them, the local time measure does remain subjective, but the global time measure – called 'cosmic time' – is an absolute measure. In this view, all 'times' may, in principle, be compared with the absolute beginning of the universe, at the beginning of this cosmic time. This view is indeed incompatible with Einstein's interpretation of the 'time measure' in physics, although it is compatible with Newton's view of an absolute origin of time.

6. Summary

Summing up, we have seen that there are many different interpretations of 'time', discussed since the ancient periods of our history. But there is, basically, a separation between the 'time' as a physical duration on the one hand and the 'time' that is taken as an abstract measure that is in our language in order to facilitate an expression of laws of matter – such as the law of a spatial trajectory of a material body. Indeed, these are the two kinds of time we see in debate between Henri Bergson [18], who defined time in the former way, and physicists who use it in the language of the laws of physics. Of course, these two kinds of time may be correlated under particular circumstances. But it is important to know that they are conceptually distinct. Otherwise, we get into logical problems, such as the 'twin paradox' of a relativity theory. The 'time' involved in the description of irreversible processes, such as physical aging of a biological organism (such as a human body), or the radioactive decay of unstable nuclei, or the quantum mechanical measuring process, is a different concept than the abstract physical language system that entails the 'word', we call 'time', though the latter as a measure may be correlated with the former [19].

In relativity theory, wherein forces propagate from one interacting component to another, one may correlate a time measure with cause (earlier) and effect (later). It is important to note with this example, however, that the relativity of time measure, as identified with cause and effect, means that cause and effect must also be relative to the observer - i.e. what is a cause and an effect to one observer, say one who is in the frame of reference of an emitter of a signal, would be an effect and a cause to an observer in the reference frame of the previously named absorber, now becoming the emitter, and vice versa. That is, with the identification of the time measure with cause and effect, its relativity implies the relativity of cause and effect themselves. This implies that if A emits a signal to B, B must simultaneously emit a signal to A, according to the requirement of symmetry of the theory of relativity [20].

We have also seen that a more general expression of the time parameter in the description as a spatial trajectory is in terms of the noncommuntative quaternion algebra, wherein 'time' is a four-parameter set rather than the one-parameter set of the classical view or of the standard view in general relativity. With this expression, the geodesic equation predicts a breakdown of the 'cosmological principle' – that the matter of a material system, such as the universe as a whole, must be isotropically and homogeneously distributed. It also predicts a natural rotational as well as translational motion (an extension of Galileo's 'principle of inertia'), implying, for example, the rotations of the galaxies, [15] and it predicts, as a useful approximation, the Hubble law in cosmology [17].

References

1. For Aristotle's views of time, see his Metaphysics, Book XII, Ch. 8, in McKeon, R.: 1941, *The Basic Works of Aristotle* (Random House)
2. St. Augustine's views of time are expressed in his Confessions, Book XI. For a translation, see: Hyman, A. and Walsh, J. J., editors: 1973, *Philosophy of the Middle Ages* (Hackett)
3. Maimonides, M., transl. Pines, S.: 1963, *The Guide of the Perplexed* (Chicago), Part II, Ch. 17
4. Galilei, G., transl. Drake, S.: 1967, *Dialogue Concerning the Two Chief World Systems* (California), p. 114
5. Newton, I. transl. Cajori, B.: 1962, *Principia*, Volume I, *The Motion of Bodies*, (California), in Scholium, p. 6
6. Reichenbach, H.: 1982, *The Direction of Time* (California)
7. Noether's theorem is discussed in: Lanczos, C.: 1966, *The Variational Principles of Mechanics* (Toronto), third Edition, p. 357
8. Einstein summarizes his views in: "Autobiographical Notes", in Schilpp, P. A., editor: 1949, *Albert Einstein – Philosopher-Scientist* (Open Court)
9. Sachs, M.: 1993, *Relativity in Our Time* (Taylor and Francis), Ch. 7

10. A mathematical as well as a logical argument against the validity of the 'twin paradox' is given in: Sachs, M.: 1971, *Physics Today*, **24**, 23. See also ref. 9, Ch. 12
11. Goedel, K.: 1949, *Rev. Mod. Phys.*, **21**, 447
12. Sachs, M.: 1976, *British Journal for the Philosophy of Science*, **27**, 225
13. The details of this derivation are given in: Sachs, M.: 1982, *General Relativity and Matter* (Reidel), Ch. 7
14. Halberstam H. and Ingram, R. E.: 1967, *The Mathematical Papers of Sir William Rowan Hamilton, Volume III, Algebra* (Cambridge)
15. Sachs, M.: 1994, *Physics Essays*, **7**, 490
16. Sachs, M.: 1989, *Annales de la Fondation Louis de Broglie*, **14**, 115. Also see ref. 9, Ch. 7
17. Sachs, M.: 1975, *International Journal of Theoretical Physics*, **14**, 115
18. Bergson, H.: 1965, *Duration and Simultaneity* (Bobs-Merrill)
19. I have compared the different types of 'time' in: Sachs, M.: 1978, *La Recherche*, **9**, 104
20. I have shown that the relativistic basis of cause and effect is a necessary concept in the 'delayed-action-at-a-distance' stand of Wheeler and Feynman, in: Sachs, M.: 1996, in *The Present Status of the Quantum Theory of Light*, A Symposium in Honour of Jean-Pierre Vigier (Kluwer)

ON THE TRANSMUTATION AND ANNIHILATION OF PENCIL-GENERATED SPACETIME DIMENSIONS

METOD SANIGA

Astronomical Institute of the Slovak Academy of Sciences
SK-059 60 Tatranská Lomnica, The Slovak Republic

Abstract. A spacetime manifold generated by the pencil of conics defined by two distinct pairs of complex-conjugated lines and a pair of real lines is considered. The manifold, originally endowed with two spatial and two temporal dimensions, is shown to substantially change its properties as we change the affine properties of the pencil. Two kinds of transformation are of particular interest. A dimensionality-preserving process, characterized by the transmutation of a temporal coordinate into a spatial one and leading to familiar $(3 + 1)$D spacetime, and a dimensionality-reducing scenario, featuring simultaneous 'annihilation' of one temporal and one spatial dimension and ending up with a $(1 + 1)$D spacetime. A striking difference between the nature of temporal and spatial is revealed; whereas we find purely spatial manifolds, those comprising exclusively temporal dimensions do *not* exist.

1. Introduction

A number of experiments show that the observed physical world can be, at least at the classical level, well approximated as a four-dimensional pseudo-Riemannian manifold (see, e.g. Misner *et al.*, 1973, or Weinberg, 1972). Although this concept proved to be extremely fruitful, especially when putting gravity down to the curvature of the spacetime, it tells us nothing (or very little) about why the world is just *four*-dimensional nor does it explain the internal structure of the time coordinate known as its *arrow*, i.e. its structuralization into three distinct domains – the past, present, and future. With the aim of addressing these two items we recently put forward a qualitatively new model of spacetime based on a specially selected and affinized pencil of conics in the projective plane (Saniga, 1996; henceforth

Astrophysics and Space Science **244**:283-290,1996.

referred to as Paper I). Although being quite condensed, the exposition of the fundamental features of the model given there nevertheless allow us to see that not only the properties of the spacetime, but also its dimensionality, are very sensitive on the way of how such an affinization of the pencil is done. The aim of the present paper is to inquire into this feature in more detail.

2. The Pencil of Conics and its Basic Properties

To this end we will consider the pencil of conics defined as

$$\Omega_{\breve{x}\breve{x}}^{\vartheta} \equiv a_{ij}(\vartheta)\breve{x}^i\breve{x}^j = \vartheta_1\breve{x}_1^2 + (\vartheta_2 + \vartheta_1)\breve{x}_2^2 - \vartheta_2\breve{x}_3^2 = 0 \qquad (1)$$

where, following the symbols and notation of Paper I, $\breve{x}_i$ ($i = 1, 2, 3$) are homogeneous coordinates of the projective plane, $\vartheta_{1,2}$ (($\vartheta_1, \vartheta_2) \neq (0, 0)$) stand for real-valued parameters, and a summation over the repeated indices is assumed. [1]

In order to examine the structure of the pencil let us first find its *degenerate* objects, i.e. the conics for which $A(\vartheta) \equiv \det a_{ij}(\vartheta)$ vanishes. From eq. (1) it is quite straightforward to see that there are just three degenerates, corresponding to $\vartheta \equiv \vartheta_2/\vartheta_1 = 0, -1$, and $\pm\infty$; while the first two represent a pair of *complex*–conjugated lines

$$(\breve{x}_1 + i\breve{x}_2)(\breve{x}_1 - i\breve{x}_2) = 0 \quad \text{and} \quad (\breve{x}_1 + i\breve{x}_3)(\breve{x}_1 - i\breve{x}_3) = 0 \qquad (2)$$

respectively, the last one comprises a pair of *real* lines,

$$(\breve{x}_2 + \breve{x}_3)(\breve{x}_2 - \breve{x}_3) = 0. \qquad (3)$$

These degenerates thus separate the set of regular conics into three distinct families, namely

$$-\infty < \vartheta < -1 \qquad \text{(1st family)}, \qquad (4)$$

$$-1 < \vartheta < 0 \qquad \text{(2nd family)}, \qquad (5)$$

and

$$0 < \vartheta < \infty \qquad \text{(3rd family)}. \qquad (6)$$

It is, however, important to note that these families are not equivalent among themselves as the second one consists of *imaginary* conics only; really, for ϑ within the range given by eq. (5) all non-zero a_{ij}'s of eq. (1) are positive-valued, which implies that the only real solution to the latter is

$$\breve{x}_1 = \breve{x}_2 = \breve{x}_3 = 0 \qquad (7)$$

---

[1]The reader who is interested in in-depth acquaintance with the properties of a projective plane is recommended to consult e.g. Klein (1928).

that, by definition, does not represent any point of a projective plane.

The third crucial characteristic of any pencil of conics is the multiplicity and character of its base (i.e. common to all the conics) points. In this respect our pencil is very simple, for it does not possess any real base point; to see this it is sufficient to realize that the only real $\vartheta - independent$ solution to eq. (1) is that given by eq. (7).

Pencil of conics (1) thus differs profoundly in its structure from the pencil dealt with in Paper I; hence, the properties of the spacetime manifolds generated by the two pencils must differ crucially from each other, as we will indeed find in what follows.

3. 'Most Regular' Spacetime Generated by the Pencil

In accordance with what was postulated in Paper I, in order to make manifest the multiplicity and character of temporo-spatial dimensions borne by a pencil of conics it is necessary to affinize the projective plane. This means singling out, or deleting, from this plane one line, usually termed as the 'ideal line' or 'line at infinity', and studying the intersection properties of the individual conics with it. Because any pencil of conics necessarily contains degenerate conics and, so, singular points, [2] it is reasonable to begin our discussion with the case where the ideal line does not meet any of the latter.

To follow this strategy we look back at eqs. (2–3) to find out that the degenerates contain one singular point each, namely $\breve{x}_1 = \breve{x}_2 = 0$ $(\vartheta = 0)$, $\breve{x}_1 = \breve{x}_3 = 0$ $(\vartheta = -1)$, and $\breve{x}_2 = \breve{x}_3 = 0$ $(\vartheta = \pm\infty)$, henceforth referred to as S_1, S_2, and S_3, respectively. It represents no difficulty to verify that the equation of the ideal line which avoids all of them can in the most general form be chosen as

$$\breve{x}_1 - m\breve{x}_2 - n\breve{x}_3 = 0 \tag{8}$$

with both m and n being non–zero real numbers.

Let us now insert the last equation into eq. (1). Denoting $\zeta \equiv \breve{x}_3/\breve{x}_2$ we obtain

$$\left(n^2 - \vartheta\right)\zeta^2 + 2mn\zeta + \vartheta + m^2 + 1 = 0, \tag{9}$$

which is a quadratic equation with the roots

$$\zeta_\pm = -\frac{mn \pm \sqrt{\Theta(\vartheta)}}{n^2 - \vartheta}, \tag{10}$$

$$\Theta(\vartheta) \equiv m^2 n^2 + \left(\vartheta - n^2\right)\left(\vartheta + m^2 + 1\right), \tag{11}$$

[2] A singular point of a planar curve is a point at which there exists no tangent line to the curve.

that, m and n being kept fixed, tell us which of the conics are hyperbolae ($\Theta(\vartheta) > 0$), parabolae ($\Theta(\vartheta) = 0$), and/or ellipses ($\Theta(\vartheta) < 0$), i.e. give us the information of what the intrinsic structure of the induced *temporal* dimension(s) looks like. Notice that we have intentionally put the letter 's' in brackets because here – and this is perphaps the most pronounced departure from the model discussed in Paper I – we do have something to do with *two* distinct temporal dimensions, rather than with a single one! This immediately follows from the fact that not only does the equation $\Theta(\vartheta) = 0$, being quadratic in ϑ, have two distinct real roots

$$\vartheta_\pm = -\frac{1}{2}\left(m^2 - n^2 + 1\right)\left(1 \mp \frac{|m^2 - n^2 + 1|}{m^2 - n^2 + 1}\sqrt{1 + \frac{4n^2}{(m^2 - n^2 + 1)^2}}\right),$$
(12)

but the corresponding parabolae are shared out by the two real (i.e. 1st and 3rd) families of conics; in fact, a brief inspection of eq. (12) shows that irrespective of the sign of term $(m^2 - n^2 + 1)$ one root is always positive (a conic of the 1st family) and the other always negative (a conic of the 3rd family). Each family thus gives rise to a single temporal dimension whose properties – assuming, for simplicity's sake, the term $(m^2 - n^2 + 1)$ to be positive – are listed in Table 1.

TABLE 1. The intrinsic structure of the two temporal dimensions.

Temporal Domain	1st Family	3rd Family
past	$-\infty < \vartheta < \vartheta_-$	$\vartheta_+ < \vartheta < +\infty$
present	$\vartheta = \vartheta_-$	$\vartheta = \vartheta_+$
future	$\vartheta_- < \vartheta < -1$	$0 < \vartheta < \vartheta_+$

This situation is depicted in Fig. 1a, where the conics of the 1st and 3rd families occupy the sectors denoted by the '–' and '+' signs, respectively; in both cases we can clearly see the domain of past (hyperbolae – shaded area) as well as the domain of future (ellipses – dotted area), the two being separated from each other by a single moment of present (the parabola – a heavy-drawn curve).

   To complete this section it remains to look at what the situation is as for *spatial* dimensions. Since the only real points lying on the $\vartheta = 0$ and $\vartheta = -1$ degenerates are, respectively, points S_1 and S_2, and these – as assumed – do not lie on the ideal line, $\vartheta = \pm\infty$ is the only degenerate having with this line real points in common. And there are just *two* such

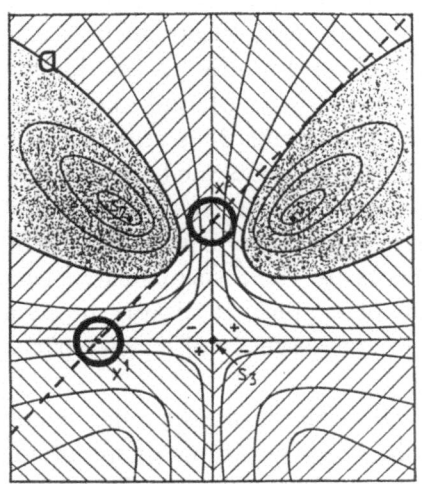 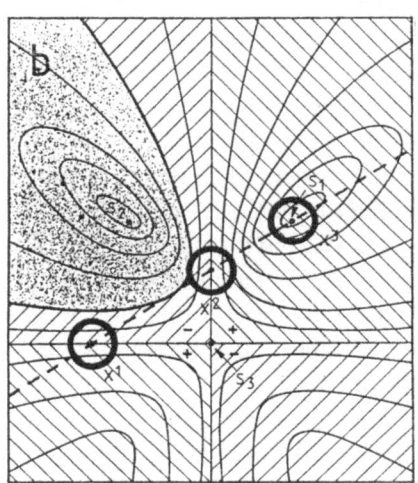

Figure 1.   (a) – A sketch of the structure of the (2 + 2)D spacetime generated by a 'regular' affine image of pencil (1), and – (b) – its transformation into the ordinary (3 + 1)D manifold, when the ideal line (dashed) incorporates singular point S_1. For more details see the text.

points as implied by eqs. (10)–(11) in the limit of $\vartheta \to \pm\infty$. Hence, there are just two observable spatial dimensions, those generated by the pencils of lines having these points as carriers. This is again visualised in Fig. 1a, where both pencils are represented by heavy circles and the corresponding dimensions denoted as x^1 and x^2.

4. 'Transmutation' of Temporal into Spatial

The (2 + 2)D spacetime manifold described in the previous section can be viewed as the 'most regular' one since the line at infinity, given by eq. (8), does not meet any of the singular points S_1 – S_3. It is therefore natural to turn our attention now to some less regular cases, where the ideal line incorporates some singular point(s).

To this end let us start shifting the ideal line from its original position, uniquely specified by the fixed non-zero values of m and n, and see what happens when it comes in contact with one of the singular points, say S_1. Since the coordinates of S_1 satisfy eq. (8) with $n = 0$, the whole story thus reduces to finding the corresponding limits of eq. (12), which read

$$\lim_{n \to 0} \vartheta_+ = 0 \quad \text{and} \quad \lim_{n \to 0} \vartheta_- = -(m^2 + 1). \tag{13}$$

While $\vartheta = \vartheta_- < -1$ is still a regular conic, the conic $\vartheta = \vartheta_+ = 0$ is degenerate. This simply means that in the limit the concept of both the future and present collapses for the conics of the 3rd family (as the domain of ellipses shrinks to a point (S_1) – see Table 1 and Fig. 1b), and so does the notion of temporal. One observes, however, that point S_1 is real and of a degenerate and the pencil of lines it carries thus represents the basis of a new *spatial* dimension – as illustrated in Fig. 1b, where this dimension is denoted as x^3. We have thus revealed a very intriguing scenario of what might be called a dimensionality-preserving metamorphosis; the arrow-like structure of the temporal dimension is gradually distorted until it disappears as a whole, this being accompanied by the sudden emergence of a new spatial coordinate!

The scenario just described is a typical example of the phenomena which could collectively be described as the transmutations of manifolds and which are inherent in the innumerable ways in which the affinizations of pencil (1) can be performed.

5. 'Annihilation' of Temporal and Spatial

Another, perhaps even more striking and bizarre aspect is the transformation in which one temporal and one spatial dimension 'cancel' each other, reducing thereby the dimensionality of the original manifold by two.

To handle this case quantitatively we start again with the $(2 + 2)$D spacetime configuration shown in Fig. 1a, but translate the ideal line in a different way. In particular, being kept all the time to be a tangent to one and the same conic of the 3rd family (and namely to that corresponding to $\vartheta = \vartheta_+$ given by eq. (12)), this line now approaches singular point S_3 until the two objects get in touch – this involving once again an abrupt, phase-transition-like change in the structure of the original configuration. A mathematically rigorous description of such a change consists of noticing that an ideal line that passes via S_3, but contains neither S_1 nor S_2, has generally the equation

$$\breve{x}_3 = \kappa \breve{x}_2, \quad 0 < \kappa^2 < \infty, \tag{14}$$

and inserting this equation into eq. (1) that yields ($\zeta \equiv \breve{x}_1 / \breve{x}_2$)

$$\zeta^2 + \vartheta \left(1 - \kappa^2\right) + 1 = 0. \tag{15}$$

The roots of the last equation,

$$\zeta_\pm = \pm\sqrt{\vartheta\left(\kappa^2 - 1\right) - 1},\tag{16}$$

then tell us what the character is of the points in which our ideal line intersects a given conic of pencil (1). Now, as we assume the ideal line to be a tangent to the conic of the 3rd family, which as a whole is characterized by positive ϑ's (see eq. (6)), we must take $\kappa^2 > 1$ in order to maintain that eq. (16) is consistent with this assumption (i.e., that it yields the double root for $\vartheta > 0$). This, however, implies that for negative ϑ's both roots are purely imaginary and, in the light of eq. (4), that all the conics of the 1st family are *ellipses*. It is thus the time dimension generated by the 1st family that now sinks completely, but here due to the dissolution of the domains of the past and present and with the simultaneous vanishing of a spatial coordinate; the last part of this statement is quite obvious as for $\kappa^2 \neq 1$ point S_3 is the only real intersection of the degenerate $\vartheta = \pm\infty$ with the ideal line given by eq. (14).

At this point the theory has been developed to such an extent that it allows us to spot and realize a fundamental difference between the nature of the temporal and spatial. Since an ideal line is real by definition and the $\vartheta = \pm\infty$ degenerate consists of real lines too, and because the intersection of any two distinct *real* lines is always a *real* point, it is obvious that *no* further reduction in the number of spatial coordinates is possible here. However, this is not the case with regard to temporal dimensions for these can be completely eliminated by the appropriately selected ideal line – namely that passing via any couple of singular points. Hence, *space can exist on its own, but time cannot.*

6. Conclusion

The theory exposed here suggests that we could gain some important physical insights into the nature of the observed physical world by making further studies of the concept of what we have termed a pencil-based manifold. The chief lesson to be learned from such a concept is a very intimate and intriguing coupling between time and space that fairly surpasses the footing which these two notions were given in the framework of general relativity. Even leaving aside the fact that our theory (at least qualitatively) accounts for the internal structure of the temporal coordinate, i.e. for its arrow, there are many other features which our spacetime is endowed with that represent a considerable departure from the currently adopted view; in particular, we have found that pencil-generated temporal and spatial coordinates can be continuously transformed into each other, and that this can be done in a way that either preserves (transmutation) or changes (annihilation) the dimensionality of the original manifold. Although there is still much that is unsettled and uncertain at this stage of the development of the theory, it already provides invaluable guidance for our scientific imagination, as we strive to decipher and appreciate the laws of Nature.

I would like to thank Mr. P. Bendík for drawing the figures. This work was supported in part by the grants # 2/506/93 of the Slovak Academy of Sciences and # 303404 of the Academy of Sciences of the Czech Republic.

References

Klein, F.: 1928, *Vorlesungen über Nicht-Euklidische Geometrie*, Springer-Verlag, Berlin
Misner, C. W., Thorne, K. S., and Wheeler, J. A.: 1973, *Gravitation*, Freeman, New York
Saniga, M.: 1996, 'Arrow of Time and Spatial Dimensions', in K. Sato, T. Suginohara, and N. Sugiyama (eds.), *The Cosmological Constant and the Evolution of the Universe*, Universal Academy Press, Inc., Tokyo, 283–284
Weinberg, S.: 1972, *Gravitation and Cosmology*, John Wiley and Sons, New York

NUCLEAR AND PARTICLE PHYSICS
MATHEMATICAL MODELS AND METHODS

DAY 3

Steward Observatory, University of Arizona

Figure 1.   Sketch art courtesy Janet A. Tifft

PARTICLES, SPACE, AND TIME

VINCENT ICKE
Sterrewacht Leiden and
Universiteit van Amsterdam

Abstract. Our Universe consistes of particles, space and time. Ever since Descartes we have known that true emptiness cannot exist; ever since Einstein we have known that space and time are part of the stuff of our world. Efforts to determine the structure of particles go in parallel with the search for the structure of spacetime. Einstein gave us a geometrical answer regarding the structure of spacetime: a distance recipe (Lorentz-Minkowski) suffices. The theory boils down to a patching together of local Lorentz frames into a global whole, which gives it the form of a gauge field theory based on local Lorentz symmetry. On large scales, the Einstein Equation seems to work well. The structure of particles is described by a gauge field, too. On small scales the 'Standard Model' seems to work very well.

However, we know from Newtonian gravity that the presence of particles must be related to the structure of spacetime. Einstein made a conjecture for the form of this connection using the Newtonian limit of small speeds and weak fields. The right hand side of his equation for the bulk theory of matter (the energy-momentum tensor), is equated to the Einstein tensor from non-Euclidian geometry.

But that connection is wrong. The structure of spacetime cannot be equated to the density of particles if we include the Standard Model in the matter tensor. In field theory a potential is not something that can be freely changed by adding an arbitrary scalar term; due to the local (as opposed to global) character of the fields, a potential becomes an entity in itself. Einstein's conjecture runs into profound trouble because the reality of potentials implies that the zero point energy of the vacuum must be included in the Einstein equation. The net result is the appearance of a term equivalent to a cosmological constant Λ of stupendous size, some 10^{118} times the critical cosmic density.

The crisis due to the zero point fluctuations in the energy-momentum tensor is a clash of titans: Einstein's geometrical ideas on spacetime structure vs the behaviour of particles and the vacuum discribed by Dirac and

Astrophysics and Space Science **244**:293-311,1996.
© 1996 *Kluwer Academic Publishers.*

followers. Someone, or everyone, is wrong. In my opinion the straightforward quantization of spacetime will always be impossible because the usual
particle symmetries (U(1), SU(2), SU(3) and relatives) connect fermions
and bosons, whereas relativistic analogies of these symmetries (the Lorentz
symmetry) says something about spacetime and not about particles.

1. Clash of Titans

When an extraterrestrial or a child asks me what we have *really* learned
about the nature of the Universe, I reply: *our Universe consists of particles,
space and time.* Even for adult terrestrials, this does not seem to be too bad
a summary, so I will use it as a starting point for a brief overview. I will try
to show that this seemingly innocuous statement harbours a grave problem,
in the following sense: the theory of particles (called somewhat preemptively
the *Standard Model*) and the theory of space and time (General Relativity,
GRT) are extremely good in their own realm of application, the world of
the very small and the very large, respectively. Taken together, however,
we get a theory that does not conform in the least to the world as we know
it.

This shows up most clearly in the fact that the Standard Model predicts
that the energy density of the vacuum is not zero, in the sense that vacuum
fluctuations are always present in it. Oversimplifying somewhat we may
say that in relativistic quantum mechanics the uncertainty relations imply
that one can never be quite sure that a cubic metre of spacetime is empty.
Particle number is no longer a constant of the motion.

These vacuum fluctuations contribute to the effective mass density of
the Universe. Suppose that we have a field with quanta of mass m. Then
integration over momentum space shows that the energy density e_V of the
vacuum is

$$e_V = (2\pi)^{-3}\hbar^{-3} \int\limits_0^K 4\pi k^2 \frac{1}{2}\sqrt{k^2c^2 + m^2c^4}\; dk \simeq \frac{cK^4}{16\pi^2\hbar^3} \quad \mathrm{J\,m^{-3}} \qquad (1)$$

in which $K \gg m$ is the upper cutoff of the wave number. By itself this
need not be a meaningful number; we could transform it away by a renormalization procedure. However, in GRT this cannot be done and we must
consider the gravitational effects of this vacuum energy density. In the case
of GRT we expect that K is inversely proportional to the coupling constant

$8\pi G/c^2$, and so

$$K \approx \sqrt{\frac{\hbar c^3}{8\pi G}} \tag{2}$$

$$e_V = \frac{c^7}{1024\pi^3 \hbar G^2} = 1.46 \times 10^{109} \quad \text{J m}^{-3}. \tag{3}$$

Considering that the energy density equivalent in an Einstein-De Sitter universe would be

$$e_0 = \rho_0 c^2 = \frac{3H_0^2 c^2}{8\pi G} = 4.24 \times 10^{-10} \quad \text{J m}^{-3} \tag{4}$$

for a Hubble parameter of $H_0 = 50 \text{ km s}^{-1} \text{ Mpc}^{-1}$ (which is $1.62 \times 10^{-18} \text{ s}^{-1}$) we conclude that the clash between Einstein on the one hand and Dirac and Feynman and their cohorts on the other, is a clash of titans indeed: they differ by a factor of 10^{118}! Even if we only took K from quantum electrodynamics we would expect that it would be roughly equal to $m_e c$, and we would still get

$$e_V = \frac{m_e^4 c^5}{16\pi^2 \hbar^3} = 9.00 \times 10^{21} \quad \text{J m}^{-3} \tag{5}$$

still at least a factor 10^{31} larger than observations appear to allow. It is manifestly false that $\Lambda = 10^{118}$. If it isn't anywhere near, it might as well be zero. *Why doesn't space weigh anything?*

Such a massive pile of problems is a cairn along the winding path to a region where good new physics can be found. Nobody has yet returned alive from that land; let us see what difficulties we expect to find there.

2. Spacetime as Real Stuff

First, consider space and time. Ever since Descartes we have known that true emptiness cannot exist; he conjectured that force comes about through the direct physical contact between objects. In the second part of his *Principes de la Philosophie*, Article 16, Descartes wrote:

> *Concerning emptiness, in the sense given to that word by the philosophers, namely a space containing no substance, it is obvious that there is no such space in the universe, because the extent of space around or enclosed by an object is not different from the extent of that object. And even as from the sole fact that an object is extended in length, height and depth, we have reason to conclude that it is a substance, because we suppose that it is not possible for anything to have no extent, we must conclude the same about space which is supposedly empty: namely that, because it possesses spatial extent, it also has substance.*

In other words: space has physical attributes, namely its three dimensions, so it must be regarded as real stuff. This powerful notion lay hidden for three hundred years, until it was rediscovered independently by Einstein.

3. From Global to Local Lorentz Symmetry

Ever since Einstein we have known that space and time are part of the stuff of our world, and not invisible graph paper. Thus, efforts to determine the structure of particles go in parallel with the search for the structure of spacetime. Einstein gave us a geometrical answer for the latter: a distance recipe (Lorentz-Minkowski) suffices. The theory boils down to the patching together of local Lorentz frames into a global whole, which gives it the form of a gauge field theory based on local Lorentz symmetry.

Special relativity starts from the invariance of the speed of light. The equation of motion for light is the equation for a sphere in 3-space: $s^2 = c^2 t^2 - r^2$ where $s = 0$ for a light ray. Now we note that the above leads to a maximum value $v \leq c$ for all speeds v. That means that the above global Lorentz symmetry cannot be maintained. To stay consistent, we may require *local* symmetry only. After all, if signals travel with a finite speed, how would our colleagues at Arcturus know that we have just performed some Lorentz transformation here? We must restrict ourselves to the infinitesimal patch around the origin of our arbitrarily chosen standpoint:

$$ds^2 = c^2 dt^2 - dr^2 \tag{6}$$

Local Lorentz symmetry means that, wherever you are, you can always find coordinates such that the above holds ('freely falling coordinates'). Thus, your neighbours in spacetime will also be able to do this. However, it isn't guaranteed that you will agree with the neighbours that their coordinate system $\{x^\mu\}$ is the same as yours. The best you can hope for is a patch-up between the two of you, by means of a bilinear form in the infinitesimal coordinates of some common coordinate system $\{\xi^\mu\}$ which you've both agreed to use, and for which local Lorentz symmetry holds too:

$$ds^2 = g_{\mu\nu} d\xi^\mu d\xi^\nu \tag{7}$$

The equation of motion corresponding to the spacetime structure encoded in $g_{\mu\nu}$ can be found by an argument similar to what one uses in classical mechanics. First, consider free motion only, the kinematics of motion without force. Classical motion is subject to certain symmetries, namely Galilei symmetry and time reversal invariance. These symmetries generate conserved quantities: momentum and energy. That is to say, the state of rest is equivalent to motion with constant energy and momentum: $dv^i/dt = d^2 x^i/dt^2 = 0$. Deviations from inertial motion are attributed to an external force, $dv^i/dt = F/m$. This was basically the method that Galilei and Huygens used in their quest for the equations of motion of classical mechanics.

In special relativity we use Lorentz symmetry instead of Galilei; in GRT we extend this to local Lorentz symmetry and find that the structure of spacetime is given by $g_{\mu\nu}$. Thus the equivalent of motion under the influence of a force is force-free motion in curved spacetime. Loosely speaking: "curved spacetime gives curved paths".

Now let us proceed to the algebraical expression of that statement. A classical free particle moves according to the law of inertia: $dv^i/dt = d^2x^i/dt^2 = 0$. If an external force is present, say due to a gravitational potential Φ, we have $d^2x^i/dt^2 = -\partial\Phi/\partial x_i$. In GRT the analogue of classical free motion must be written as $du^\mu/ds = 0$, where $u^\mu \equiv dx^\mu/ds$ is the four-velocity, having changed from using the time derivative to a derivation with respect to the interval s. If we arrange with 'the neighbours' elsewhere in the Universe to refer all descriptions to a global but otherwise arbitrary coordinate system $\{\xi^\mu\}$ (much as at a conference one usually agrees to speak English), we find

$$u^\mu = \frac{dx^\mu}{ds} = \frac{\partial x^\mu}{\partial \xi^\alpha}\frac{d\xi^\alpha}{ds} = v^\alpha \frac{\partial x^\mu}{\partial \xi^\alpha} \tag{8}$$

Here v^α is the *local* four-velocity, the 'free-fall motion'. This produces an expression for the way in which the four-velocity changes with respect to interval:

$$0 = \frac{du^\alpha}{ds} = \frac{d^2\xi^\alpha}{ds^2} = \frac{d}{ds}\left(\frac{\partial\xi^\alpha}{\partial x^\mu}\frac{dx^\mu}{ds}\right) \tag{9}$$

and, after working out the differentiations, one uses the metric tensor to obtain

$$\frac{d^2x^\lambda}{ds^2} + \Gamma^\lambda_{\mu\nu}\frac{dx^\mu}{ds}\frac{dx^\nu}{ds} = 0 \tag{10}$$

$$\Gamma^\lambda_{\mu\nu} \equiv \frac{\partial x^\lambda}{\partial \xi^\alpha}\frac{\partial^2\xi^\alpha}{\partial x^\mu \partial x^\nu} \tag{11}$$

The above equation of motion can be interpreted as "curved spacetime gives curved orbits": the four-acceleration is no longer zero but proportional to Γ, in which we have dumped all the garbage due to the 'mismatch' in spacetime caused by the introduction of local Lorentz symmetry. Thus, although we can always *locally* transform away the curvature of spacetime ("freely falling coordinates"), we cannot do so globally.

4. Space, Time and Particles Connected

Einstein made a conjecture that encompasses the slogan 'Nature is made of particles, space and time' in a particular way. He did this by using, on the left hand side of his famous field equation, non-Euclidean geometry to describe the structure of spacetime via a local generalization of the global Minkowski distance recipe. On the right hand side he used a description of matter based on two approximations. First, he used a continuum representation of matter, averaging over the individual particles to obtain densities. Second, he connected the presence of matter to the structure of spacetime by demanding correspondence with the Newtonian equations in the limit for weak fields and small velocities. Accordingly, his equation reads something like

$$\left\{ \begin{array}{c} \text{structure of} \\ \text{space and time} \end{array} \right\} = \left\{ \begin{array}{c} \text{distribution of} \\ \text{particles and fields} \end{array} \right\}$$

If we read the equals sign as implying an interaction, this says 'particles interact with spacetime'. Of course this interpretation immediately shows that the equation is incomplete, because the back-reaction of the structure of spacetime on the properties of the particles is not included, other than by their trajectories as pointlike test particles in the field $g_{\mu\nu}$.

How are matter and spacetime curvature connected algebraically? In other words, what connects matter and the distance recipe in spacetime? When we compare the Newtonian and the Einsteinian equations for the trajectory of a particle, we notice immediately that Φ and $\Gamma^\lambda_{\mu\nu}$ are apparently related. Of course it cannot be that $\Gamma = \Phi$ or something as simple as that, because then Newtonian theory would be relativistic already! We must incorporate a physical ingredient in our theory. The one that Einstein used is: in the limit for small velocities and small curvatures in a static spacetime we must recover the classical equations.

In that case, the only components that remain are those related to the time-indices: everything is zero except Γ^μ_{00}. Using the above calculations one may see that this is closely related to the fact that the limit of the four-velocity u^μ for small three-velocities is $u^\mu = (c, 0, 0, 0)$. The zero-component does not vanish, just like the 0-component of the energy-momentum does not vanish for $v \to 0$ (this is the famous mc^2-term). Furthermore, from the definition of the metric we know that $ds^2 \simeq g_{00}dt^2$ (the other components of $g_{\mu\nu}$ vanish in the limit for small curvature), and therefore the low-velocity limit of Eq.(10) is

$$\frac{d^2x^\mu}{dt^2} + \Gamma^\mu_{00} = 0, \tag{12}$$

from which we finally conclude that the desired correspondence with classical gravitation can be obtained if we relate Γ to the gradient of the gravitational potential:

$$\Gamma^{\mu}_{00} \Longleftrightarrow \frac{\partial \Phi}{\partial x_{\mu}}. \tag{13}$$

In order to connect with the classical Newtonian case, we felt obliged to equate Γ with a coordinate derivative of the classical potential. Then Γ is related to a linear combination of first derivatives of $g_{\mu\nu}$:

$$\Gamma^{\kappa}_{\lambda\mu} = \frac{1}{2} g^{\nu\kappa} \left\{ \frac{\partial g_{\mu\nu}}{\partial x^{\lambda}} + \frac{\partial g_{\lambda\nu}}{\partial x^{\mu}} - \frac{\partial g_{\mu\lambda}}{\partial x^{\nu}} \right\}, \tag{14}$$

and we finally conclude that *the role which the potential Φ plays in classical Newtonian gravity is taken over by the metric tensor $g_{\mu\nu}$ in general relativity.*

The classical potential is related to the presence of matter by means of the Poisson equation

$$\Delta \Phi = 4\pi G \rho, \tag{15}$$

which should give us a hint about how the presence of matter can be connected to the structure of spacetime. The point here is that the density field ρ could not possibly be used, because it is in no way Lorentz invariant. In fact, we can see immediately that a Lorentz transformation of ρ should go as

$$\rho' \propto \gamma^2 \rho. \tag{16}$$

One Lorentz factor γ comes from the change of the effective mass via $E = \gamma mc^2$; the other one comes from the fact that a volume seen in motion decreases by one factor γ because of its Lorentz-FitzGerald contraction. Accordingly we suspect that ρ should be part of a tensor because a Lorentz scalar transforms with γ^0, a vector with γ^1 and a tensor with γ^2. Since ρ is classically a scalar field we also suspect that it is the 00-component of a tensor.

In the limit of small velocities without external forces we may take that tensor, which we'll call $T_{\mu\nu}$, to be diagonal. Then ρ is placed in the top-left corner. What will we have on the remaining three places of the diagonal? Since ρ is a mass density, and since in relativity we have to take mass and energy as equivalent, it seems natural to use an energy density. The mass density is the mass of a collection of particles per unit volume. The corresponding energy per unit volume we know as the *pressure P* of the collection of particles. The entries elsewhere in the tensor can be found by Lorentz transformation of T; the complete form is the *energy-momentum tensor*

$$T_{\mu\nu} = P\eta_{\mu\nu} + (\rho c^2 + P)u_{\mu}u_{\nu}. \tag{17}$$

The final order of business now is to construct a tensor $G_{\mu\nu}$ from $g_{\mu\nu}$ and its derivatives that has the same transformation properties as $T_{\mu\nu}$. The most obvious choice, taking T simply proportional to g, is not sufficient because it would not include Newtonian gravity; in order to obtain that, as we had seen above, we must include the derivatives of g. In particular, in the Newtonian limit we retain the 00-components only, namely

$$\frac{\partial^2 g_{00}}{\partial x^\alpha \partial x_\alpha} \propto T_{00}. \tag{18}$$

Apparently, the desired tensor must contain at least second derivatives of g. This implies that *a fourth-rank tensor must be involved!* Einstein showed that the correct expression is related to the monstrous *Riemann-Christoffel curvature tensor*

$$R^\lambda_{\mu\nu\kappa} \equiv \frac{\partial \Gamma^\lambda_{\mu\nu}}{\partial x^\kappa} - \frac{\partial \Gamma^\lambda_{\mu\kappa}}{\partial x^\nu} + \Gamma^\alpha_{\mu\nu}\Gamma^\lambda_{\kappa\alpha} - \Gamma^\beta_{\mu\kappa}\Gamma^\lambda_{\nu\beta}. \tag{19}$$

This animal must be reduced to second rank before we can equate it to $T_{\mu\nu}$. This is done by contracting it over one index. The most general expression for the required tensor is then

$$R^\lambda_{\mu\lambda\nu} - \frac{1}{2}g_{\mu\nu}g^{\alpha\beta}R_{\alpha\beta} + \Lambda g_{\mu\nu} = R_{\mu\nu} - \frac{1}{2}g_{\mu\nu}R + \Lambda g_{\mu\nu}, \tag{20}$$

and in units where the gravitational constant G is retained explicitly, correspondence with the Poisson equation shows that

$$R_{\mu\nu} - \frac{1}{2}Rg_{\mu\nu} + \Lambda g_{\mu\nu} = -\frac{8\pi G}{c^2}T_{\mu\nu}. \tag{21}$$

This is a physical choice; it does not have the mathematical necessity of a gauge theory because of the way in which $T_{\mu\nu}$ was put in by hand. It is Einstein's guess, based on correspondence with Newtonian mechanics, and it works very well on large scales: black holes, relativistic stars, the Universe. But it works not at all on small, atomic, quantum-mechanical scales. The hassle is that we are obliged to include the vacuum zero-point energy in $T_{\mu\nu}$. As we saw in Sec.1, it shows up in the form of a finite value $\Lambda \approx 10^{118}\rho_0 c^2$ of the cosmological constant which is totally excluded by cosmological observations.

Possibly there is another guess one could make by searching for an expression that corresponds with (say) Schrödinger's Equation instead of Poisson's Equation, but nobody has yet succeeded in doing so.

Note that the rather peculiar and counter-intuitive behaviour of Λ, and the related possibility of inflation ("you get something out of nothing") is due to our initial assumption for the connection between $g_{\mu\nu}$ and the potential Φ, and via Φ and the Poisson equation to the (thermo)dynamic mass- and energy densities ρ and P. The connection between matter and spacetime curvature is still a conjecture, since we do not have a quantum gravity theory.

One immediate cause for worry is that it seems like 'double dipping' to introduce spacetime in $g_{\mu\nu}$ as well as in Λ. Ought we not to include the Planck-scale fluctuations in some (possibly extended) form of $g_{\mu\nu}$ rather than in the potential term that produces Λ? After all, a potential is a classical beast that would be wiped out by second quantization.

5. Twists and Wrinkles

Having described the structure and behaviour of space and time as they appear in our summary expression 'the Universe is made of particles, space and time', let us consider the particles. The trick of getting a field $g_{\mu\nu}$ that corrects for the consequences of using a *local* Lorentz symmetry instead of a global one is common to all current theories of interaction. First let me try to explain in somewhat pedestrian terms what happens here.

The similarity between the symmetries of Nature and simple rotations enables us to understand how a symmetry can produce a force field. Take before you, on a smooth table, a small tablecloth, or something similar (e.g. a large piece of aluminium foil). The material must be a uniform colour, without any patterns. Make sure it is quite smooth. We are not looking so closely that the individual fibres are visible, and we will pretend that the material extends to infinity: our tablecloth is a small piece of an unbounded model universe. Now rotate the whole cloth through an arbitrary angle. Any piece of the surface, when inspected individually, appears the same as before: the cloth is *invariant under global rotations*.

But it would be impossible, even in principle, to do something like this with the real Universe. Imagine that we want to perform a global symmetry transformation. Then we would have to let the symmetry act in all of space at exactly the same time. But this is impossible to do in reality because no signal can propagate faster than the speed of light. We must *accept only local symmetry rotations*, that is, a symmetry where the amount of rotation *differs* from event to event in spacetime.

Return to the tablecloth before you. Put your finger on a point near the centre and give the cloth an arbitrary twist, keeping the edges of the cloth in place. When you remove your finger, you notice that the piece of the surface you have just rotated still appears the same; at that one point, the cloth is invariant under local rotations. But in the vicinity of the twisted point, something has happened: *a spray of wrinkles radiates outward from it*. The local twist cannot be connected smoothly with the undisturbed cloth at large distances: the difference must be patched up. Because of relativity, all symmetries must be local, and *any local symmetry creates a field*. The wrinkles are related to the 'field lines'.

6. Gauge Twists and Velpons

Elementary particles belong to certain families. Within such a family the particles treat each other as equals, at least in the ideal case. That is to say that they are, in some sense, interchangeable: because of the perfect equality one would not notice such a swap. This operation is a symmetry. If one were to subject the whole Universe at once to such a *global* symmetry, nothing would change at all.

If we pick a fundamental fermion multiplet with N members we expect that the symmetry group that acts on this N-plet should behave as a rotation in some abstract N-dimensional space. The rotation of a particle over the *mixing angle* θ literally makes the particle 'turn into' a different one! Some symmetries are actually connected with rotation in space (which creates the angular momentum of a particle) or rotation in spacetime (Lorentz symmetry). Other symmetries behave like rotations too, but not in ordinary space; rather, these symmetries are rotations about other directions than the axes of space and time. Apparently, the vacuum possesses more possible directions than those of spacetime.

A symmetry can be responsible for generating a field. Now in the quantum picture a field is built up from field quanta; and the exchange of a quantum produces a force. Thus, *any local symmetry creates a force*. The quanta of such a field are gauge quanta, or, more precisely, *gauge bosons*, because Lorentz invariance and exact gauge symmetry combined demand that the particle have mass zero and integral spin. The force that corresponds to the exchange of gauge bosons can be considered as a binding agent, a kind of glue between the particles to which the bosons are coupled.

Each force has its own set of glue quanta. A generic name (other than the insipid 'gauge boson') for these does not exist in the professional literature. Therefore, I will succumb to the temptation to name something, and use the generic name *velpon*, after one of the most common brands of glue in my home country.

All known forces are due to gauge symmetries. A local gauge twist causes wrinkles in the vacuum, because the mismatch between the twisted space and the unperturbed vacuum in the distance must be patched up somewhere in between. In the non-quantum picture, of which our tablecloth is a model, the difference is made good by wrinkles, 'field lines' that radiate from the twisted spot. But in a quantum world, where interaction is all-or-nothing, *the twist must be taken away by a single quantum*. The exchanged quantum, the velpon, then becomes the carrier of the vacuum wrinkle caused by the local gauge twist.

7. The Lagrangian in QED and in Gravity

The above explanation is only an analogy and needs to be made more precise for practical purposes. This I will do by briefly showing how the 'wrinkle' picture of fields can be cast in algebraic form. It all begins by guessing a proper symmetry from some notion of similarity between particles, in particular fermions. For example, one may note that a proton is really rather like a neutron, with the exception of a small difference in mass and a difference in electric charge. It takes a little faith to overlook these differences, but one may with some justification surmise that the mass difference will ultimately appear to be due to the charge or something like that. Or one may note that in a sense an electron is rather like a neutrino, in that they always appear together in weak decays.

Start with such a guessed global symmetry. Then one notes that using this symmetry globally is contrary to the spirit of relativity. But using a *local* symmetry is not possible unless one inserts a new field to counteract the mismatch caused by locality. The quanta of this mismatch field transmit a force. In this way, a local symmetry of a basic multiplet of fermions produces bosons that couple to the fermions in a way that is prescribed by the symmetry.

Suppose that our mechanical system is described by a Lagrangian $\mathcal{L}$, which is a function over spacetime $\{x_\mu\}$ of a generalized coordinate vector q and its corresponding momentum $q_{,\mu}$ (we use the abbreviation $q_{,\mu} \equiv \partial q/\partial x^\mu$). The action corresponding to this is found by integrating the Lagrangian density over an arbitrary four-volume Ω:

$$S = \int_\Omega \mathcal{L}(q, q_{,\mu}) dx_\nu, \tag{22}$$

where the dynamical variables q and $q_{,\mu}$ are to be seen as functions of x_μ. Because Ω is arbitrary, the requirement $\delta S = 0$ implies

$$\int_\Omega \frac{\partial \mathcal{L}}{\partial q} \delta q + \frac{\partial \mathcal{L}}{\partial q_{,\mu}} \delta q_{,\mu} \, dx_\nu = \int_\Omega \left(\frac{\partial \mathcal{L}}{\partial q} - \frac{\partial}{\partial x^\mu} \frac{\partial \mathcal{L}}{\partial q_{,\mu}} \right) \delta q \, dx_\nu + \oint_\Omega \frac{\partial \mathcal{L}}{\partial q_{,\mu}} \delta q \, dx_\mu = 0, \tag{23}$$

from which the Lagrangian equations of motion follow directly because the surface integral is zero. Note that we are allowed to subject $\mathcal{L}$ to the same symmetry under which q is supposed to be symmetric, because $\mathcal{L} = f(q, q_{,\mu})$.

Now let us change q infinitesimally by some symmetry **L**. A global symmetry does not change the equations of motion, because **L** commutes with δ. However, if **L** is a *local* symmetry, then $\mathbf{L} = \mathbf{L}(x_\mu)$, and therefore

$$q \to q + \delta q \quad \text{and} \quad \delta q = \epsilon(x_\mu)q. \tag{24}$$

It follows immediately that if $q_{,\mu} \to q_{,\mu} + \delta q_{,\mu}$, we get

$$\delta q_{,\mu} = \epsilon_{,\mu}q + \epsilon(x_\mu)q_{,\mu}, \tag{25}$$

so that the integrand of δS becomes

$$\delta \mathcal{L} = \left(\frac{\partial \mathcal{L}}{\partial q}q + \frac{\partial \mathcal{L}}{\partial q_{,\mu}}q_{,\mu} \right) \epsilon + \frac{\partial \mathcal{L}}{\partial q_{,\mu}}\epsilon_{,\mu}q. \tag{26}$$

The term in brackets drops out because of the equations of motion, and we conclude that $\delta \mathcal{L} \neq 0$ because of the derivative $\epsilon_{,\mu}$: the *local* character of the transformation **L** spoils the proper extremum behaviour of $\mathcal{L}$, and no good equations of motion result!

In other words, the fact that **L** changes from event to event in spacetime produces a *mismatch* between $\mathbf{L}\mathcal{L}$ at one event and the $\mathbf{L}\mathcal{L}$ elsewhere. The key idea now is, to patch this up by adding extra terms to the Lagrangian to correct the mismatch. It is by no means obvious that this can be done successfully!

Because the culprit is a vector $\epsilon_{,\mu}$, we try to patch up $\mathcal{L}$ by adding a vector field to it. For the moment, let us call this field A', and the corresponding Lagrangian is

$$\mathcal{L}' = \mathcal{L}'(q, q_{,\mu}, A'), \tag{27}$$

of which we will now rigorously require that $\delta \mathcal{L}' = 0$. This requirement prescribes a functional dependence of $\mathcal{L}'$ on its arguments, as follows. The δq and $\delta q_{,\mu}$ are found as before; the variation $\delta A'$ is, of course, a linear combination of ϵ and $\epsilon_{,\mu}$ (for infinitesimal transformations). The most general form for $\delta A'$ is then

$$\delta A' = U A'\epsilon + C^\mu \epsilon_{,\mu} \tag{28}$$

with constant scalar U and vector C^μ, to be determined afterwards. To get a proper equation of motion from the patched-up Lagrangian, we require

$$\delta \mathcal{L}' = \frac{\partial \mathcal{L}'}{\partial q}\delta q + \frac{\partial \mathcal{L}'}{\partial q_{,\mu}}\delta q_{,\mu} + \frac{\partial \mathcal{L}'}{\partial A'}\delta A' = 0. \tag{29}$$

Inserting the expressions for δq and so forth yields a linear equation in ϵ and $\epsilon_{,\mu}$. Because the magnitude of ϵ is arbitrary (provided it is infinitesimal), each coefficient of ϵ and $\epsilon_{,\mu}$ must vanish independently. This gives

$$\frac{\partial \mathcal{L}'}{\partial q} q + \frac{\partial \mathcal{L}'}{\partial q_{,\mu}} q_{,\mu} + \frac{\partial \mathcal{L}'}{\partial A'} U A' = 0 \tag{30}$$

$$\frac{\partial \mathcal{L}'}{\partial q_{,\mu}} q + \frac{\partial \mathcal{L}'}{\partial A'} C^{\mu} = 0 \tag{31}$$

with the consistency requirement that

$$C^{\mu} C_{\mu} = 1. \tag{32}$$

The latter means that C_{μ} has an inverse; if it did not, then some of the above equations would be linearly dependent and the system could not be solved. Now define the vector field A_{μ} as

$$A_{\mu} \equiv C_{\mu} A' \tag{33}$$

to find that

$$\frac{\partial \mathcal{L}'}{\partial q_{,\mu}} q + \frac{\partial \mathcal{L}'}{\partial A_{\mu}} = 0. \tag{34}$$

The remarkable thing is, that *this equation is in fact a prescription for the way in which the Lagrangian must depend on its arguments.* We see directly that it requires that the vector field A_{μ}, which was introduced to patch up the mismatch created by the locality of **L** (i.e. the dependency $\mathbf{L} = \mathbf{L}(x_{\mu})$) occurs in $\mathcal{L}'$ *only* through the combination

$$q_{;\mu} \equiv q_{,\mu} - q A_{\mu}, \tag{35}$$

the *covariant derivative* of q. The form $\mathcal{L}' = \mathcal{L}'(q, q_{,\mu}, A')$ allows us only one way to insert $q_{;\mu}$ into $\mathcal{L}'$, namely in exactly the same way as $\mathcal{L}$ depends on $q_{,\mu}$. This must be so because $q_{;\mu}$ contains a term that is linear in $q_{,\mu}$. and another term that can be made zero by letting **L** equal the identity. Thus we get

$$\mathcal{L}' = \mathcal{L}(q, q_{;\mu}), \tag{36}$$

and from now on we use this form.

Note that in the covariant derivative the local symmetry prescribes that q and A_{μ} couple by means of the product $q A_{\mu}$; in quantum electrodynamics this appears in the form where the dynamical variables q and $q_{,\mu}$ are replaced by the derivative ∂ and a constant factor ie:

$$q_{,\mu} - q A_{\mu} \rightarrow \partial - ieA \rightarrow i\psi^{*}(\gamma \cdot \partial - ie\gamma \cdot A)\psi, \tag{37}$$

which is the famous 'minimal coupling' term in the Dirac equation (the γ's are Dirac matrices).

Having now found that there is only one functional form of the Lagrangian which allows us to patch up the mismatch due to the local symmetry, it remains to determine the constants U and C^μ. First, we note that

$$\delta A_\mu = C_\mu \delta A' = C_\mu U \epsilon(x_\alpha) A' + \epsilon_{,\mu} = C_\mu C^\nu U \epsilon A_\nu + \epsilon_{,\mu}. \qquad (38)$$

Second, we recall the expressions for the variations $\delta \mathcal{L}$ and $\delta \mathcal{L}'$, which lead directly to

$$\frac{\partial \mathcal{L}'}{\partial q} = \left.\frac{\partial \mathcal{L}}{\partial q}\right|_{q;\mu} - \left.\frac{\partial \mathcal{L}}{\partial q_{;\mu}}\right|_q A_\mu \qquad (39)$$

$$\frac{\partial \mathcal{L}'}{\partial q_{,\mu}} = \left.\frac{\partial \mathcal{L}}{\partial q_{;\mu}}\right|_q \qquad (40)$$

$$\frac{\partial \mathcal{L}'}{\partial A'} = -\left.\frac{\partial \mathcal{L}}{\partial q_{;\nu}}\right|_q C_\nu q \qquad (41)$$

Inserting these into the equation resulting from $\delta \mathcal{L}' = 0$, we find

$$\left(\frac{\partial \mathcal{L}}{\partial q}q + \frac{\partial \mathcal{L}}{\partial q_{;\mu}}q_{;\mu}\right) - \frac{\partial \mathcal{L}}{\partial q_{;\nu}}qU A_\nu = 0. \qquad (42)$$

The term in brackets vanishes because of the equation of motion for $\mathcal{L}$, and because we had $\mathcal{L}' = \mathcal{L}(q, q_{;\mu})$. It follows immediately that $U = 0$. The definition of A_μ then gives, by means of the expression for $\delta A'$, that

$$\delta A_\mu = \epsilon_{,\mu}. \qquad (43)$$

This demonstrates quite clearly how the vector field A_μ comes in because of the *local* character of the symmetry: if **L** were independent of x_α, we would have $\epsilon_{,\mu} = 0$!

One further point remains to be settled. By patching up the Lagrangian, we have let a genie out of a bottle, namely the field A_μ. We are now obliged to take this field seriously, and to identify it with an actual particle. In that case, we must of course allow A_μ to occur in the Lagrangian as a free field (i.e. as more than just an entity which couples to the q-field by qA_μ). It may be a trifle much to ask, but can the locality of **L** prescribe the form of the occurrence of this free field too?

The patch-up vector field A_μ may occur itself in the Lagrangian as a dynamical variable, together with its spacetime derivative $A_{\mu,\nu}$, in the same way that we had a dependence on the dynamical variables of the q-field. Because $\mathcal{L}$ is linear, we can insert extra terms by simple addition, so we can restrict ourselves to finding the sub-part $\mathcal{L}''$ that depends on the A's only, and then add it to what we had already (note that the coupling term has already been disposed of!) We use the same variational form:

$$\delta \mathcal{L}'' = \frac{\partial \mathcal{L}''}{\partial A_\mu} \delta A_\mu + \frac{\partial \mathcal{L}''}{\partial A_{\mu,\nu}} \delta A_{\mu,\nu}. \qquad (44)$$

As usual, we insert δA_μ and require that each coefficient of ϵ and $\epsilon_{,\mu}$ vanish independently. This yields the equations

$$\frac{\partial \mathcal{L}''}{\partial A_\mu} = 0 \qquad (45)$$

$$\frac{\partial \mathcal{L}''}{\partial A_{\mu,\nu}} + \frac{\partial \mathcal{L}''}{\partial A_{\nu,\mu}} = 0 \qquad (46)$$

Accordingly, we find that the patch-up field *cannot itself occur in the Lagrangian.* Consequently, the field A_μ *is not an observable;* but it *can* couple to the dynamical variable q by means of the term qA_μ. In Feynman terms: the A_μ can only occur between vertices, it is an intermediary, a *virtual particle.* The above shows that the new field can occur in $\mathcal{L}$ only through the combination

$$F_{\mu\nu} = A_{\nu,\mu} - A_{\mu,\nu}. \qquad (47)$$

That is to say, *the curl of the field is an observable!* This should of course look very familiar to aficionados of Maxwell's Equations.

This completes the demonstration that the requirement of local symmetry of the Lagrangian is so severe, that not only the way in which the patch-up field A_μ couples to the q-field, but also the way in which it must occur in the Lagrangian is prescribed entirely. This almost total lack of arbitrariness is what makes the local symmetry concept so compelling.

It can be shown that the same kind of construction works for vector fields q^a (in fact, this is what the original Yang-Mills paper was all about). In that case, it can be shown that the strictness and cleanness with which the form of the Lagrangian is prescribed is due to the group structure of the symmetry.

We have four such cases in Nature:

(1) the case of a phase-rotation symmetry $U(1)$, (i.e. the multiplication with a complex scalar function as treated above), which produces electromagnetism;

(2) the case of the "isospin symmetry" $SU(2)$ (i.e. multiplication with a factor derived from a 2×2 symmetry via $\exp(\frac{1}{2}ig\tau \cdot \omega)$, where τ are Pauli matrices and ω is an arbitrary smooth function over spacetime), which produces the weak interaction;

(3) the group $SU(3)$, leading to the colour interaction;

(4) Lorentz symmetry, which gives rise to the gravitational interaction (General Relativity).

In the Yang-Mills case, the group has nonzero structure constants f^a_{bc}, and following exactly the same line of resoning one may show that the "wrinkle" or "patch-up" field A^a can occur in the Lagrangian only through the combination

$$F^a_{\mu\nu} = A^a_{\nu,\mu} - A^a_{\mu,\nu} - \frac{1}{2}f^a_{bc}\left(A^b_\mu A^c_\nu - A^b_\nu A^c_\mu\right),\qquad(48)$$

which clearly shows the occurrence of nonlinear terms due to the non-Abelian character of the group. The range of the index a depends on the group dimension; for SU(N), it is $N^2 - 1$.

If the local symmetry is Lorentz symmetry, one may show in precisely the same way – though with much more effort – that the patch-up fields (which in gravity theory are traditionally called Γ instead if A) can occur in the Lagrangian only through the combination

$$R^\lambda_{\mu\nu\kappa} \equiv \Gamma^\lambda_{\mu\nu,\kappa} - \Gamma^\lambda_{\mu\kappa,\nu} + \Gamma^\eta_{\mu\nu}\Gamma^\lambda_{\kappa\eta} - \Gamma^\eta_{\mu\kappa}\Gamma^\lambda_{\nu\eta},\qquad(49)$$

which is the Riemann-Christoffel tensor.

8. Clash

The remarkable fact is that both GRT and the Standard Model describe fundamental interactions by means of a gauge field. On small scales, these are quantum fields due to the symmetries U(1), SU(2) and SU(3). On large scales, the gauge field is the Christoffel object $\Gamma^\lambda_{\mu\nu}$ due to local Lorentz symmetry.

It would be great if these similarities allowed us to bring all known forces together in one formalism. Then the expression 'the Universe is made of particles, space and time' would get a truly compelling uniformity, so that the field equations would read something like

$$\left\{\begin{array}{c}\text{structure of}\\\text{space, time, and}\\\text{particles}\end{array}\right\} = 0$$

Alas, we are disappointed. The structure of spacetime cannot be included with the density of particles if we use the Standard Model in the matter tensor. In field theory a potential is no longer something that can be freely changed by adding an arbitrary scalar term; due to the local (as opposed to global) character of the fields, a potential becomes an entity in itself, witness for example the occurrence of A_μ. In electrodynamics this is merely the vector potential, in quantum electrodynamics it stands for a real particle, the photon, which cannot be transformed away. Einstein's conjecture runs into profound trouble because the reality of potentials implies that the zero point energy of the vacuum must be included in the Einstein equation.

9. Pigs in Space

So the theory of forces and matter tells us that in the vacuum, this space 'supposed to be empty', spontaneous particle-antiparticle pairs arise. And that means that this apparently empty vacuum plays an active role in Nature: it has properties that have a profound influence on the behaviour of particles and their interaction. This is apparent, among other things, in the anomalous magnetic moment of the electron and in the Casimir-Polder force, in which the zero-point energy corresponding to the spontaneous pairs exerts a measurable influence on the force between conductors.

If one introduces matter into space in this manner, things simply won't fit. We cannot have pigs in space, because their attendant vacuum fluctuations would ruin the Universe. Einstein built his theory of spacetime expressly in such a way that Newtonian mechanics was recovered in the limit for small speeds and weak fields. But Newtonian theory is all wrong on a small scale, so it would be a stupendous marvel if the Einstein equation gave the right result all the way. If only he had believed in quanta, maybe he could have forged a correspondence with the Schrödinger or Dirac equation!

There are several ways out. People have tried to cancel the vacuum fluctuations against each other. That is in itself not so bizarre: a force is the net result of a quantum sum over *all* Feynman diagrams, and counterdiagrams might be dreamed up, as in the case when the c-quark was predicted from the absence of the decay of the kaon into a pair of muons. But the cancellation would have to be so extraordinarily perfect that it is contrived in the extreme. Only unbroken supersymmetry seems to help. In these theories, there is a symmetry that connects fermions and bosons. Quite a desperate move, because bosons and fermions are as un-alike as possible! However, such un-kosher combinations can be made, using generators Q_α, four-momentum p^μ, Pauli matrices τ_μ and the anticommutation rule

$$\left\{Q_\alpha, Q_\beta^\dagger\right\} = (\tau_\mu)_{\alpha\beta}\, p^\mu. \tag{50}$$

The vacuum is the state which has

$$Q_\alpha|0\rangle = Q_\alpha^\dagger|0\rangle = 0, \tag{51}$$

so that the anticommutator $\{\}$ immediately produces

$$\langle 0|p^\mu|0\rangle = 0; \tag{52}$$

that is to say, the energy-momentum density of the vacuum is zero: $\Lambda = 0$. For this mechanism to work, each fermion should have a bosonic counterpart and vice versa. Our world doesn't look like that in the least, so we are not much further along.

Who's gotta yield, the left- or the right hand side of the Einstein equation? Both, in a sense, but the left more than the right. The right hand side, $T_{\mu\nu}$, is a continuum average and doesn't contain individual particles. That must be modified: no more spacetime averages, no densities, because in a quantum formulation we should not expect space and time to be continuous in the conventional sense. Averages and derivatives $\partial/\partial x^\mu$ would lose their meaning.

The left hand side, the Einstein tensor, at first appears to be the strongest fortress because it is purely mathematical. However, its description of the structure of spacetime in terms of a distance recipe (via $g_{\mu\nu}$) can probably not be quantized. After all, if one were to interpret g in terms of a collection of particles instead of a classical field, we get a paradox: *if spacetime is made of particles, how could such a particle move through space and time?* Stated somewhat differently, the usual particle symmetries (U(1), SU(2), SU(3) and relatives) connect fermions and bosons, whereas the Lorentz symmetry says something about spacetime and not about particles. Furthermore, it is my prejudice that quantum behaviour is a much more strongly established physical effect than the Einsteinian variant of gravity. Anyone can see the Balmer series with minimal equipment. Compared with that loud-and-clear demonstration of quantization, the usual relativistic tests (Mercury, gravitational lensing) seem weak and indirect. A black hole would be a convincing thing, analogous to the hydrogen atom, but its properties are still inferred only indirectly.

So how can we make particles out of g? When matching $g_{\mu\nu}$ to quantum degrees of freedom we need not assign every component to a dynamical field. In QED, the classical vector potential A_μ reappears as the photon; in GRT we have a correspondence between the Newtonian potential Φ and $g_{\mu\nu}$. But part of the covariance with respect to g could be purely a coordinate effect, i.e. there might be one specific 'nature-given' set of coordinates where some fields vanish. The large-scale homogeneity and isotropy of the Universe is a case in point: spherical coordinates might be a preferred reference frame, in which case Lorentz covariance is no longer guaranteed. Having decided which components of g should appear as quantum degrees of freedom, we must decide how to assign these to observable particles. Quantum superposition allows us to construct states by linear combination, as in the case of the U(1)⊗SU(2) unification: the photon γ and the neutral vector boson Z^0 are superpositions of the U(1) and SU(2) velpons. Compare also the Englert-Brout-Higgs mechanism for making particles massive: the degrees of freedom of the scalar particles are used to generate two extra degrees of freedom for the W and the Z, which – if they were massless – would have only two helicity states each, instead of the four of a massive particle.

One could make $\Lambda = 0$ at one specific point in time, but the expansion of the Universe would shift us away from that point and we'd be just as badly off. Similarly, if we inflate the Universe by using some sort of value K derived from grand unification or a similar theory, we should expect that today we'd still be not too far away from an $n = 1$ state in the bottom of the potential of the GUT, and thus have a substantial fraction of the GUT potential still around.

Currently I am trying to do away with the problem more radically by just stating that, by fiat, gravitons do not interact with vacuum fluctuations. The attractiveness is that this removes the need for renormalization of gravity: there'd be no more loops in the graviton propagator. But taken literally that should also exclude one-loop diagrams from the interaction between gravitons; gravity would no longer be nonlinear, against all the evidence.

At this point, I'm talking pie-in-the-sky. It is like Bohr's treatment of the hydrogen atom. Bohr knew perfectly well that an accelerated electron in an atom ought to radiate like crazy. But he pretended it doesn't, just to see what happens. But I'm not Bohr, and I don't know what the equivalent of the hydrogen atom is. The Schwarzschild black hole may fit the bill. But the effects I've calculated are nowhere near observable, and probably all false. Nobody, so far, has found an escape route.

10. *Envoi*

What I have discussed here is known and unknown. So well known that I'm quite embarrassed to talk about it, and so utterly unknown that I feel similarly embarrassed. In between we may be lucky to find fertile ground where we can practice what Medawar called *the art of the soluble*.

References

Icke, V.: 1995, *The Force of Symmetry*, Cambridge University Press, Cambridge
Utiyama, R.: 1956, *Phys. Rev.* **101**, 1597-1607
Weinberg, S.: 1989, *Revs. Mod. Phys.* **61**, 1-23

TIME ANISOTROPY AND QUANTUM MEASUREMENT: CLUES FOR TRANSCENDING THE GEOMETRIC PICTURE OF TIME

AVSHALOM C. ELITZUR

School of Physics and Astronomy
The Raymond and Beverly Sackler Faculty of Exact Sciences
Tel-Aviv University, Israel

Abstract. The yet-unknown origin of the various manifest time asymmetries seems to be related to another persistent problem, namely, the lack of a satisfactory explanation for the presumed "collapse of the wave function" in quantum mechanics. An experiment is proposed to test some hypotheses concerning both quantum collapse and the origin of time asymmetries.

1. Introduction

Various processes that are not invariant under time reversal are known from thermodynamics, gravitational physics, cosmology, and other domains of modern physics. These anisotropies have no trace in the perfectly symmetric laws of physics. Whence this incompatibility? Is there some "master asymmetry" from which all these asymmetries originate? The diversity of answers proposed so far (Davies 1974; Halliwell et al. 1994; Landsberg 1989; Penrose 1979; Price 1996; Zeh 1989) attests to the acuity of the problem, regarded by many as one of modern physics' greatest puzzles.

Oddly, quantum mechanics is seldom considered as instructive for studying time asymmetry. One reason for this is that QM is plagued with its own paradoxes, such as the non-local effects of measurement and their incompatibility with relativity theory. Many competing interpretations of QM have been proposed to tackle these difficulties, yet nearly none of them yields experimental predictions that enable proving or disproving it. In this respect, the interpretation of QM still belongs to philosophy rather than to physics.

Astrophysics and Space Science **244**:313-319,1996.

However, a closer analysis of all these interpretations reveals an implicit assumption about the nature of time that divides them into two groups. About half of them presume that the evolution of a quantum system is, in principle, reversible, while the remaining ones insist that "collapse of the wave function" is a real phenomenon that cannot be reversed by whatever means. It therefore follows that once this reversibility is put to empirical test, it will be capable of eliminating at least some of these interpretations and yielding support for the remaining ones. Moreover, such a test is bound to disclose new insights about the nature of time in itself. Consider, then, the following experiment.

2. Can Collapse be Time-Reversed?

Let a single particle go through an interferometer (Fig. 1). Following the wave-function's impinging on a beam-splitter, only one half of it (on the right arm of the interferometer) undergoes a measurement while the other remains unmeasured. The measurement on the right arm takes place in a small and isolated system such that it never gets entangled with the surrounding environment. Then, while the half wave function is still inside the isolated system, the measurement process undergoes a complete reversal. Finally, both halves of the wave function are reunited by a reverse beam-splitter so as to enable measuring an interference effect.

QM holds that the outcome of the measurement of one half of the wave function instantaneously affects the other half of the wave function, such that if a particle has been detected on the right arm, the wave function vanishes from the left arm and vice versa. In both cases, interference will not appear. It is at this point that a hitherto unnoticed question is imposed by our experiment: *Would the measurement's undoing exert a non-local effect, like that exerted by the measurement itself?*

First, let us dispose with two possible objections to this experimental setting. One is von-Neumann's and Wigner's claim that no measurement has taken place until a conscious observer has read the measurement's result. Now, of all scientific conferences, I think that in this conference I am exempt from the need to show that this interpretation is nothing but metaphysics. For the astrophysicist and the cosmologist, it is utterly ridiculous to believe that the universe has been in a superposition, like a giant Schrödinger cat, for billions of years, until conscious beings evolved and observed it! If collapse of the wave function is a real process, it should be independent of the presence or absence of conscious beings.

Another possible objection is that no measurement has been performed on the wave function until the measuring instrument has interacted with the entire universe. Again, it is the astrophysicist who is in the position to

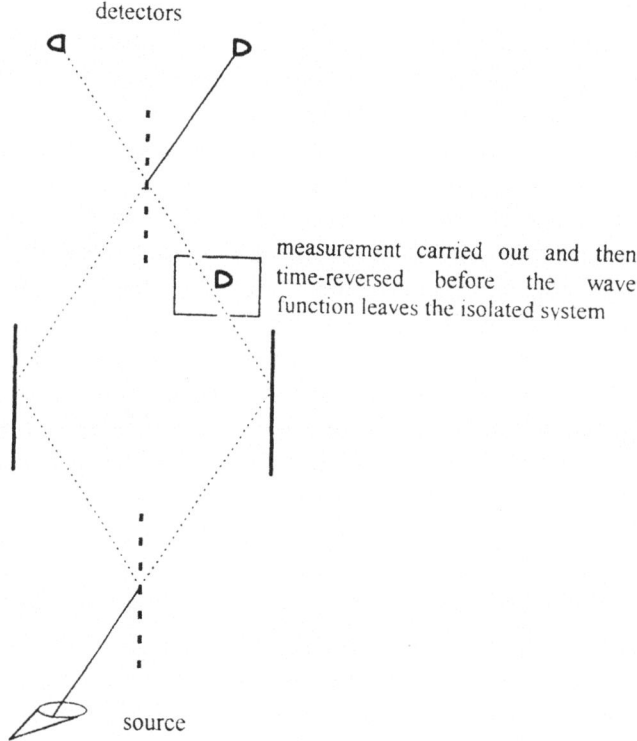

detectors

measurement carried out and then
time-reversed before the wave
function leaves the isolated system

source

Figure 1.

dispense with this objection on pure physical grounds. By the relativistic
prohibition on velocities greater than c, none of the events normally called
"measurements" meets this criterion either, since it would take billions of
years until the entire universe is entangled with the measurement's result.
That leaves us, therefore, with the loose term "environment" as the necessary counterpart in measurement. But then, there is no known physical
difference between an entire laboratory, a human observer, or the small, isolated system employed in our experiment. The question therefore persists:
Would the measurement's undoing in the interferometer's right arm lead
to non-local undoing of the measurement's earlier effects on the left arm?
Whatever the answer turns out to be, it would have nontrivial bearings on
the nature of time.

If interference fails to show up, we shall conclude that non-locality operates only upon measurement and not upon the time-reversed process.
That would mean that quantum mechanics gives rise to a microscopic time
asymmetry still absent in the present formalism. This, in turn, would render quantum interactions the source of the other time-arrows, somewhat in

the spirit of Penrose (1979).

If, however, interference does show up, then "collapse of the wave function" is a false notion. "Reversible collapse" is a self-contradictory term. Hence, if the apparent collapse has been reversed, we must conclude that no collapse has occurred in the first place. To see why, consider the case in which, due to the measurement on the interferometer's right arm, no particle is detected. If collapse has indeed taken place, it must have forced the particle to assume a definite position in the other arm. In this case, the undoing of the measurement cannot work, because, assuming a true collapse, the wave function has totally vanished from the right arm and no undoing can be carried out. We would therefore expect interference to vanish in 50% of the cases. But then, if interference does show up in the remaining 50%, that would violate the uncertainty principle, because one would be able to infer in these cases that the particle has been detected on the right arm. However, observing interference and at the same time knowing which path the particle has taken is prohibited by the uncertainty principle. Conclusion: if the collapse is a real event, it must be irreversible. Conversely, if measurement can be reversed, no collapse ever takes place.

Time asymmetry is not the only problematic bearing that the notion of collapse has on time. A more familiar problem is that collapse of the wave function obliges absolute simultaneity. I would like to show that this effect too shows that the present relativistic picture of time is inadequate.

The incompatibility between quantum non-locality and special relativity has been noted long ago in the vast literature discussing the Einstein-Podolsky-Rosen experiment, but I wish to bring the problem to the extreme by considering a special case. Consider (Fig. 2) a particle split by a beam-splitter. The two halves of the wave function i) travel far away from one another, then, ii) with appropriate mirrors, their motions are reversed and they travel towards one another, and then again, iii) another pair of mirrors reflects them back such that they travel again away from one another. Let us denote the wave function's two halves by A and B, and their parallel stages, respectively, by A_i, B_i, etc. By special relativity, events at stage (ii) of each half of the wave function are simultaneous with the events at stage (iii) of the other half. This is because the two halves of the wave function constitute the same reference frame during these stages. Now suppose that a measurement is carried out during stage A_{iii}. Suppose that no particle has been detected. This is an "interaction free measurement" (Elitzur & Vaidman 1993) that collapses the wave function such that the particle resides in the wave function's other half. However, due to the above relativistic definition of simultaneity, this effect of the collapse occurs at stage B_{ii} (which, for stationary observers, took place earlier). Consequently, this must effect at B_{iii} by ordinary causality. But then again, by the same relativistic

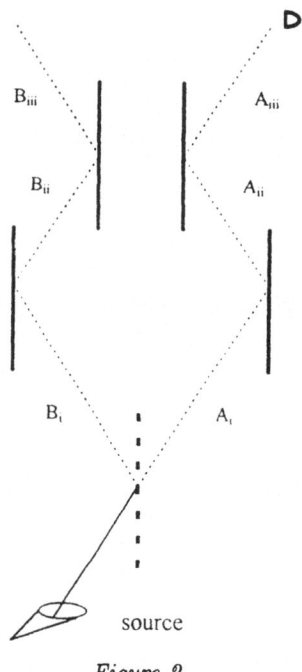

Figure 2.

reasoning, stage B_{iii} is simultaneous with A_{ii}, which means that the wave function must vanish at A_{ii} too. But then, no interaction-free measurement could have taken place at A_{iii}! The only way to avoid such causal loops is to assume some privileged frame in which spacelike separated events are "earlier" or "later" than one another in an absolute sense. Such a universal frame of reference has been proposed by Rosen (1980, 1990) on cosmological grounds. If collapse is an objective event, it must occur in such an absolute reference frame.

This problem will plague even the less radical possible consequences of the above experiment. Let us consider again the no-collapse interpretations of QM, such as the "guide wave" or the "many worlds." These interpretations deny that nature possesses a time-arrow at the microscopic level. Unfortunately, they require absolute time just like the collapse theories do. In fact, they make this requirement even stronger by their strong realistic claim that the guide wave or the other worlds are not mere mathematical constructs but have full objective existence.

Finally, let us examine an interpretation that avoids both notions of collapse and absolute time. This is the transactional interpretation, based on Wheeler & Feynmann's (1945) absorber theory. The transactional model asserts that non-local effects are mediated by retarded-plus-advanced waves via a spacetime zigzag that connects the particles and their source (Cramer 1986; Elitzur 1991). This way, backward causation becomes a natural aspect of any quantum interaction and no absolute time is required to explain the correlation between arms A and B. However, if the undoing of measurement has non-local effects, this would impose on the transactional interpretation an odd conclusion: *The same path in spacetime must be traversed more than once.* First, the measurement is supposed to emit an advanced wave that goes backwards in time to the source and then back to the other half of the wave function, and later the measurement's undoing sends another advanced wave that goes over the first one and overrides it. Such a bizarre process in which the time dimension is traversed more than once entails a higher evolution parameter that ascribes evolution to the Minkowskian spacetime itself (Elitzur 1995; Horwitz et al.1988).

3. Prospects and Conclusions

How practical is the undoing experiment proposed above? Reversal of measurement is approaching technical feasibility due to the works of Cirac & Zoller (1995) and Herzog et al. (1995) involving reversible quantum computation and quantum erasers. Notice, however, that in these experiments the process of measurement and its reversal are carried out on the entire wave function, whereas this article shows that it would be much more interesting to do it on only part of the wave function. If quantum mechanics is correct, even such an experiment must suffice to restore the interference pattern. In the future, as technological advances enable the reversal procedure to be carried out on larger and larger measuring devices, it will be instructive to see whether there is any physical magnitude where non-locality fails, indicating an objective limit between classical and quantum physics.

Another candidate, although still impractical, for the time-reversal of measurement is the "time machine" proposed by Aharonov et al. (1990; Vaidman 1991). It involves a unique combination of quantum mechanical and gravitational effects that can make an isolated system's evolution to be reversed. In view of the fact that the measurement proposed in this article constitutes a tiny process (involving, say, only a few atoms), it would be the first candidate for trying this time machine.

To summarize, it has been shown that any interpretation of QM necessitates going beyond the present description of time. While numerous works have discussed the problem of quantum-mechanical measurement, nearly none has considered the time-reversed process. Yet it is this case that conceals some novel insights. For it proves that quantum non-locality entails either *i*) some fundamental microscopic time-arrow, or *ii*) absolute, reference-frame-independent time, or *iii*) a higher evolution parameter that parametrizes spacetime itself. (Worse, QM might entail more than one of these outrages!) I believe that these are only few of the novel questions that arise once we consider the hitherto-neglected possibility of time-reversed measurement.

References

Aharonov, Y., Anandan, J., Popescu, S., & Vaidman, L.: 1990, *Phys. Lett.*, **64**, 2965

Cirac, J. I., & Zoller, P.: 1995, *Phys. Rev. Lett.*, **74**, 4091

Cramer, J. G.: 1986, *Rev. Mod. Phys.*, **58**, 647

Davies, P. C. W.: 1974, *The Physics of Time-Asymmetry*. University of California Press

Elitzur, A. C.: 1991, *Found. Phys. Lett.*, **3**, 525

Elitzur, A. C.: 1995, "Time Transience and Entropy Increase: An Enigmatic Link". Paper read at the 9th Conference of The International Society for the Study of Time, St.Adèle, Quebéc, Canada

Elitzur, A. C., & Vaidman, L.: 1993, *Found. Phys.*, **23**, 987

Halliwell, J. J., Pérez-Mercader, J., & Zurek, W. H. (Eds.): 1994, *Physical Origins of Time-Asymmetry*. Cambridge: Cambridge University Press

Herzog, T. J., Kwiat, P. G., Weinfurter, H., & Zeilinger, A.: 1995, *Phys. Rev. Lett.*, **7**, 3034

Horwitz, L. P., Arshansky, R. I., & Elitzur, A. C.: 1988, *Found. Phys.*, **18**, 1159

Landsberg, P. T.: 1989, In Sarlemijn, A., & Sparnaay, M.J. (Eds.) *Physics in the Making. Essays on Developments in 20th Century Physics.*, pg 131. Amsterdam: Elsevier Science Publishers B. V.

Penrose, R.: 1979, In Hawking, S. W., & Israel, W. (Eds.): *General Relativity: An Einstein Centenary Survey*, pg 581. Cambridge: Cambridge University Press

Price, H.: 1996, *Time's Arrow and Archimedes' Point*. Oxford: Oxford University Press

Rosen, N.: 1980, *Found. Phys.*, **10**, 673

Rosen, N.: 1990, *Found. Phys.*, **21**, 459

Vaidman, L.: 1991, *Found. Phys.*, **22**, 947

Wheeler, J. A. & Feynman, R. P.: 1945, *Rev. Mod. Phys.*, **17**, 157

Zeh, H. D.: 1989, *The Physical Basis of the Direction of Time*. Berlin: Springer-Verlag

3-D PERIOD DOUBLING
AND MAGNETIC MOMENTS OF PARTICLES

ARI LEHTO

University of Helsinki
Department of Physics
c/o P.O. Box 11012, 02044 VTT, Finland

Abstract. Several invariant properties of matter seem to fit remarkably well to quantized values obtained from Planck domain units by period doubling in three dimensions. The units are defined using natural constants only. Such properties include the elementary electric charge, electron and proton rest energies and others. It has been shown by W. G. Tifft that the quantized redshifts of galaxies fit the pattern, too. Magnetic moments of the electron, proton, neutron, muon and lambda particles also seem to obey the doubling rule. The same number of doublings seems to yield the electron and proton rest masses and magnetic moments correspondingly. The proton rest mass is $E_o \times 2^{-64}$ and the magnetic moment $\mu_o \times 2^{64}$, where E_o and μ_o are the Planck domain units. The corresponding exponent of 2 for the electron is 75.66. The units for the proton and the electron differ by a factor of $\sqrt{\pi}$. The magnetic moment of the muon suggests a close relationship between the magnetic structures of the muon and the proton, whereas the lambda particle seems to be related to the neutron. Reasons for the supposed existence of quantized 3-d time are represented.

1. Introduction

Continuous space-time can be divided into infinitesimally small intervals dx and dt. It is then, in principle, possible to construct a potential well of width dx, where the ground state wavelength will go towards zero by continuously narrowing the well. This, in turn, means that even the ground state energy will go to infinity, not to mention the excited states. Thus, in a continuum model, infinite local energies can be obtained.

Astrophysics and Space Science **244**:321-328,1996.

In a time-continuum this means that the period of oscillation will go to zero and the frequency to infinity, which, according to the Planck relation E=hν, also means infinite energies. The infinities of continuum models have been criticised e.g. by W. Pauli and V. Weisskopf (Pauli and Weisskopf 1934).

An alternative to the space- time continuum is a discrete space-time with a suitably chosen unit cell. We would like to make a distinction between the "external space-time" and "internal space-time" as experienced by matter. By the "external space-time" we mean our ordinary 4-d space-time, where matter is able to move. The "internal space-time" means those quantized degrees of freedom, which characterize the intrinsic properties of matter, e.g. electron rest mass and electric charge.

The Planck cell is a natural choice for the unit cell of the internal space-time. The edge lengths of the cell are the Planck length (10^{-35} m) and time (10^{-43} s) respectively. The maximum (local) energy is now the Planck energy (defined with h) and there will be no infinities.

The Planck energy is huge, 10^{19} GeV, as compared to the proton rest energy of 1 GeV. A mechanism to decrease the Planck energy to the levels of our everyday life is therefore needed.

Non-linear oscillators show period doubling, which provides the route towards chaos. Different kinds of non-linear oscillators behave in a universal way, as discussed by M. Feigenbaum (Feigenbaum 1980).

Generally, conditions necessary for period doubling to occur are non-linearity and iteration. The physical world is seldom linear and periodic motion represents iteration, since the final condition, f_n, of period n equals the initial condition i_{n+1} of period n+1. For these reasons period doubling will be used for reducing the Planck energy and quantizing the levels at the same time.

If period doubling occurs in its simplest form, then

$$\text{Observable} = \text{Unit} \cdot 2^{\pm D}, \qquad (1)$$

where D is the number of doublings.

When several ratios of commensurate properties of matter were examined, it was found that the decimal parts of the exponents D in equation (1) concentrate near 1/3 integer values (Lehto 1990). This means that D is of the form $\pm N/3$, or

$$\text{Observable} = \text{Unit} \cdot 2^{\pm N/3}. \qquad (2)$$

This was interpreted to mean that the observables are actually cubic roots and thus originate from a 3-d system. The ratios examined contained ratios of lengths. If a spatial volume is given, then the characteristic length is a cubic root of this volume.

The ratios examined include energies, which are related to *time* (= period) t by the Planck relation E=h/t. These seem to be cubic roots as well. It was therefore concluded that the temporal part of the internal space-time of matter is three dimensional, too.

The periodic time is thus related to the invariant properties of matter; it is not a generalization of the normal flowing 1-d time. The periodic 3-d time can be considered as three internal degrees of freedom, which define the energy levels, among other things.

The period doubling process yields for instance the following fits (Lehto 1990):

1. 21 cm H-line: $\lambda(21cm) = 2^{112} \cdot l_o$, l_o = Planck length.

2. Electron rest mass m_e: $m_e = 2^{-75.67} \cdot m_o$, m_o = Planck mass.

3. CBR temperature T_{3K}: $T_{3K} = 2^{-106.67} \cdot T_o$, T_o is the Planck temperature.

4. Elementary electric charge (force quantization): $e^2 = 2^{-9.75} \cdot q_o^2$, q_o is the Planck domain unit charge.

5. Fine structure constant (from definition): $\alpha = 2\pi \cdot 2^{-9.75}$.

6. Velocity can be defined as: $v = \Delta l/\Delta t$: $v = c \cdot 2^{-n/3}$.

$c = l_o/t_o$ is the speed of light (t_o = Planck time). Sometimes even finer details are seen, for example in galactic redshifts, where the 9'th root appears (Tifft 1996). The apparent reason for this is that the natural constants are perceived quantities as well, which exhibit an internal cubic root structure, too.

2. Magnetic Moment μ

The magnetic moment μ of a particle is traditionally considered to be a result of its charge distribution and spin S:

$$\mu = g(e/2m)S, \tag{3}$$

where g is the so called spin gyromagnetic ratio characteristic to the complex internal structure of the particle. The experimental values of g for the electron, proton and neutron are -2, 5.586 and -3.826 respectively. There is no *simple* explanation for these values.

Let us now treat the magnetic moments as classical current loops the size of the Planck cell. The current is obtained by dividing the elementary charge e by the period of orbital rotation. Two different loops, shown in Figure 1, are defined:

a) e-loop (electron related) with the Planck length as the circumference

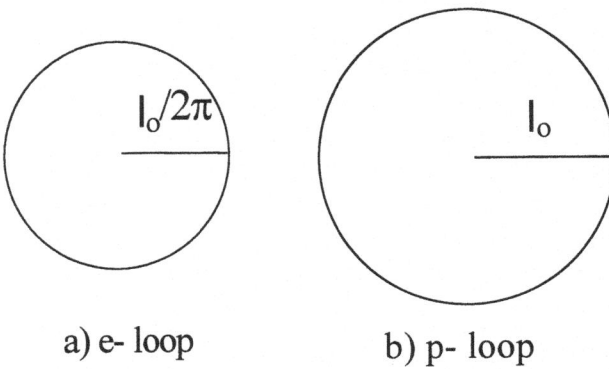

a) e- loop   b) p- loop

Figure 1.   Definition of the current loops.

b) p-loop (proton related) with the Planck length as the radius.

The unit magnetic moment μ_{oe} for the e-loop may be defined as

$$\mu_{oe} = \text{current} \cdot \text{loop area} = (e/t_o) \cdot \pi \cdot (l_o/2\pi)^2. \tag{4}$$

The numeric value of μ_{oe} is $1.5483 \cdot 10^{-46}$ Am^2.

The unit magnetic moment for the p-loop is correspondingly

$$\mu_{op} = \text{current} \cdot \text{loop area} = e/(8t_o) \cdot \pi \cdot l_o^2. \tag{5}$$

The numeric value of μ_{op} is $7.6407 \cdot 10^{-46}$ Am^2. The factor 8 in the denominator of the p-loop unit magnetic moment follows from the assumption that the speed of light cannot be exceeded. If the radius of the loop is l_o, then the circumference is $2\pi l_o$. Therefore $2\pi t_o$ is needed for one period. In the period doubling system the nearest allowed number to 2π is $2^3 = 8$ and the shortest period is thus $8t_o$. The unit current is thus $e/(8t_o)$ and the velocity of the charge is somewhat less than c.

2.1. THE ELECTRON MAGNETIC MOMENT

The electron magnetic moment $\mu_e = 9.2848 \cdot 10^{-24}$ Am^2 is obtained from the unit magnetic moment μ_{oe}:

$$\mu_e = \mu_{oe} \cdot 2^{75.67}. \tag{6}$$

The result shows that the electron magnetic moment has doubled 75.67 times ($3 \cdot 75.67 = 227$ times in three dimensions).

2.2. THE PROTON MAGNETIC MOMENT

The proton magnetic moment $\mu_p = 1.4106 \cdot 10^{-26}$ Am^2 is obtained from μ_{op}:

$$\mu_p = \mu_{op} \cdot 2^{64.00}, \tag{7}$$

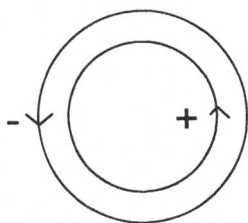

Figure 2. The neutron current loops.

showing that the magnetic moment has doubled 64 times.

2.3. THE NEUTRON MAGNETIC MOMENT

The neutron charge distribution is known to be formed of a positively charged inner layer and a negatively charged outer layer. There are thus two current loops of opposite magnetic moments. The negative loop has a larger magnetic moment. The simplest case is now assumed: The larger loop is identical to that of the proton (save the charge) and a concentric smaller positive loop μ_x is added, see figure 2.

The neutron magnetic moment is $-9.6624 \cdot 10^{-27}$ Am^2. μ_x can now be calculated from $\mu_x = \mu_p + \mu_n$, thus $\mu_x = 2^{62.33} \cdot \mu_{op}$ or $\mu_x = 2^{-1.67} \cdot \mu_p$. This shows that the magnitude of μ_x results from period doubling in 3-d, too.

$$\mu_n = (2^{62.33} - 2^{64.00})\mu_{op}. \tag{8}$$

2.4. THE LAMBDA MAGNETIC MOMENT

The experimental value of the lambda particle's magnetic moment is somewhat uncertain, but

$$\mu_\Lambda = \mu_n \cdot 2^{-1.64}. \tag{9}$$

This result suggests that the magnetic structure of the lambda particle is a scaled-down version of the neutron.

2.5. THE MUON MAGNETIC MOMENT

The muon magnetic moment is $4.4905 \cdot 10^{-26}$ Am^2 . Comparison to the proton magnetic moment yields:

$$\mu_\mu = 2^{1.67} \cdot \mu_p = 2^{65.67} \cdot \mu_{op}. \tag{10}$$

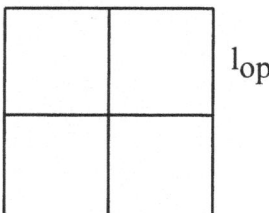

Figure 3. Magnetic moments of the proton close-packed.

The muon's magnetic structure seems to be related to the proton, not the electron.

3. The Proton Rest Energy

We have now related several basic properties of particles to period doubling in 3-d. One might expect the proton rest energy to be related to the Planck energy in a simple manner, too. However, the proton rest mass is $2^{-64.82}m_o$, not $2^{-64}m_o$ as might be assumed from the magnetic moment of the proton.

The space-related behavioral difference between the electron and the proton is that the electrons avoid dense packaging, whereas the protons (and neutrons) prefer being packed closely due to the strong force. What can be done is to make a close-packed arrangement of the unit magnetic moments μ_{op} conserving the area and period (i.e the value of the magnetic moment), as shown in figure 3:

Thus $\mu_{op} = e/(8t_o) \cdot \pi \cdot l_o^2 = e/(8to) \cdot l_{op}^2$. Then $l_{op} = l_o \cdot \sqrt{\pi}$. The Planck energy for this length is $E_{op} = E_o/\sqrt{\pi}$, where E_o is the original Planck energy. The proton rest energy will be now:

$$E_p = 2^{-64} \cdot E_{op}. \tag{11}$$

Furthermore 2^{64} times l_{op} is $1.3 \cdot 10^{-15}$ m, close to the proton "size".

Robert Ehrlich in 1978 published results of his work on the possible existence of an elementary length and an elementary time associated with the strong interaction (Ehrlich 1978). Ehrlich suggests that there should exist a length $l_o = 0.66$ fm (this l_o is Ehrlich's notation, not the Planck length) and a time $\tau_o = 0.66$ fm/c. The Planck length l_{op} is $7.178 \cdot 10^{-35}$ m. The Ehrlich length is $2^{62.995} \cdot l_{op}$. Thus the Ehrlich length fits very closely to the length obtained from the Planck length l_{op} doubled 63 times. The same applies to the Ehrlich time τ_o. We may note that the 63 level corresponds to the energy of the proton-antiproton pair.

4. Consistency of Results for the Electron and Proton

Table 1 shows that the same number of period doublings determine the rest energy and magnetic moment for the electron and the proton.

TABLE 1. Electron and Proton Properties

Property	Relationship
Electron rest energy	$E_e = 2^{-75.67} E_o$
Electron magnetic moment	$\mu_e = 2^{75.67} \mu_o$
Proton rest energy	$E_p = 2^{-64.00} E_{op}$
Proton magnetic moment	$\mu_p = 2^{64.00} \mu_{op}$
Proton "size"	$r_p = 2^{64} l_{op}$

5. Discussion

The Planck domain units are defined here with four natural constants, namely h, c, ϵ_o and G. The gravitational constant G is by far the most inaccurate of these constants. This may have an effect on the value of the Planck mass, reflecting upon the mass- energies of the electron and the proton, which do not fit as well as their electric properties. Table 2 shows the inaccuracies of the fits. The values of the natural constants used are given in Cohen and Taylor (1991). Inaccuracy is defined as the absolute value of (calculated − measured)/(measured).

As W. G. Tifft has shown (Tifft 1996), the period doubling scheme also gives excellent fits to the observed quantized redshifts.

6. Summary

It has been suggested earlier (Lehto 1990) that several invariant properties of matter may be related to some kind of an internal (3+3)-d quantized space-time structure. The unit cell of the periodic space-time is the Planck cell; period doubling in three dimensions is used to quantize the periodic space-time and to scale the Planck domain units to levels observed in our everyday world.

TABLE 2. Inaccuracies of the Fits

Quantity	Inaccuracy
Electron rest energy	0.1%
Proton rest energy	0.2%
Elementary charge	0.003%
Electron magnetic moment	0.007%
Proton magnetic moment	0.08%
Neutron magnetic moment	0.07%
Muon magnetic moment	0.3%
Fine structure constant	0.006%

In a periodic (3+3)-dimensional space-time several properties of matter seem to obey

$$(\text{Observable}) = (\text{Planck domain unit}) \cdot 2^{\pm N/3}, \qquad (12)$$

where N is the total number of doublings.

The magnetic moments of the particles discussed here seem to originate from simple Planck domain current loops. The values agree with the measured values without introducing any additional factors, like gyromagnetic ratios. The rest energy and the magnetic moment are determined by the same number of period doublings for the electron and the proton. The Planck domain unit energy for the proton is determined by the close-packed structure of unit Planck domain magnetic moments.

7. Acknowledgements

The author is greatly indebted to Prof. W. G. Tifft, Dr. J. Cocke, Dr. C. DeVito and Dr. A. Pitucco for many interesting and fruitful conversations.

References

Cohen E. R. and Taylor B. N.: 1991, *Physics Today*, August
Ehrlich R.: 1978, *Phys. Rev. D*, **18**, 320
Feigenbaum M. J.: 1980, *Los Alamos Science* **1**, 4
Lehto A.: 1990, *Chin. J. Phys.* **28**, 215
Pauli W. and Weisskopf V.: *Helv. Phys.* Acta **7**, 709
Tifft W. G.: 1996, *Astroph. J.*, **468**, 491

RELICS OF THE PRIMORDIAL ORIGIN
OF SPACE AND TIME
IN THE LOW ENERGY WORLD

C. WOLF
Department of Physics
North Adams State College
North Adams, MA 01247, U.S.A.

Abstract.
In the earliest stage of cosmological evolution due to high matter densities space and time most likely admitted a very complex geometrical and topological structure. After themalization, statistical averaging and cooling, flat Minkowski space developed but statistical fluctuations from this "averaged out space-time" may still exist in the low energy world. In the following, we explore the consequences of these fluctuations in the low energy world based on a "microscopic uncertainty principle for time". Phenomena such as spin polarization precession, spectral shifts, spin flips, C.P. violating phenomena and neutron interferometry may all be influenced by these fluctuations and we discuss just how the conventional theory of these temporal phenomena could be affected by fluctuations away from Minkowski space-time. We also discuss the experimental limits on the discrete time interval setting the scale of these fluctuations along with possible temporal changes of the discrete time interval over cosmological time scales in the spirit of Dirac's "Large number hypothesis".

1. Introduction

If we assume that the present theory of cosmological evolution is correct, it is only natural to conjecture that the present low energy structure of space-time had it origins in very energetic quantum processes that may involve geometrical and topological notions foreign to our present way of thinking. Within the present structure of quantum gravity, the early universe may have contained wormholes [1], instantons [2], strong black holes [3] and

Astrophysics and Space Science **244**:329-346,1996.

dimensions in addition to the usual four space-time dimensions we perceive [4]. In fact Sakharov suggested that even the present universe might contain domains of varying dimensionality, varying topology and varying signature [5]. These possibilities still lie within the general domain of continuum physics where general field theoretical methods can be applied. There is an alternate school of thought in physics made popular by Wheeler [6, 7], but originally suggested by Heisenberg [8] and others [9, 10], namely that the world has an intrinsic discrete countable nature associated with it. Wheeler pictures "events" as endowed with a certain Yes-No outcome and when combined in a combinatoric sequence the physical measureables of the world we perceive are born. What the fundamental events are is left an open question, the achievement of such a "construct" is that it asserts that geometry and field theory emerge from a sequence of binomial choices connected in a combinatoric fashion. In an unpublished work [11] we have discussed a pre-geometric picture of the early universe based on a binomial distribution for a two fold quantum variable in the spirit of the problem of a random walk. In much the same spirit, Finkelstein [12] has erected a theory of a "quantum net" with space-time and field theory following after an averaging process and Wootters [13] has constructed a theory of how space and time develop using "spin" as the "primitive variable", in his picture, space and time intervals are the result of generic quantum correlations between spin variables at different points. Other authors including Evako [14], Antonsen [15], Kull et. al. [16] and Bombelli et. al. [17] have considered discrete space-time using graph theory and topology to predict notions of curvature, dimensionality and signature in the continuous limit. These attempts are representative of the general ideas originally put forth by Wheeler [6], the problem is that it is difficult to test any of these theories in the low energy world. It is for this reason that we direct our attention toward phenomena and "theoretical structures" where discreteness can be studied by simple modifications of the corresponding continuous theory. Admittedly to derive the principles of Quantum Mechanics and the structure of space-time from such primitive notions as a discrete topological set or a fundamental Yes-No choice would be an incredible accomplishment, however at this point we really have no experimental information to guide us in this direction. The original attempts at constructing discrete space-time theories were not so much motivated by the desire to uncover the true microstructure of the continuum but rather were intended to render quantum electrodynamics and quantum gravity finite by introducing a fundamental length into the theory [18, 19]. In this same spirit, Lee [20] introduces a discrete time into quantum mechanics to facilitate the calculation of path integrals and make the measure a more well defined quantity. Following Wilson's historic paper on lattice gauge theory [21], there was a vast avalanche of work proving the equivalence of a Euclidian lattice gauge theory and Hamiltonian lattice theory when only space is discrete [22, 23]. Again these works are important within the general context of discrete space-time, but their aim was directed toward the calculation of Q.C.D. confinement rather than uncovering certain generic properties of the space-time lattice. Other results of lattice studies hint at why fermions must break chiral invariance [24] on a lattice (a property of the electroweak theory) and why certain dimensionalities exhibit phase transitions and others don't for "SU(N) lattice gauge theory [25].

One remarkable property of lattice theory is that Lorentz invariance can never be restored [26], this suggests that if the world did begin in a discrete manner, Lorentz invariance is a low energy symmetry when the lattice passes to the continuum limit. If we put the Schrodinger equation on a space-time lattice certain chaotic fluctuations occur in wave packet propagations that may be observable if the particles De Broglie wavelength is in the same order of magnitude as the lattice spacing, this would be the case for very heavy particles [27]. For a non-linear Schrodinger equation various authors have discussed soliton-soliton interaction and single soliton propagation when the space coordinate alone is discretized [28, 29]. If two times are introduced into a quantum theory (one continuous and one discrete) we have pointed out that certain peculiar features occur in quantum phenomena somewhat related to chaos and the second time could also be looked at as a hidden variable [30].

The above mentioned works on lattice gauge theory (21 – 26) suggest how to incorporate a discrete time and perhaps a discrete spatial variable into a fundamental quantum mechanical equation, namely replace derivatives by finite differences in both space and time on a discrete space-time lattice and solve the resulting difference equation. In what follows, we will illustrate the techniques in three situations, the first is the solution of the potential free one dimensional Schrodinger equation [31], the second is the application of a purely discrete time theory to discrete time spin polarization precession [32], the third is the application of a quantum formalism when there are two times, one of the times being continuous and the second discrete [30, 33].

The second approach to the discretization of time goes back to the historic work of Caldirola [34, 35]. Caldirola simply replaced time derivatives by finite differences with a discrete time interval τ, that is

$$\frac{\partial \Psi}{\partial \tau} \to \frac{\Psi\left(t + \frac{\tau}{2}\right) - \Psi\left(t - \frac{\tau}{2}\right)}{\tau}$$

There are alternates to this replacement but they simply involve the above replacement over a continuum of values for τ [36]. The interpretation of this replacement is the main "physical motivation" of this study, it emerges from the belief that in the early universe there was no global notion of space and time. Only individual particles could be identified, this can be somewhat related to the fact that at high matter densities the formation of strong black holes (elementary particles) prevents communication between particles because of the formation of horizons [37, 38].

Through thermalization and quantum correlations and in the correspondence limit an averaged out notion of space and time was born which we know as Minkowski space-time, however when quantum mechanics is applied to individual particles because fluctuations from this averaged out time still exist the quantum dynamical equations must reflect the presence of these fluctuations by replacing the time derivatives by finite differences where the width of the finite difference measures the discrete time interval τ [39]. Said in another way, the exterior Hamiltonian which is in a sense an empirical expression of all the local information we have on dynamical systems, acts on the particle's wave function, the wave function of the particle then responds over a finite time interval rather than at a specific time mirroring the fact that there is an uncertainty in tuning between the particles sense of time and the averaged out time of Minkowski's space-time. In other words, the wave function remembers its pre-geometric origin, the Hamiltonian doesn't. In a separate paper we have shown that this "fundamental uncertainty principle in time" leads to the upper limit for the mass of an elementary particle [40]. Other phenomena that we have studied using this approach include electron spin resonance [41], electron spin polarization precession [42], spectral shifts in the spectrum of hydrogen [43], spin polarization precession of composite leptons [44], spin flip phenomena in high magnetic fields of confined particles [45] and composite particles (leading to the study of the internal hidden quantum numbers) [46], spin polarization precession of composite gauge bosons leading to definite signatures for composite gauge bosons [47, 48], and spin flip transitions of composite particles with electric and magnetic charge degrees of freedom in high magnetic fields [49]. We have also applied discrete time difference notions to the propagation of a double wave-packet [50], to time asymmetric (CP violating) decays of the K_L, K_S system [51] and E.P.R. correlations in ϕ decays [52]. It is also interesting that the above discrete time difference theory can be proven to be equivalent to non-local hidden variable theory [53] for particles emitted in opposite directions from a static source, and lastly discrete time difference Q.M. can be identified with both energy dependent width phenomena in particle decays [54] and a cosmological source of time reversal violation [55] in addition to the usual source ascribed to phases in particle couplings. If instead of descretizing time differences in this above sense we discretize space we have shown that the modified quantum theory leads to modifications in neutron interferometry predictions [56] and if we discretize both space and time differences simultaneously, definite signatures for wave packet propagation arise due to the modified quantum theory [57]. Looking for actual experimental tests of the above discrete space-time difference Q.M. we have pointed out that definite shifts in the width of the central diffraction pattern for coherent particles arise for both a discrete time difference Quantum Theory [58] and a discrete time – discrete space quantum theory [59]. Also the famous angle slow down effect for the 4π spinor rotation in a magnetic field can be correctly accounted for by discrete time difference Q.M. [60]. A variety of experiments can be used to set limits on the discrete time and space interval (some including the above references) and we have summarized the experimental tests in [36].

The third area of exploration involving discrete time effects involves the notion of a Markov jump process where due to environmental influences a quantum system behaves like a Brownian particle responding both to the usual Hamiltonian effects plus Markov environmental statistical effects [61]. Just recently Nanopolous has pointed out that the "Procrustean principle" which considers quantum systems as open systems has deep foundations within the structure of string theories [62]. Along the same lines of reasoning Itzykson and Drouffe [63] have shown that a Schrodinger-like equation results for a random-walk on a hypercubical lattice. If a spin system is indeed an open system it would seem that because of its simplicity, Markov effects could most easily be applied to phenomena involving spin flips. We have applied these ideas to single leptons without internal structure [32], to gauge bosons with and without internal structure for a variety of models [64, 65], and to leptons with internal structure [66] providing us with potential probes to the compositeness of "thought to be elementary particles". The time interval between Markov steps might or might not be linearly related to the usual Minkowski time and this is a question that can only be settled by experiment. Unexpected deviations from the theoretical predictions in particle experiments are seldom ascribed to discrete time effects because of the "present disposition of modern day particle theory". However it could very well be that many phenomena such as particle widths, and corrections to anomalous moments have their origin in discrete space and time effects of the above mentioned nature. We hope the experimental community becomes increasingly aware of this with an emphasis on placing more accurate limits on discrete time parameters.

A last point to mention in the introduction is that if a discrete time interval (τ) did emerge from pre-geometric motions we might expect it to vary over cosmological time scales [67], this suggests that spectral shifts induced by discrete time difference quantum theory might be different for nearby galaxies than for those at the sphere of last scattering due to a "Dirac-like cosmological time variation" [68, 69], of the fundamental discrete time interval. If spectral lines used in red-shift analyses attain an unexplained shift or broadening we might use this both as a tool to put limits on τ and its variation over cosmological time scales.

2. Quantum Theory Admitting Discrete Time and Discrete Time Differences

To begin our review of discrete time Q.M. we first classify the possible theoretical structures that can arise from generalizations of the known structure of Quantum Theory.

I. Pure Discrete Time Quantum Theory

- a) Applications to a Free Particle;
- b) Influence on Spin Polarization Precession;
- c) Enhanced Wave Packet Spreading and Catalysis of Chaos;
- d) Theories with Two Times Present (one continuous and one discrete); and
- e) Pure Discrete Spatial Quantum Theory and the Development of Negative Temperatures in the Universe.

II. Discrete Time Difference Quantum Theory

- a) Spin Polarization Precession as a Probe to the Compositeness of Elementary Particles;
- b) Spin-Flip Spectra and Atomic Transitions as a Probe to Discrete Time Q.M. and the Composite Structure of Elementary Particles;
- c) Discrete Time Difference Q.M., CP Violating Phenomena and EPR Phenomena;
- d) Upper Limit to Elementary Particle Masses Brought About by Discrete Time Theory;
- e) Resolving the "angle slow down effect" in Rauch's Experiment using Discrete Time Difference Quantum Theory; and
- f) Cosmological Implications of Discrete Time Difference Quantum Theory.

III. Environmental Discrete Time Markov Processes as a Probe to the Compositeness of Elementary Particles

- a) Composite Models of Gauge Bosons; and
- b) Composite Models of Leptons.

The Schrodinger equation on a space-time lattice in one spatial dimension can be found by making the following replacement [31]

$$\frac{\partial \Psi}{\partial x} \rightarrow \frac{\Psi\left(x + \frac{L_0}{2}\right) - \Psi\left(x - \frac{L_0}{2}\right)}{L_0}$$

$$\frac{\partial \Psi}{\partial t} \rightarrow \frac{\Psi\left(t + \frac{\tau}{2}\right) - \Psi\left(t - \frac{\tau}{2}\right)}{\tau} \tag{1}$$

(L_0, τ are discrete space and time intervals respectively.) The resulting discretized Schrodinger equation reads

$$-\frac{\hbar^2}{2mL_0^2}\left(\Psi(x + L_0, t) + \Psi(x - L_0, t) - 2\Psi(x, t)\right)$$

$$= \frac{i\hbar}{\tau}\left[\Psi\left(x, t + \frac{\tau}{2}\right) - \Psi\left(x, t - \frac{\tau}{2}\right)\right] \tag{2}$$

We set $x = kL_0, t = \frac{n\tau}{2}$ (choosing units such that $L_0 = 1, \frac{\tau}{2} = 1$) and we find that Eq. (2) gives

$$-\frac{\hbar^2}{2m}[\Psi_{k+1,n} + \Psi_{k-1,n} - 2\Psi_{kn}]$$

$$= \frac{i\hbar}{2}[\Psi_{k,n+1} - \Psi_{k,n-1}] \tag{3}$$

Eq. (3) can be solved by the generating function method [70] to give (here we assume $\Psi_{0,n} = \Psi_{1,n} = 0$ for all n).

$$\Psi_{k,2} = \frac{i\hbar}{m}\Psi_{k+1,1} + \frac{i\hbar}{m}\Psi_{k-1,1} - \frac{2i\hbar}{m}\Psi_{k,1} + \Psi_{k,0} \tag{4}$$

Eq. (4) demonstrates that we need the distribution $\Psi_{k,n}$ at two temporal lattice points to generate the wave function at later times. Where the initial data comes from is unknown, perhaps from pre- geometric notion expressed in the language of quantum correlations. We also find that for the propagation of a wave packet (for heavy particles) when the de Brogolie wave length is of the same order of magnitude as the lattice spacing, that chaotic fluctuations occur away from continuous wave packet propagation that would be a signature of the underlying discrete space-time quantum theory.

As a second application of a pure discrete time quantum theory we consider electron-spin polarization precession in a z component magnetic field [32], for the Hamiltonian we have

$$H = \left(\frac{e}{2m}\right)(2)S_z B$$

and for the discrete time amended Schrodinger equation

$$H\Psi = i\hbar\frac{\left[\Psi\left(t + \frac{\tau}{2}\right) - \Psi\left(t - \frac{\tau}{2}\right)\right]}{\tau} \tag{5}$$

Letting $t = \frac{n\tau}{2}$ and scaling such that $\frac{\tau}{2} = 1$ we have for the wave function

$$\Psi = \begin{pmatrix} a_1 \\ a_2 \end{pmatrix} U_n \tag{6}$$

the eigenvalues are $E_+ = \frac{e\hbar B}{2m}$ and

$$E_+ = \frac{e\hbar B}{2n}, \begin{pmatrix} a_1 \\ a_2 \end{pmatrix} = \begin{pmatrix} 1 \\ 0 \end{pmatrix}, E_- = -\frac{e\hbar B}{2m}, \begin{pmatrix} a_1 \\ a_2 \end{pmatrix} = \begin{pmatrix} 0 \\ 1 \end{pmatrix}$$

For the discrete temporal function U_n we have

$$E U_n = i\hbar \frac{[U_{n+1} - U_{n-1}]}{2} \tag{7}$$

We find two solutions corresponding to each $(E)(E\pm)$ in Eq. (7), one solution corresponds to the usual $e^{-\frac{iEt}{\hbar}}$, the other corresponds to $e^{\frac{iEt}{\hbar}}$. When we allow for a small coefficient of the positive exponential term in addition to the usual term $\left(e^{-\frac{iEt}{\hbar}} \right)$ and assume $\langle S_x \rangle_{n=0} = \frac{\hbar}{2}$ we find

$$\langle S_x \rangle = \frac{\frac{\hbar}{2}(1 - \epsilon_1 - \epsilon_2) \cos\left(\frac{eB}{m}\right)\left(\frac{n\tau}{2}\right) + (\epsilon_1 + \epsilon_2)\frac{\hbar}{2}(-1)^n}{(1 - \epsilon_1 - \epsilon_2) + (-1)^n(\epsilon_1 + \epsilon_2) \cos\left(\frac{eB}{m}\right)\left(\frac{n\tau}{2}\right)} \tag{8}$$

Eq. (8) demonstrates that for $\epsilon_1, \epsilon_2 \neq 0$ that the spin polarization exhibits small chaotic fluctuations away from the usual value of

$$\langle S_x \rangle = \frac{\hbar}{2} \cos\left(\frac{eB}{m}\right) t \quad \left(t = \frac{n\tau}{2} \right)$$

However small ϵ_1, ϵ_2 are these fluctuations might be measurable if other perturbing effects could be filtered out.

If we consider that the above theory (Eq. 5, with $t = \frac{n\tau}{2}$) governs the propagation of a wave packet, we have found that the wave packet spreads faster than a normal wave packet and because time is a discrete number we observe a zig-zag deviation from a smooth gaussian [71]. If a theory with two times governs the propagation of a wave packet (one continuous and one discrete) we have again chaotic effects on wave packet propagation which could be related to the existence of a hidden variable [30].

If instead of time being "discretized", space is discretized [72] and the Schrodinger Equation is generalized to accommodate discrete spatial effects we find that a bound particle will have a finite number of bound states. In the early universe or in the present universe where domains of discrete space exist this could lead to the development of negative temperatures which are actually hotter than colder temperatures [73, 74]. The annihilation of such regions could produce C.M.B. anisotropies as well as being a candidate for γ ray burst phenomena [75].

As mentioned earlier in the introduction, if instead of a pure discrete time theory we have a discrete time difference quantum theory in a background of continuous time (motivated by a microscopic uncertainty principle for time) we may write

$$H\Psi = i\hbar \frac{\left[\Psi\left(t + \frac{\tau}{2}\right) - \Psi\left(t - \frac{\tau}{2}\right)\right]}{\tau} \tag{9}$$

(t is continuous)

For a spinning electron we have the eigenstates

$$\Psi = \begin{pmatrix} a_1 \\ a_2 \end{pmatrix} T(t)$$

Using the Hamiltonian $H = \frac{e}{m} S_z B$ we find $E_\pm = \pm \frac{e\hbar B}{2m}$. The temporal function is for each eigenvalue

$$T_\pm = e^{-\frac{2}{\tau}\left(\sin^{-1}\frac{E\tau}{2\hbar}\right)it} \tag{10}$$

For a state initially polarized in the x direction we have

$$\Psi = \begin{bmatrix} \frac{1}{\sqrt{2}} e^{-\frac{2}{\tau}\sin^{-1}\left(\frac{E_+\tau}{2\hbar}\right)it} \\ \frac{1}{\sqrt{2}} e^{-\frac{2}{\tau}\sin^{-1}\left(\frac{E_-\tau}{2\hbar}\right)it} \end{bmatrix} \tag{11}$$

When we evaluate $\langle S_x \rangle$ we find

$$\langle S_x \rangle = \frac{\hbar}{2} \cos\left(\frac{4}{\tau} \sin^{-1} \frac{eB\tau}{4m}\right)$$

and

$$\omega = \frac{4}{\tau} \sin^{-1}\left(\frac{eB\tau}{4m}\right)$$

Thus for small τ,

$$\omega \propto \frac{eB}{m} + \frac{e^3 B^3 \tau^2}{96m^3} \tag{12}$$

Our first observation in Eq. (12) is that for high B, ω receives corrections of order B^3, secondly if the electron is a composite particle [44] and the discrete time interval is a random variable generated from the "primitive unknown of pre-geometry", we might expect that if the electrons contain (preons) that $\tau = n\tau_0$ or $\tau = \sqrt{n}\tau_0$ depending on whether τ is the average over a random distribution or τ^2 is the width squared of a random distribution. Here we have applied the central limit theorem for the n preons comprising the electron, also $\tau_0 =$ discrete time interval for a single preon. Then Eq. (12) would read

$$\omega \simeq \frac{eB}{m} + \frac{e^3 B^3 n^2 \tau_0^2}{96m^3} \tag{13}$$

or

$$\omega \simeq \frac{eB}{m} + \frac{e^3 B^3 n \tau_0^2}{96m^3} \tag{14}$$

Eq. (13), Eq. (14) could be used to probe the generation structure of charged leptons in that ω now depends on the number of preons in each generation. If the above ideas are applied to a composite spin 1 gauge boson (ω^-) [47] with two preons comprising the internal structure the following formula results for the x spin polarization

$$\langle S_x \rangle = \frac{\hbar}{2} \cos(a_1 - a_3)t + \frac{\hbar}{2} \cos(a_3 - a_2)t \tag{15}$$

here the Hamiltonian of the two spin 1/2 preons is ($q_1 = q_2 = -e, e = \frac{e_e}{2}, e_e =$ electronic charge)

$$H = M_0 C^2 + \frac{P_1^2}{2m} + \frac{P_2^2}{2m} + \frac{e}{m}(S_{z_1} + S_{z_2})B + g\vec{S}_1 \cdot \vec{S}_2 \tag{16}$$

with the three eigenstates

$$\begin{aligned}
S_z &= 1, a_1 = \frac{2}{\tau} \sin^{-1}\left(\frac{E_+ \tau}{2\hbar}\right) \\
S_z &= -1, a_2 = \frac{2}{\tau} \sin^{-1}\left(\frac{E_- \tau}{2\hbar}\right) \\
S_z &= 0, a_3 = \frac{2}{\tau} \sin^{-1}\left(\frac{E_0 \tau}{2\hbar}\right)
\end{aligned} \tag{17}$$

We see that the eigenvalues of Eq. (16) have the same internal spatial state

$$\begin{aligned}
E_\pm &= M_0 C^2 + \frac{(n_1^2 + n_2^2)h^2}{8mL^2} + \frac{g\hbar^2}{4} \pm \frac{e\hbar B}{m} \\
E_0 &= M_0 C^2 + \frac{(n_1^2 + n_2^2)h^2}{8mL^2} + \frac{g\hbar^2}{4}
\end{aligned} \tag{18}$$

($M_0 C^2 =$ rest mass parameter, $g =$ spin-spin coupling constant, $L =$ confinement scale, $m =$ heavy preon mass).

Eq. (15) suggests that the spin polarization will precess with two sinusoidal functions, for small τ a Doppler like effect will result for $\langle S_x \rangle$.

For the same reason that anomalous behavior appears in spin polarization precession, it also appears in the transition between different energy states of atomic systems, nuclei and spin-flip transitions of elementary particles. The transition frequency according to time dependent perturbations theory becomes altered to read

$$\omega = \frac{2}{\tau} \sin^{-1} \left(\frac{E_F \tau}{2\hbar} \right) - \frac{2}{\tau} \sin^{-1} \left(\frac{E_1 \tau}{2\hbar} \right) \tag{19}$$

When Eq. (19) is applied to atomic transitions [43], spin flips of confined electrons [45], and spin flips of particles with internal structure [46, 49], spectral shifts result which are dependent on both the discrete time interval and the individual state which characterizes the composite structure. This is essentially because the transition frequency (Eq. (19)) is non-linearly related to the difference between the two eigenstates.

Another application of the above ideas is in the probability of CP or T violation. As is well known CP violation results because of complex phases in the Yukawa couplings, the pure Higgs-Higgs coupling and the vacuum expectation values of the Higgs field (soft CP violation) [76]. If however there is explicit T violation in the equation of quantum theory, we may amend Eq. (9) to read [51]

$$H\Psi = i\hbar \frac{\left[\Psi \left(t + \frac{\tau}{2} - \epsilon \right) - \Psi \left(t - \frac{\tau}{2} - \epsilon \right) \right]}{\tau} \tag{20}$$

(ϵ = time asymmetry parameter).

When Eq. (20) is applied to the K_L, K_S system in the $\left(\begin{smallmatrix} K_0 \\ \bar{K}_0 \end{smallmatrix} \right)$ basis (H = Hamiltonian) we have the following equation for $\Psi = \left(\begin{smallmatrix} a_1 \\ a_2 \end{smallmatrix} \right) T(T)$

$$H \left(\begin{matrix} a_1 \\ a_2 \end{matrix} \right) T(t) = E \left(\begin{matrix} a_1 \\ a_2 \end{matrix} \right) T(t)$$

$$= i\hbar \left(\begin{matrix} a_1 \\ a_2 \end{matrix} \right) \left[\frac{T \left(t + \frac{\tau}{2} - \epsilon \right) - T \left(t - \frac{\tau}{2} - \epsilon \right)}{\tau} \right] \tag{21}$$

The solution to order ϵ is

$$\Psi = \left(\begin{matrix} a_1 \\ a_2 \end{matrix} \right) e^{-\frac{2}{\tau} \sin^{-1} \left(\frac{E\tau}{2\hbar} \right) it} e^{-\frac{2r_0}{\tau} \left(\tan \frac{r_0 \tau}{2} \right) \epsilon t}$$

$$\left(r_0 = -\frac{2}{\tau} \sin^{-1} \left(\frac{E\tau}{2\hbar} \right) \right) \tag{22}$$

$\left(\begin{pmatrix} a_1 \\ a_2 \end{pmatrix}\right)$ are different for K_L and K_S).

If E contains an imaginary component (as it must in the $K_0, \bar{K}_0$ system) the last factor gives an additional contribution to the lifetime of K_L, K_S. The hope here is that when all the phenomenological input to H is given by particle theory, additional corrections to the experimental lifetime of K_L, K_S may be used to put limits on ϵ (parameter representing explicit T violations). We have also applied discrete time difference Q.M. to the decay of ϕ into K_L, K_S [77]. If we consider a ϕ to decay to a right moving K meson (a) and left moving K meson (b), we write for the CP odd state at $t = 0$

$$\Psi = \frac{1}{\sqrt{2}}(|K_S>_a |K_L>_b -|K_L>_a |K_S>_b) \tag{23}$$

If the temporal part of the wave function is included in Eq. (23) and we go back to the $K_0, \bar{K}_0$ basis

$$|K_L> = \frac{1}{\sqrt{2}}(|K_0> -|\bar{K}_0>)$$

$$|K_S> = \frac{1}{\sqrt{2}}(|K_0> +|\bar{K}_0>)$$

we obtain the following formula for the coefficient of $|K_0>_a |K_0>_b$

$$C_{K_{0_a},K_{0_b}} = \frac{1}{2\sqrt{2}}\left[\begin{array}{c} -e^{-\frac{2}{\tau}\sin^{-1}\left(\frac{E_S\tau}{2\hbar}\right)it_a - \frac{2}{\tau}\sin^{-1}\left(\frac{E_L\tau}{2\hbar}\right)it_b} \\ +(t_a \rightleftharpoons t_b) \end{array} \right) \tag{24}$$

When measurements on the two beams at t_a (right) and t_b (left) are made we find that Eq. (24) gives a probability below that of normal Q.M. and increases the difference in probability predicted by Q.M. and that of local realism [78].

If we now turn our attention to bound state phenomena we observe from the form of the solution in Eq. (10) that if $\frac{E\tau}{2\hbar} > 1$ the function $\sin^{-1}\left(\frac{E\tau}{2\hbar}\right)$ attains an imaginary component [40], this suggests that if elementary particles are comprised of preons and the total composite system interacts with the surrounding space-time so as to include discrete time difference quantum effects then there should be an upper limit to the bound state energy. In fact, if we write the Hamiltonian of two preons as

$$H = M_0C^2 + \frac{P_1^2}{2m_1} + \frac{P_2^2}{2m_2} + Kr + g\vec{S}_1 \cdot \vec{S}_2 \tag{25}$$

where

$$g(\vec{S}_1 \cdot \vec{S}_2) = \text{spin} - \text{spin interaction},$$

$$Kr = \text{hypercolor confining potential,}$$
$$M_0 C^2 = \text{rest mass energy}$$

and the eigenvalue problem is

$$H\Phi = E\Phi = i\hbar \left[\frac{\Phi\left(t + \frac{\tau}{2}\right) - \Phi\left(t - \frac{\tau}{2}\right)}{\tau} \right]$$

$$\Phi = \varphi(\vec{r}, s)T(t), \tag{26}$$

then the temporal solution is $T(t) = e^{-\frac{2}{\tau}\sin^{-1}\left(\frac{E\tau}{2\hbar}\right)it}$ and for $\frac{E\tau}{2\hbar} > 0$ we have

$$T(t) = e^{-\frac{2}{\tau}i\left(\frac{\pi}{2} + 2\pi n\right)t} e^{-\frac{2}{\tau}\ln_e\left[\frac{E\tau}{2\hbar} + \sqrt{\frac{E^2\tau^2}{4\hbar^2} - 1}\right]t} \tag{27}$$

Eq. (27) represents a decaying solution with lifetime

$$T_L = \frac{\tau}{4\ln_e\left[\frac{E\tau}{2\hbar} + \sqrt{\frac{E^2\tau^2}{4\hbar^2} - 1}\right]} \tag{28}$$

If we estimate the discrete time interval to be $\tau \approx 10^{-26}$ sec then the most energetic stable bound state appears at

$$\frac{E\tau}{2\hbar} = 1, \quad E = \frac{2\hbar}{\tau}(E = 90 GeV)$$

This is about the mass of heaviest particle discovered to date $(\omega^{\pm})$. We note from Eq. (28) that heavy particles would decay so quickly that they would be much more short lived than the shortest lived resonance known $(T \simeq 10^{-23} \text{ sec})$.

Two other applications of discrete time difference quantum theory that are worth mentioning are the resolution of the angle slow down effect in Rauch's experiment [79] and the effect on cosmological dynamics in the early universe produced by a discrete time derived equation of state.

Rauch's experiment consists of superimposing two neutron beams, one traversing a region with a magnetic field in the z direction, the other with no magnetic field, constructive interference should result when the beam with the magnetic field precesses 4π about the B field. The experimental results demonstrate that for constructive interference there is a consistently smaller value of the angle of precession (smaller than 4π) using the known flight time in the interferometer and the precession frequency $\omega = \frac{eB}{m}$. Since discrete time difference Q.M. increases ω (Eq. (12)), it can account for this difference and generate the necessary 4π [60]. When statistical mechanics is applied to particles admitting discrete time difference Q.M. we find that in the relativistic limit a stiff equation of state $P = \epsilon$ results rather than the usual $P = \frac{\epsilon}{3}$ for radiation for highly relativistic particles. This would slow the expansion rate of the universe during the radiation era and change the perturbation spectra leading to C.M.B. anisotropies [80].

The third approach to incorporating discrete time effects into quantum theory involves the following "construct", consider a quantum system with all known "Hamiltonian effects" represented by the Schrodinger time evolution of the system, if in addition there are statistical effects that cannot be put in Hamiltonian form, then we might expect the wave function to zeroeth order to be described by the Schrodinger evolution with first order statistical effects leading to modifications in the form of the wave functions. We choose to study spin systems in high magnetic fields and alter the zeroeth order wave functions by inserting Markov probabilities for coefficients of the spin function components. For composite particles this leads to both modifications of the spin polarization precession amplitude which is different for different composite structures and to possible connections between the "ordinary time" and the Markov "jump time" interval. This also provides us with an excellent probe to compositeness totally foreign to any existing techniques. We illustrate the general idea using a spin 1 gauge boson Hamiltonian composed of two similar preons ($q_1 = q_2 = -e$) [64], the Hamiltonian can be written according to Eq. (16) as

$$H = M_0 C^2 + \frac{P_1^2}{2m} + \frac{P_2^2}{2m} + \frac{e}{m}(S_{z_1} + S_{z_2})B + g(\vec{S}_1 \cdot \vec{S}_2)$$

The total wave function in a $S = 1$ triplet state that gives $\langle S_x \rangle = \hbar$ at $t = 0$ is

$$\Psi = \left[\frac{\alpha\alpha}{2}e^{-\frac{iE_+t}{\hbar}} + \frac{\beta\beta}{2}e^{-\frac{iE_-}{\hbar}t} + \frac{1}{2}(\alpha\beta + \beta\alpha)e^{-\frac{iE_0t}{\hbar}} \right] U(x_1, x_2) \qquad (29)$$

here E_+, E_-, E_0 are given by Eq. (18) with $U(x_1, x_2)$ the same for each S_z. Also $\alpha, \beta =$ spin up and spin down functions respectively. We now consider a two step Markov process for each preon, calling $P_n(+)$ the probability of spin up after n steps, $P_n(-)$ the probability of spin down after n steps we have after n steps [61]

$$P_n(+) = \frac{p}{p+q} + (1-p-q)^n \left(\frac{1}{2} - \frac{p}{p+q} \right)$$

$$P_n(-) = \frac{q}{p+q} + (1-p-q)^n \left(\frac{1}{2} - \frac{q}{p+q} \right) \qquad (30)$$

p = probability of spin flip from down to up, q = probability of flip up to down in external magnetic field $B_z = B$. We now modify Eq. (29) to read

$$\Psi = \begin{array}{l} \sqrt{P_n(+)P_n(+)}\alpha\alpha e^{-\frac{iE_+ t}{\hbar}} + \sqrt{P_n(-)P_n(-)}\beta\beta e^{-\frac{iE_- t}{\hbar}} \\ + \left(\sqrt{P_n(+)P_n(-)}\alpha\beta + \sqrt{P_n(-)P_n(+)}\beta\alpha\right) e^{-\frac{iE_0}{\hbar}t} \end{array} \tag{31}$$

here $P_n(+), P_n(-)$ are the same for each preon. If we calculate

$$\langle S_x \rangle = \Psi^+ (S_{x_1} + S_{x_2})\Psi$$

we obtain

$$\langle S_x \rangle = 2\hbar \cos \frac{e_e B}{M_\omega} t \left(\sqrt{P_n(+)^3 P_n(-)} + \sqrt{P_n(-)^3 P_n(+)} \right) \tag{32}$$

here we set $m = \frac{M_\omega}{2}, e = \frac{e_e}{2}$ (e_e = electronic charge) in Eq. (31).
In Eq. (31) we have omitted the spatial function since it integrates out in calculating $\langle S_x \rangle$ in Eq. (31). Eq. (32) demonstrates that the amplitude varies with the Markov probabilities of the individual preons. We have applied these ideas to composite gauge bosons as mentioned above, to non-composite gauge bosons [64], to composite gauge bosons with different preons [65] and to composite charged leptons [66] composed of three fermions or a fermion-boson pair. The general results of these studies indicate that small time variations of $\langle S_x \rangle$ can probe the probabilities $P_n(+), P_n(-)$ and ascertain the relationship between the discrete Markov jump time and the normal time of Minkowski space.

3. Experimental Tests

The above theoretical discussions suggest various experimental tests to discrete time Quantum theory and discrete spatial Quantum theory that for the most part help to establish limits on the discrete space and time interval rather than provide a clear test for the theory. Some of the phenomena sensitive to a discrete time interval include the diffraction of heavy particles giving rise to a central maximum (sensitive to discrete time effects) [58, 81], the shift in the spin precession frequency of precessing particles in a magnetic field [36, 81], spectral shifts in Hydrogen transitions (sensitive to discrete time effects) [43, 81], energy dependent width phenomena in particle decays [54], shifts in the x neutron spin polarization in neutron interferometry experiments [56], Rauch's experiment involving the 4π rotation of spinors [79], wave packet spreading phenomena [50] and possible cosmological spectral shifts due to cosmological variations of the discrete time interval. Discrete time quantum theory also predicts a cut-off in the mass spectrum of a group of similar particles [40] which should be a measurable as higher energies are attained in accelerator experiments. The same ideas predict that charged particles should decay in high magnetic fields even if the particle is stable without the field. In [40] a "Meissner like" effect is induced by discrete time effects for particles themselves rather than for "Couper pairs" as in the B.C.S. theory [82]. Probably the best qualitative signal of a discrete time quantum theory would be a spin precession frequency varying with B^3 in high magnetic fields (Eq. 12).

4. Conclusion

Admittingly, the above approach to discrete time Q.M. and discrete time difference Q.M. is open to an avalanche of criticism in that a Hamiltonian is assumed that emerges from continuum physics and then acts on a wave function that responds in a discrete manner. My response to this criticism is that the Hamiltonian is constructed from a knowledge of local physics and is really the only empirical information we have to work with. The wave function's response to zeroeth order would be a time derivative, but because of the "relic" fluctuations from a "smooth flowing time variable" due to the generic pre-geometric origin of time we must admit discrete time differences specified by τ. Similar ideas exist in the "theory of irreversible processes" [83] where a "correlation time exists" which is bigger than the molecular interaction time and smaller than the characteristic time for the decay of a fluctuation. Perhaps the discrete time interval we refer to is bigger than the preon- preon interaction time and of the same order of magnitude or less than the time for the decay of a fluctuation from a Minkowski background. Thus τ represents the characteristic time for the interaction of the particle with the background, the Hamiltonian inducing the interaction and all knowledge of the particle (contained in the wave function) responds within a width τ of t. In the above discussion we have seen that these ideas naturally lead to an independent probe for compositeness of particles and as an independent source of CP violation over and above that offered by local quantum field theory. In addition to the above motivation to believe in the existence of a discrete time interval, the existence of virtual wormholes in a mini-superspace model might also induce an effective discrete time interval in the space-time foam that leaves its remnants in the low energy world. [84]

5. Acknowledgement

I'd like to thank the Physics Departments at Williams College and Harvard University for the use of their facilities.

References

1. Harris, E. G.: 1993, *Am. J. of Phys.*, **61**, 1140
2. Gonzales-Diaz, P. F.: 1991, *Il Nuovo Cimento B*, **106**, 335
3. Salam, A. and Strathdee, J.: 1978, *Phys. Rev. D*, **8**, 4598
4. Klein, O.: 1926, *Z. Phys.*, **27**, 895
5. Sakharov, A. D.: 1984, *Sov. Phys. J.E.T.P.* (Eng. translation), **60**, 214
6. Wheeler, J.: 1980, *Chap. in Quantum Theory and Gravitation*, Proc. of a symposium held at Loyola Univ., New Orleans, May 23-26, 1979 (Academic, NY)
7. Wheeler, J.: 1990, *Information, Physics, Quantum (The Search for Links)* Preprint, Princeton Univ., Feb.
8. Heisenberg, W.: 1930, *Z. Phys.*, **65**, 4
9. Beck, G.: 1929, *Z. Phys.*, **53**, 675
10. Carazza, B. and Kragh, H.: 1995, *Am. J. Phys.*, **63** (7), 595
11. Wolf, C.: 1987, Unpublished
12. Finkelstein, D.: 1985, *Int. J. of Theoretical Phys.*, **27**, 473
13. Wootters, W.: 1984, *Int. J. of Theoretical Phys.*, **23**, 701
14. Evako, A. V.: 1994, *Int. J. of Theoretical Phys.*, **33**, 1553
15. Antonsen, F.: 1994, *Int. J. of Theoretical Phys.*, **33**, 1189
16. Kull A. and Treumann, R. A.: 1994, *Int. J. of Theoretical Phys.*, **34**, 435
17. Bombelli, L., Lee, J., Meyer, D. and Sorkin, R.: 1987, Preprint IA-SSNS-HEP 87/23 Inst. of Advanced Study, Princeton, NJ
18. Snyder, H. S.: 1974, *Phys. Rev.*, **71**, 38
19. t'Hooft, G.: 1984, *Phys. Rep.*, **104**, 133
20. Lee, T. D.: 1983, *Phys. Lett. B*, **122**, 218
21. Wilson, K. G.: 1974, *Phys. Rev D.*, **10**, 2445
22. Christ, N. H., Friedberg, R. and Lee, T. D.: 1982, *Nucl. Phys. B*, **210**, 310
23. Creutz, M.: 1977, *Phys. Rev. D*, **15**, 1128
24. Nielson, H. B. and Ninomiya, M.: 1981, *Nucl. Phys. B*, **185**, 20
25. Creutz, M.: 1979, *Phys. Rev. Lett.*, **43**, 533
26. Rebbi, G.: 1983, *Lattice Gauge Theories and Monte Carlo Simulation* (World Scientific Publ. Co., Singapore)
27. Wolf, C.: 1993, Submitted to *Annalen der Physik*
28. Maslov, E. M.: 1990, *Phys. Lett. A*, **151**, 47
29. Tamga, J. M., Remoissenet, M. and Pouget, J.: 1995, *Phy. Rev. Lett.*, **75**, 357
30. Wolf, C.: 1994, Unpublished
31. Wolf, C.: 1990, Proc. of Fifth International Conference on Hadronic Mechanics and Non-potential Interactions, Aug. 13-17, 1990, Univ. of N. Iowa, Cedar Falls, IA
32. Wolf, C.: 1994, *Fizika B*, **3**, 9
33. Wolf, C.: 1990, Fourth Workshop on Hadronic Mechanics and Non-Potential Interactions; 22-26 Aug. 1988, Skopje, Yugoslavia (Nova Sci. Pub., NY), p. 301
34. Caldirola, P.: 1956, *Nuovo Cimento*, **3**, 97
35. Caldirola, P.: 1976, *Lett. Nuovo Cimento*, **16**, 151
36. Wolf, C.: 1992, *Hadronic J.*, **15**, 371
37. Sinha K. P. and Sivaram, C.: 1979, *Phys. Rep.*, **51**, 3
38. Sijacki D. and Ne'eman, Y.: 1990, *Phys. Lett. B*, **247**, 571
39. Recami, E.: 1990, Personal Communication at fourth workshop on Hadronic Mechanics and Non-potential Interactions, 22-26 Aug. 1988, Skopje, Yugoslavia (Nova. Sci. Pub., NY)
40. Wolf, C.: 1994, *Il Nuovo Cimento B*, **109**, 213
41. Wolf, C.: 1987, *Phys. Lett. A*, **123**, 208
42. Wolf, C.: 1987, *Il Nuovo Cimento Note Brevi B*, **100**, 431
43. Wolf, C.: 1987, *Eur. J. of Phys.*, **10**, 197
44. Wolf, C.: 1990, *Hadronic J.*, **13**, 22
45. Wolf, C.: 1990, *Il Nuovo Cimento B*, **105**, 805

46. Wolf, C.: 1991, *Hadronic J.*, **14**, 321
47. Wolf, C.: 1993, *Annales de le Foundation Louis de Broglie*, **18**, 403
48. Wolf, C.: 1996, To appear in *Annales de la Foundation Louis de Broglie*
49. Wolf, C.: 1990, *Annales de la Foundation Louis de Broglie*, **15**, 487
50. Wolf, C.: 1989, *Hadronic J.*, **12**, 45
51. Wolf, C.: 1988, *Hadronic J.*, **11**, 227
52. Wolf, C.: 1994, Unpublished
53. Wolf, C.: 1995, *Phys. Education* (India), Apr.-June, 44
54. Wolf, C.: 1990, Proc. of fourth Workshop on Hadronic Mechanics and Non-Potential Interactions, Aug. 22-26, 1988, Skopje, Yugoslavia (Novo Sci. Pub., NY), p. 298
55. Wolf, C.: 1989, Unpublished
56. Wolf, C.: 1990, *Foundations of Phys.*, **20**, 133
57. Wolf, C.: 1990, *Annales de la Foundation Louis de Broglie*, **15**, 189
58. Wolf, C.: 1989, *Il Nuovo Cimento B Note Brevi*, **103**, 649
59. Wolf, C.: 1992, Unpublished
60. Wolf, C.: 1993, Unpublished
61. Hoel, P. G., Post, S. C. and Stone, C. J.: 1972, *Intro. to Stochastic Processes* (Houghton Mifflin Comp., Boston)
62. Nanopoulos, D. V.: 1994, CERN-TH-7620/94, CTP-TAMU-20/94, ACT - 07/94.
63. Itzykson, C. and Drouffe, J. M.: 1989, *Statistical Field Theory*, Vol. 1 (Cambridge Univ. Press, Cambridge) , p. 3
64. Wolf, C.: 1995, *Fizika B*, **4**, 167
65. Wolf, C.: 1995, Proc. of IRB Int. Workshop, Aug. 8-12, 1995, Monteroduni, Molise, Italy
66. Wolf, C.: 1996, *Fizika B*, **5**, 49
67. Wolf, C.: 1995, Unpublished
68. Beesham, A.: 1994, *Int. J. of Theoretical Phys.*, **33**, 1383
69. Beesham, A.: 1994, *Int. J. of Theoretical Phys.*, **33**, 1935
70. Hildebrand, F. B.: 1968, *Finite Difference Equations and Simulations* (Prentice Hall, Englewood Cliffs, NJ)
71. Wolf, C.: 1994, Submitted to *Indian J. of Pure and Applied Phys.*
72. Wolf, C.: 1996, To appear in *Foundations of Phys. Letters*
73. Ramsey, N. F.: 1956, *Phys. Rev.*, **103**, 20
74. Purcell, E. M. and Pound, R. V.: 1951, *Phys. Rev.*, **81**, 279
75. Paczynski, B.: 1992, *Comments on Astrophysics*, **16**, 241
76. Wolfenstein, L.: 1986, *Annual Review of Nucl. and Part. Science*, **36**, 137
77. Wolf, C.: 1994, Unpublished
78. Previtera, P. and Selleri, E.: 1992, *Phys. Lett. B*, **296**, 261
79. Rauch, H., Wolfing, A., Bauspiess. W. and Bonse, V.: 1978, *Z. Physik B*, **29**, 281
80. Wolf, C.: 1995, *Il Nuovo Cimento B*, **110**, 967
81. Wolf, C.: 1994, *Frontiers of Fundamental Physics*, Proc. of International Conference, Olympia, Greece, 27-30 Sept. 1993 (Plenum, NY, NY), p. 449
82. Bardeen, J., Cooper, L. N. and Schrieffer, J. R.: 1957, *Phys. Rev.*, **108**, 1175
83. De Groot, S. R.: 1951, *Thermodynamics of Irreversible Processes* (North Holland Pub. Comp., Amsterdam)
84. Redmount, I. H. and Waimo, S.: 1994, *Phys. Rev. D*, **49**, 5199

UNEXPLAINED EMPIRICAL RELATIONS
AMONG CERTAIN SCATTERINGS

L.W. MORROW
P.O. Box 11639
Pittsburgh, PA 15228

Abstract. When the differential cross-section data of π^+-proton scattering are compared to the data of proton-alpha scattering at momenta related by $4p_4 = p_\pi$ (where p_4 is the momentum of p-alpha scattering and p_π is the momentum of π^+-p scattering) a correlation is observed between $p_\pi = 125$ MeV/c and 1100 MeV/c. The correlation is most pronounced in the region of the lowest energy resonance of both scatterings ($p_\pi \sim 210$ MeV/c). A less pronounced correlation (in the lower energy regions) among four scatterings is observed when the differential cross-section data of π^+-p and p-alpha scatterings and the data of p-^{3}He and proton-deuteron scatterings are compared using relations $4p_4 = 3p_3 = 2p_2 = p_\pi$ (where p_3 is the p-^{3}He momentum and p_2 is the proton-deuteron momentum). The facts strongly suggest something of physical significance but no theoretical explanation is known. Indeed, an explanation seems to be beyond the scope of existing theory.

The following is a condensation of a paper published in Il Nuovo Cimento, Part A in November 1992 (Morrow 1992). The paper is concerned with unexplained empirical relations in the data sets of four elastic scatterings, with primary attention to a comparison of the differential cross-section data of π^+-p scattering and p-alpha scattering. The finding of the relations in these comparisons was accidental. No theoretical reason for considering such comparisons is known. The only case that can be made for considering such comparisons is found after the fact, from the results.

In all but one of the figures the quantity compared is $k^2 d\sigma/d\Omega$, where $k = p/\hbar$. This will be referred to as "the quantity". Only experimental data are used; the curves are used to facilitate the comparison and have no theoretical significance. The first comparison is for p-alpha and π^+-p scatterings. Both of these are spin 1/2 on spin 0. In the low energy region

Astrophysics and Space Science **244**:347-351,1996.

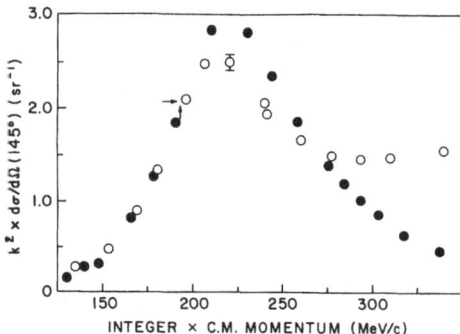

Figure 1. Comparison of the quantity at 145° for π^+-p and p-^{4}He elastic scatterings.
The abscissa is the product Np. For, π^+-p, N = 1 and for p-^{4}He, N = 4. Filled circles
are π^+-p points. Open circles are p-^{4}He points.

to be considered first, accurate cross-section measurements and good phase
shift analyses have been available for both scatterings for over 40 years.
Both scatterings have a $^2P_{3/2}$ resonance as lowest resonance. Considering
the quantity for large back angles in these resonance regions, a maximum
in the quantity for π^+-p occurs near a momentum of 220 MeV/c and in
the p-alpha the maximum occurs near 55 MeV/c. This suggests making
a comparison of the quantities in these regions using momenta related by
$p_\pi/p_4 = 4$. It is convenient to introduce a variable Np where N = 1 for
π^+-p and N = 4 for p-alpha. Figure 1 shows a comparison of the quantity
at 145°. The abscissa is Np in MeV/c. The quantity is, of course, dimensionless; sr^{-1} is reciprocal steradians. The π^+-p points are shown by filled
circles; the p-alpha points are shown by open circles. The arrows indicate a
filled circle that has been omitted to avoid overlapping an open circle. The
uncertainties in the cross-section data are mostly less than the radii of the
circles. Note that there is no scaling on the vertical axis. Thus the only constant used in this comparison is 4, the ratio of momenta. Below 200 MeV/c
the two quantities are indistinguishable. Between ~200 and ~240 MeV/c
the quantities show a general similarity. Figure 2 shows the comparison at
166°.

To get an idea of the uncertainty in the momentum ratio, comparisons
were made at ratios of 4 ± 5%, that is, at ratios of 3.8 and 4.2. The results
are shown in Figure 3. This is the only figure for which the abscissa is
not Np. The solid curve is a smooth curve through the π^+-p data with
constant = 1. The p-alpha data are shown by half-filled circles for the
constants 3.8 and 4.2.

Although the emphasis is on directly measured quantities, it is of some
interest to compare phase shifts near 200 MeV/c. For both scatterings,
only S and P wave phase shifts are of appreciable size. Figure 4 shows a

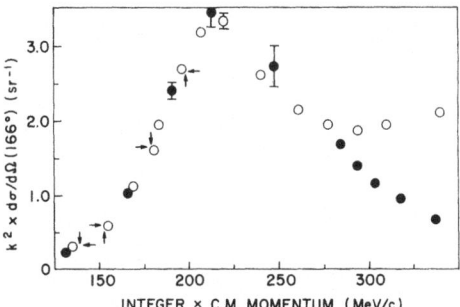

Figure 2.   Comparison of the quantity at 166° for π^+-p and p-^{4}He elastic scatterings.

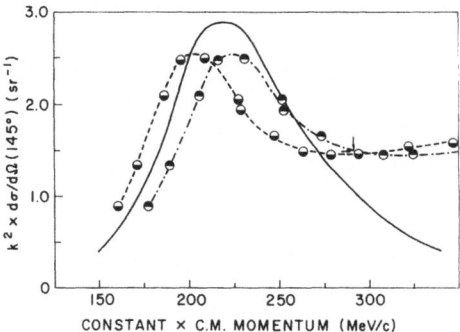

Figure 3.   Comparison of the quantity at 145° for π^+-p and p-^{4}He. Note the different variable on the abscissa. The solid curve is a smooth curve through π^+-p data with constant = 1. The p-^{4}He data are shown by half-filled circles for which the constants are 3.8 and 4.2.

comparison of the resonance phase shift, the $^2P_{3/2}$ one. It is plotted in degrees against Np. The curve is a smooth curve for p-alpha and the circles are the π^+-p points. All of the latter lie on the low side of the p-alpha curve. For the points between 175 and 225 MeV/c the actual best ratio is 3.88, 3% less than 4.

The factor 4 suggests a connection with the number of nucleons in the alpha particle. This suggests looking at the data for the scattering of protons by ^{3}He and by the hydrogen isotopes. For p-^{3}He there is one very broad peak in the low energy region. The quantities for p-^{3}He and p-alpha were compared for 17 values of the ratio p_3/p_4. These values ranged from 1/3 to 3. The largest degree of resemblance was for ratios between 1.2 and 1.6. It is plausible to select 4/3, leading to N = 3 for p-^{3}He scattering. Figure 5 shows a comparison of the quantity at 145° over approximately the same Np range considered in the earlier comparisons. The p-^{3}He points are shown by triangles and the p-alpha points by circles. For Np less than

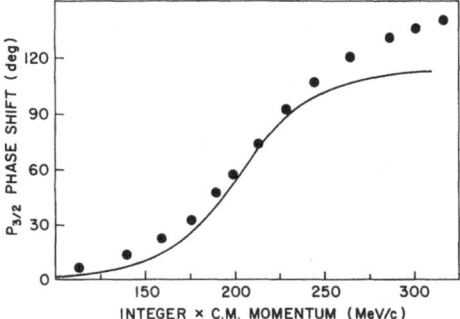

Figure 4. Comparison of the $^2P_{3/2}$ phase shift for π^+-p and p-^{4}He elastic scatterings plotted against Np. The smooth curve shows p-^{4}He scattering. The circles represent the π^+-p points.

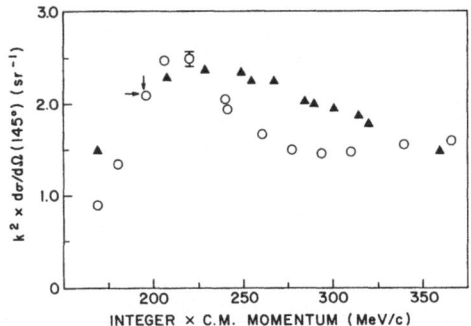

Figure 5. The quantity at 145° for p-^{3}He and p-^{4}He elastic scatterings. N = 3 for p-^{3}He and N = 4 for p-^{4}He. The p-^{3}He points are shown by triangles and the p-^{4}He points by circles.

225 MeV/c a resemblance is seen although it is not as pronounced as that seen in the comparison of π^+-p and p-alpha. A comparison of the quantities for p-deuteron and p-alpha at 15 ratios of momenta showed a weak resemblance at ratios between 1.5 and 2, suggesting N = 2 for p-D.

Comparison of a certain aspect of the asymmetry data of the four scatterings provides additional evidence for the inclusion of p-^{3}He and p-D scattering with the values of N = 3 and N = 2.

Another approach is to compare the quantities for p-alpha and π^+-p over a wider range of the variable Np. This is shown in Figure 6 where the range of Np is from 0 to 1200 MeV/c. The circles represent p-alpha points at 166°. The broken curve at the lower momenta is for π^+-p at 166°; the solid curve is for π^+-p scattering near 180°. For the present purpose the broken curve and solid curves will be considered as one curve. Before considering the comparison with the p-alpha points it is useful to look closely at this curve.

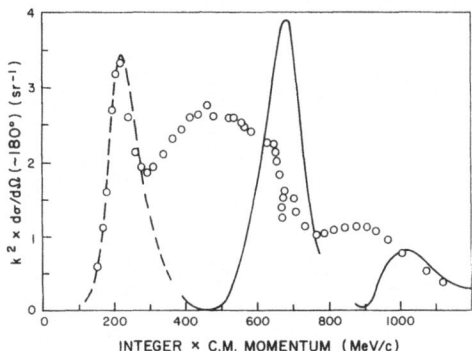

Figure 6.   Comparison of the quantity for very large back angles for π^+-p and p-^{4}He elastic scatterings. The circles represent p-^{4}He points at 166°. The broken curve shows π^+-p scattering at 166°; the solid curve shows π^+-p scattering near 180°.

Looking at the extrema one finds that the four extrema between 100 and 950 MeV/c all occur within 6 MeV/c of integral multiples of 224 MeV/c, i.e. near 224, 448, 672 and 896 MeV/c. To the best of my knowledge this regularity has not been noted in the pion-proton literature. Returning now to the comparison with p-alpha, it is evident that above the low energy peak the quantities are widely separated. However, it is found for the 3 extrema of the π^+-p curve near 448, 672 and 896 there are extrema of the opposite kind in curves through the p-alpha points at about the same locations- thus a p-alpha maximum at 461, etc. It is conceded that there are three other extrema for p-alpha in this region for which there are no corresponding extrema in π^+-p. It is also conceded that the comparison of the locations of extrema in this way is a significant change in method from the lower energy comparisons.

In conclusion, it seems likely that these facts, when considered together, form a physically significant pattern but no explanation is known. If there is indeed a physically significant pattern there must be a physical relation among these scatterings that is not at present known. I therefore believe that this matter deserves thorough study, including statistical analysis and further experimental work.

References

Morrow, L.W.: 1992, *Il Nuovo Cimento* , **105 A** , 1579

SPHERICAL ROTATION, PARTICLES AND COSMOLOGY

MARTIN KOKUS
PO Box 119
Hopewell, Penn. 16650

Abstract. There have been many models of both fundamental particles and the cosmos which have involved rotation, spin or vorticity. Most of these entail "something" spinning about a fixed axis. This paper qualitatively describes a more general rotation which shows promise in unifying the fundamental forces and explaining various cosmological phenomena.

1. Introduction

There have been many models of fundamental particles which consisted of something spinning, such as a point charge, a ring of charge, or just a piece of "space". The motion is almost always a rotation about a fixed axis. The principal drawback is that they still require postulating a "charge". Even in the case where a particle is described as a vortex or similar deformity in "space", there are still problems. First, they require equal and opposite deformities that correspond to the electrical charges because an axis can be rotated so that the direction of spin about it can correspond to any direction of spin about another axis. Second, and more subtle, is that if a piece of space is rotating continuously about an axis, there must be a discontinuity someplace between the vortex and the rest of space. Therefore, for all practical purposes, there is little difference between hypothesizing a vortex model and hypothesizing a space-particle dichotomy.

There have also been hypotheses that the Universe consists of a rotational hierarchy. As the planets revolve about the Sun and the Sun revolves about the Galaxy, the Galaxy revolves about the Local Group and the Local Group revolves about a larger ensemble. This goes on until we have the whole Universe rotating. There is no evidence in the near universe that contradicts this model, but it is usually ruled out because the distant universe

Astrophysics and Space Science **244**:353-356,1996.

is very isotropic. This would be impossible if our Universe were rotating about an axis.

The problem with all of these models can be resolved with the introduction of spherical rotation, a three dimensional vortex.

2. Spherical Rotation

The term "spherical rotation" was coined by Battey-Pratt and Racey (1980, hereafter BPR) who wrote the definitive paper on the subject. See also Wolff (1990) and Speiser (1964). BPR wanted to differentiate spherical rotation from cylindrical rotation. The easiest way to understand the distinction is to imagine a ball imbedded in a block of jello, and to suppose that we have rigged up some magnets so we can rotate the ball without otherwise disturbing the jello. Also suppose that the outside of the block of jello is anchored so that it will not move. If we just keep rotating the ball about an axis (cylindrical rotation) we will quickly build up enough stress to tear the jello all around the ball. But, if we first rotate the ball 180° about any axis perpendicular to the spin axis, we can then spin the ball indefinitely without the stress accumulating. After every two rotations the configuration of the jello returns to its original state. This is the simplest vortex that satisfies the criteria that the medium be continuous (or that the curvature and torsion of space be smooth and well behaved) everywhere including the particle itself.

Battey-Pratt and Racey (1980) have shown that this is the simplest model which has SU(2) as its motion group. SU(2) is the group used to describe leptons (electrons, positrons and neutrinos).

We can then take the entire block of jello and enclose it in a larger ball which we imbed in another block of jello. We can then take the larger ball and rotate it 180° about an axis perpendicular to the spin axis and then it can spin continuously without any stress accumulating in the jello. We can continue building a hierarchy of balls of jello to have a variety of three dimensional rotations within a continuous medium provided that the original rotation is an integer multiple of 180°. It is very interesting to note that the criteria that the medium be continuous throughout the particle, in and of itself, introduces quantization into the problem.

A good way to develop an intuitive understanding of the deformity in the medium is to pick up a ball, rotate it 180° about a horizontal axis without changing your grip on it, and then rotate it continuously about the vertical axis without letting go of it (it may take several attempts to find the right starting configuration). The arm would represent a filament from a "particle" out to infinity; or, if we are interpreting this as a deformation of space, the arm would represent a coordinate. From inspection it is obvious that the total torsion (twist) along a filament oscillates between $\pm\pi$ and the total curvature (the net rotation of a vector parallel to the filament between the particle and the external boundary) also oscillates between $\pm\pi$ with the maximums being 90° out of phase.

By formulating an equation for the phase of these waves and identifying corresponding terms with the Klein-Gordon equation for an elementary particle, BPR have shown that the particles rest mass is given by $m = h\omega/c^2$, where ω is the spin frequency. The undulations in space away from the inner area would then correspond to the Compton wave with wavelength h/mc. The core region would spin with an angular velocity ω. BPR have shown that while this frequency slows to $\omega(1-v^2/c^2)^{1/2}$ when observed from a moving reference frame, the frequency, and therefore the mass, inferred from the rate at which the observer transverses the de Broglie wave, would increase by a factor of $(1 - v^2/c^2)^{-1/2}$

What BPR failed to note is that the spherical rotation model provides an additional way for the observed mass of a particle to differ from its mass in its rest frame. If the particle possesses angular momentum in the plane of its continuous motion then its spin rate and therefore its observed mass would be altered. While this effect should be insignificant if the spin orientation is random, if we exist in a rotating hierarchy where spin orientation is correlated with the angular momentum of the body, the observed mass of a fundamental particle in a distant rotating body would be less than its mass in the observers frame. This would result in an apparent redshift that is not a result of recessional velocity.

3. Pairs of Particles

If we look at pairs of particles, there should be four possible orientations of the original 180° rotation and spin that would produce deformations where the medium is always continuous. They would correspond to all of the combinations of unlike and like charge, and parallel and antiparallel spin. The orientation of the original 180° rotation corresponds to the spin. Following the analogy, it can be seen by inspection that standing waves in the orbital plane can only exist between particles of opposite charge, and standing waves along the spin axis can only exist between opposite poles. BPR have shown that all four possibilities satisfy the Dirac equation for particles.

4. Ensembles of Particles and Particle Creation

In the original BPR model they allowed the jello to be anchored at infinity. In reality it is anchored in nearby particles. When these particles rotate about each other, they perturb the principal field in the original model, creating various unique phenomena. One phenomenon that is exclusive to ensembles of spherically rotating particles involves an oscillating torsion and curvature that occurs between pairs of rotating particles which are themselves orbiting about another pair. The oscillation is the result of the original pair reducing (or increasing) their rotation rate by one spin each orbit compared to the surrounding particles. It is the intention of future research to see if this effect is related to the weak force, gyromagnetism, geomagnetic pole switching or gravitation.

A section of a standing wave, one wavelength long, between two spherically rotating particles of opposite charge will, at certain phases, be identical to a pair of particles at certain stages of their rotation. If a very strong shear or vorticity (which could be the result of a gravitational field produced by a large dense mass) is superimposed on the standing wave, we could get a particle pair.

5. Summary

While still far from a rigorous complete theory, spherical rotation has identified a common thread through many current problems in physics.

References

Battey-Pratt, E. P. and Racey, T. J.: 1980, *Int. J. of Theoret. Phys.*, **19**, 437-475
Speiser, D.: 1964, *Theory of Compact Lie Groups and Some Applications to Elementary Particle Physics*, ed. Gursey, F., (Gordon and Breach, New York)
Wolff, M.: 1990, *Exploring the Physics of the Unknown Universe*, (Technotron Press, Manhattan Beach, Calif.), pp 76-81

A NON-LINEAR MODEL FOR TIME

CARL L. DEVITO
Mathematics Department
University of Arizona
Tucson, AZ 85721

Abstract. Some recent astronomical observations [4], and a number of experiments in particle physics, seem to cast doubt on the validity of the standard linear model for time. These results raise two (at least) questions: (1) If time is not a linear continuum (i.e., if the standard model is incorrect), then why does this model work so well in so many areas of science? (2) Whatever the "true nature" of time is, are there any advantages, to science, in replacing the standard model with a more complicated one? The purpose of this paper is to present a non-linear, mathematical model for time that enables us to answer question (1), and to partially answer question (2). Our discussion of question (2) is incomplete, but our results are intriguing. They also show promise of helping us understand some of the observations mentioned above. A rather natural extension of our model brings it into close contact with one that has been used in quantum theory ("Stochastically branching spacetime topology" by Roy Douglas [2]). These points of contact will also be discussed.

1. Introduction

We want to construct a non-linear, mathematical model for time that conforms to our intuitive temporal sense, and also conforms to the way in which time is treated in physics. If time really is non-linear, as some of our colleagues are suggesting, then there should be no logical obstructions to developing a model of this kind. However, our strong intuitive feeling that time is linear, and the great success of the linear model in physics, seem to be telling us that any successful non-linear model must incorporate, in a fundamental way, a linear component. This is true of the model we shall present here.

Astrophysics and Space Science **244**:357-369,1996.

It seems clear to the author that time is unlike anything else in our experience. Having said that, we still find it impossible to avoid using geometric language, and geometric analogies in our discussion. One must be careful, however, not to carry these analogies too far.

Geometry is the study of space. No reference to time is needed for that study. Here we shall try to model time without any reference to space. Our treatment will be axiomatic and our presentation will be somewhat like the usual expositions of elementary geometry. We have tried, however, to arrange our presentation so that readers with no interest in the mathematical details can skip these details and still follow the development. As in elementary geometry our theorems are sometimes "intuitively true". However, deducing them from a set of axioms (there are only five) adds both clarity and precision to our discussion. It is also hoped that, just as the study of Euclid's axioms (particularly his fifth postulate) led to the more sophisticated geometries of the 19th and 20th centuries, a study of our axioms might lead to more sophisticated temporal models.

Finally, I would like to thank my colleagues William Tifft, John Cocke, and Tony Pitucco for their insights into the nature of time, and for many fascinating discussions about physics and cosmology.

2. First Axioms

In our experience some events happen before (or earlier than) certain other events and hence these events can be ordered in time – we can say which one happened "first." However, some events cannot be so ordered – perhaps they happen at the "same time" or they are so remote that we cannot tell which occurred first. Thus our sense of time seems to give us a way of ordering some, but not all, of the events we experience. This kind of situation is well-known in mathematics and has led to the formal concept of a partial ordering. This will be discussed further as we proceed [see, 2, p. 23]. Our first axiom is that time itself can be partially ordered.

Axiom 1. We shall assume that time is an infinite set I whose elements will be called instants. Furthermore, we shall assume that the elements of I can be partially ordered.

We shall use the symbol $\leq$ for our partial ordering. If x, y are instants, $x \leq y$ may be read "x is no later than y." When $x \leq y$ and $x \neq y$ we shall write $x < y$; read "x is earlier than y."

Since I is only partially ordered by $\leq$ there may be instants x and y for which neither $x \leq y$ nor $y \leq x$ is true. Such instants are said to be incomparable.

If x, y, z are in I, and any two of these instants are comparable, then we shall say that y is between x and z when $x \leq y$ and $y \leq z$ or when $z \leq y$ and $y \leq x$.

Mathematical Remark 1. A partial ordering $\leq$ on a set I is a relation on the elements of I that has the three following properties: (i) $x \leq x$ for all $x \in I$; (ii) $x \leq y$ and $y \leq x$ together imply $x = y$; (iii) $x \leq y$ and $y \leq z$ together imply $x \leq z$. The pair $(I, \leq)$ is then called a partially ordered set.

An example is this: Let $\mathbf{N} = \{1, 2, 3, 4, 5, 6, \cdots\}$ and for a, b in $\mathbf{N}$ define $a\alpha b$ to mean a divides (or divides evenly into) b. Then $(\mathbf{N}, \alpha)$ is easily seen to be a partially ordered set (i.e., α has properties (i), (ii) and (iii)). Note that $2\alpha 6$ and $3\alpha 6$, but 2 and 3 are incomparable. Furthermore, note that 6 is between 2 and 12 while 7 is not.

Given two instants x and y which can be ordered we intuitively feel that we can measure the "length of time" between them. Moreover, this "length" is zero when, and only when, these instants are the same. This intuition is reflected in our next axiom: We shall assume that there is a function that, given two comparable instants x and y, assigns a number to the "length of time" between these instants.

Axiom 2. There is a function dur (short for "duration") that assigns to each pair of comparable instants x, y a non-negative, real number. Furthermore, dur $(x, y) = $ dur (y, x) for all comparable pairs x, y and dur $(x, y) = 0$, if, and only if, $x = y$.

The duration function gives us a kind of temporal "distance" between comparable instants. In our usual linear model we have, given an instant x and a time interval (say 5 minutes), two instants that are 5 minutes distant from x. One in the past, and one in the future. In a non-linear model there may be more than two instants that are 5 minutes distant from x. We want to keep this possibility open. We might also mention that even in the linear model, there is an instant, the big-bang, for which there is only one instant that is five minutes away.

Axiom 3. Given any instant x and any positive, real number ρ we assume that there is at least one instant y such that dur $(x, y) = \rho$.

We turn now to the first question raised in the abstract of this paper. We may paraphrase it as follows: Why does the linear model for time work so well in science, and have such great intuitive appeal? Our answer is that, in the set of instants I, there are certain subsets, that we shall call time tracks, which possess all of the properties usually attributed to the linear model. In this view each object in the universe "lives" on its own time track, and this track provides a kind of temporal "base line" to which all external events may be referred. There are two differences between this view and the standard one. First, in the standard model there is one time track for all objects while here different objects have different time tracks. Secondly, in our model, objects may change time tracks. This last property will be discussed more fully in a subsequent paragraph.

Our personal time tracks seem to have three properties. First, any two instants can be ordered; we cannot order all the events we observe, but we seem to be able to do so for any two instants we experience. Secondly, the duration functions seems to be additive; i.e., the time from breakfast to dinner is the time from breakfast to lunch plus the time from lunch to dinner. Finally, given an instant and a time length (again let us say 5 minutes), there are exactly two instants that are 5 minutes away from the given instant. We formalize these properties in our next definition.

Definition 1. A non-empty subset T of I will be called a time track if it has the three following properties: (a) Any two instants on T can be compared; (b) If x, y, z are on T and y is between x and z, then dur (x, z) = dur $(x, y)+$ dur (y, z); (c) Given y on T and a positive, real number ρ there are exactly two instants x, z on T such that dur $(x, y) = \rho$ and dur $(y, z) = \rho$.

In the standard, linear model time (i.e., I) is itself a time track. Here we shall assume:

Axiom 4. The set I is not a time track, but there is at least one time track in I.

We shall now prove that any time track has all the properties usually assigned to the time axis in physics. Readers may skip the proofs without any loss in continuity.

Theorem 1. Let T be a time track, let y be an instant on T and let ρ be a fixed, positive, real number. If x, z are the two instants on T such that dur $(x, y) = \rho$ and dur $(y, z) = \rho$, then y is between x and z.

Proof. Since any two elements of T can be compared we may suppose that $x \leq z$. We shall assume that $y < x$ and deduce a contradiction: By part (b) of definition 1 we may write dur (y, z) = dur $(y, x)+$ dur (x, z). However, we also have dur $(y, z) = \rho =$ dur (y, x). These two equations give dur $(x, z) = 0$, and hence $x = z$. We conclude that there can be only one instant on T that is later than y and ρ time units from it. A similar argument shows that there can be only one instant on T that is earlier than y and ρ time units from it.

The sequence of events breakfast-lunch-dinner can be thought of as happening on a given day. However, we can also imagine that they are translated through time by, say, 24 hours. This translation does not affect their order, nor does it affect the duration between any two of them. Theorem 1 enables us to formalize this process of "translation through time".

Definition 2. Let T be a time track and let ρ be a fixed real number. We define the translation function $t\rho$, mapping T to T as follows: (a) $t_0(x) = x$ for all $x \in T$; (b) If $\rho > 0, t_\rho(x) = y$ where y is the unique instant on T such that $x < y$ and dur $(x, y) = \rho$; (c) If $\rho < 0, t_\rho(x) = z$ where z is the unique instant on T such that $z < x$ and dur $(z, x) = |\rho|$.

Lemma 1. For each fixed, real number ρ the function t_ρ maps T onto T and is a one-to-one function.

Proof. This is obvious when $\rho = 0$. If $\rho \neq 0$ we need only note that $t_\rho \circ t_{-\rho})(x) = t\rho[t_{-\rho}(x)] = x$ for all $x \in T$.

Theorem 2. Let T be a time track and let ρ be a fixed real number. Then the function t_ρ from T to itself preserves order and duration. More explicitly, for any x, y on T we have: (a) $x \leq y$ if, and only if, $t_\rho(x) \leq t_\rho(y)$; (b) dur $(x, y) =$ dur $[t_\rho(x), t_\rho(y)]$.

Proof. (a) It is clear that we may assume that the given instants x, y of T satisfy $x < y$. We argue by contradiction.

Suppose $t_\rho(y) < t_\rho(x)$. Then we must have $x < y \leq t_\rho(y) < t_\rho(x)$. But then we may write $\rho =$ dur $[x, t_\rho(x)] =$ dur $(x, y)+$ dur $[y, t_\rho(y)]+$ dur $[t_\rho(y), t_\rho(x)]$. Since dur $[y, t_\rho(y)] = \rho$, this reduces to $0 =$ dur $(x, y)+$ dur $[t_\rho(x), t_\rho(y)]$. Now both terms on the right-hand side of this equation are non-negative and so they must both be zero. Thus dur $(x, y) = 0$ giving us $x = y$. This contradiction shows that $t_\rho(x) \leq t_\rho(y)$. But since t_ρ is one-to-one (by Lemma 1) we see that $t_\rho(x) < t_\rho(y)$.

Suppose now that we know that $t_\rho(x) \leq t_\rho(y)$. Then by the argument just given we must have $t_{-\rho}[t_\rho(x)] \leq t_{-\rho}[t_\rho(y)]$. Thus $x \leq y$ as claimed.

(b) The result is certainly true when $\rho = 0$. So let us suppose that $\rho > 0$ and let x, y in $T, x \leq y$ be given. By (a) we must have $t_\rho(x) \leq t_\rho(y)$. We shall consider two cases: (i) $x < t_\rho(x) \leq y$; (ii) $x \leq y \leq t_\rho(x)$.

In case (i) we may write: (*) dur $(x, y) =$ dur $[x, t_\rho(x)]+$ dur $[t_\rho(x), y] = \rho+$ dur $[t_\rho(x), y]$.

Now in this case $t_\rho(x) \leq y$ and clearly $y < t_\rho(y)$ hence (**) dur $[t_\rho(x), t_\rho(y)] =$ dur $[t_\rho(x), y]+$ dur $[y, t_\rho(y)] =$ dur $[t_\rho(x), y] + \rho$. Equations (*) and (**) have the same right-hand side. Thus dur $(x, y) =$ dur $[t_\rho(x), t_\rho(y)]$ and the theorem is proved in this case.

In case (ii) we may write: (*) dur $[x, t_\rho(x)] = \rho =$ dur $(x, y)+$ dur $[y, t_\rho(x)]$. Since we have $x \leq y \leq t_\rho(x) \leq t_\rho(y)$ in this case, we may write (**) dur $[y, t_\rho(y)] = \rho =$ dur $[y, t_\rho(x)]+$ dur $[t_\rho(x), t_\rho(y)]$. Combining equations (*) and (**) we get dur $(x, y)+$ dur $[y, t_\rho(x)] =$ dur $[y, t_\rho(x)]+$ dur $[t_\rho(x), t_\rho(y)]$. Thus dur $(x, y) =$ dur $[t_\rho(x), t_\rho(y)]$ and we are done.

At this point we have proved (b) for $t_\rho, \rho \geq 0$. Suppose $\rho > 0$ and consider $t_{-\rho}$. Given x, y in T we have, by the paragraph just above, dur

$[t_{-\rho}(x), t_{-\rho}(y)] = \text{dur } [t_\rho(t_{-\rho}(x)), t_\rho(t_\rho(y))]$. But this last term is just dur (x, y).

The translation function enables us to transform any time track T into the numerical time axis of mathematical physics. We simply choose a point x on T and let $\varphi(0) = x, \varphi(\rho) = t_\rho(x)$ for each real number ρ. The function φ so defined gives us a one-to-one correspondence between T and the set of real numbers.

The time track of the earth has a number of natural periodic functions defined on it. There is the day-night cycle, the lunar cycle, and the yearly seasonal cycle. Each of these give us a natural "unit" of time. On an arbitrary time track T we can define a function f from T into the reals to be periodic if there is some $\rho > 0$ for which $f[t_\rho(x)] = f(x)$ for all x in T. Much of the theory of periodic functions can be done in this setting. We can also develop an integration theory for functions on T. In particular, the theory of the Riemann-Stieltjes [3, p. 532] integral fits particularly well into this context.

We end this paragraph with a Theorem that tells us that time tracks are as "large" as possible.

Theorem 3. If T_1, T_2 are time tracks, and if $T_1 \subseteq T_2$, then $T_1 = T_2$.

Proof. We argue by contradiction. Suppose that there is an element x_2 in T_2 that is not in T_1. Choose an arbitrary element x_1 in T_1, note that x_1, x_2 are both in T_2 (because $T_1 \subseteq T_2$) and hence that dur (x_1, x_2) has meaning. Moreover, this number, call it ρ, must be positive.

By part (c) of definition 1, we can find exactly two instants y_1, z_1, in T_1 such that dur $(y_1, x_1) = \rho = \text{dur } (z_1, x_1)$. Furthermore, by Theorem 1, $y_1 < x_1 < z_1$; it is only a matter of notation which instant we call y_1 and which z_1 and so we can name these instants in such a way that the stated chain of inequalities is true. We now consider the four instants x_1, y_1, z_1, and x_2 on T_2. Since they all lie on a time track, they are pairwise comparable. We distinguish four cases.

(i) Suppose that $y_1 < x_2 < x_1 < z_1$. Then we may write dur $(y_1, x_1) = $ dur $(y_1, x_2) + $ dur (x_2, x_1). Since dur $(x_1, y_1) = \rho = $ dur (x_1, x_2), we see that dur $(y_1, x_2) = 0$. But then $y_1 = x_2$ which is a contradiction because x_2 is not on T_1 while y_1 is.

(ii) Suppose $y_1 < x_1 < x_2 < z_1$. The argument here is similar to that given in (i).

(iii) Consider the case $y_1 < x_1 < z_1 < x_2$. We may write dur $(x_1, x_2) = $ dur $(x_1, z_1) + $ dur (z_1, x_2). Again, since dur $(x_1, x_2) = \rho = $ dur (x_1, z_1), we see that dur $(z_1, x_2) = 0$. This gives $z_1 = x_2$ which contradicts the fact that x_2 is not in T_1.

(iv) The case $x_2 < y_1 < x_1 < z_1$ is similar to case (iii).

3. Mathematical Tools

Sequences provide a convenient tool for working out the mathematical properties of time tracks. We deduce all the results we shall need from the four axioms stated above. Readers with no interest in these detail may skip this entire paragraph.

We recall that a sequence in a time track T is simply a function φ from the natural numbers (i.e., the set $\mathbf{N} = \{1, 2, 3, 4, 5, 6, \cdots\}$) into T. It is customary to set $x_1 = \varphi(1), x_2 = \varphi(2), \cdots, x_n = \varphi(n), \cdots$ and speak of the sequence $\{x_n\}_{n=1}^{\infty}$ in T.

Definition 1. Let T be a time track and let $\{x_n\}_{n=1}^{\infty}$ be a sequence in T. We shall say that $\{x_n\}_{n=1}^{\infty}$ is convergent to the instant $x_0 \in T$, and we shall write $\lim_{n \to \infty} x_n = x_0$, if for any given $\epsilon > 0$ we can find an integer N such that dur $(x_n, x_0) < \epsilon$ whenever $n \geq N$.

It turns out that we can prove all the theorems we need by working only with monotonic sequences. These are sequences $\{x_n\}_{n=1}^{\infty}$ on a time track T that are either increasing (meaning $x_n \leq x_{n+1}$ for all n) or are decreasing (meaning $x_n \geq x_{n+1}$ for all n). These type sequences seem to make the most sense physically.

Theorem 1. Let $\{x_n\}_{n=1}^{\infty}$ be a monotonic sequence on a time track T, and suppose that $\lim x_n = x_0$ on T. Then if $\{x_n\}$ is increasing, $x_n \leq x_0$ for all n, while if $\{x_n\}$ is decreasing $x_0 \leq x_n$ for all n. Furthermore, if $\{x_n\}$ is increasing (resp. decreasing) and if $x_n \leq y$ all n, (resp. $y \leq x_n$ all n) for some y on T, then $\lim x_n = x_0 \leq y$ (resp. $y \leq x_0 = \lim x_n$).

Proof. We give the proof only for the case of an increasing sequence. We want to show that $x_n \leq x_0$ for all n. Suppose $x_0 < x_p$ for some p, hence $x_0 < x_n$ all $n \geq p$. Then $x_0 < x_p \leq x_{p+m}$ for $m \geq 0$ so we may write dur $(x_0, x_{p+m}) =$ dur $(x_0, x_p)+$ dur $(x_p, x_{p+m}) \geq$ dur $(x_0, x_p) > 0$. Since this holds for all $m \geq 0$, we see that lim dur (x_0, x_{p+m}) cannot be a zero which is a contradiction:

Next, suppose that $x_n \leq y < x_0$ for all n. Then dur $(x_n, x_0) =$ dur $(x_n, y)+$ dur $(y, x_0) \geq$ dur $(y, x_0) > 0$. Again we have a contradiction.

Corollary 1. Let $\{x_n\}_{n=1}^{\infty}$ be a monotonic sequence on the time track T. If this sequence converges to an instant on T, this limit is unique.

Proof. Suppose that $\{x_n\}$ is an increasing sequence and that lim x_n is equal to both x_0 and y_0 on T. Then $x_n \leq x_0$ for all n and $\lim x_n = y_0$, so $y_0 \leq x_0$. However, $x_n \leq y_0$ for all n and $\lim x_n = x_0$. Hence $x_0 \leq y_0$.

Theorem 2. Let $\{x_n\}_{n=1}^{\infty}$ be a monotonic sequence on the time track T and suppose $\lim_{n \to \infty} x_n = x_0$ on T. Then: (a) For any fixed ρ, $\lim_{n \to \infty} t_\rho(x_n) =$

$t_p(x_0)$; (b) For any fixed y_0 on T the sequence of real numbers $\{$dur $(x_n, y_0)\}$ converges to the number dur (x_0, y_0).

Proof. Part (a) follows immediately from Theorem 2 of §2. To prove (b) just note that if $y_0 = x_0$ there is nothing to show; for then lim dur (x_n, x_0) and dur (x_0, x_0) are both zero.

Suppose $x_n \leq x_0 < y_0$. Then dur $(x_n, y_0) =$ dur $(x_n, x_0)+$ dur (x_0, y_0). Since lim dur $(x_n, x_0) = 0$ we see that lim dur (x_n, y_0) exists and is equal to dur (x_0, y_0).

Now suppose that $y_0 < x_0$. Then $y_0 < x_n$ for all n beyond some integer; say all $n \geq p$. Hence dur $(y_0, x_n)+$ dur $(x_n, x_0) =$ dur (y_0, x_0). For all $n \geq p$. Again the result follows by taking limits.

Finally, we have

Theorem 3. Let $\{x_n\}_{n=1}^{\infty}$ be an increasing sequence on a time track T, and suppose that for some y on T we have $x_n \leq y$ for all n. Then there is an x_0 on T such that $\lim x_n = x_0$.

Proof. Since $x_1 \leq x_2 \leq x_3 \leq \cdots \leq x_n \leq \cdots \leq y$ we must have dur $(x_1, y) \geq$ dur $(x_2, y) \geq \cdots \geq$ dur $(x_n, y) \geq \cdots \geq 0$. It follows from a well-known property of the real numbers that lim dur $(x_n, y) = \beta$ for some real number β. If $\beta = 0$ we see that $\{x_n\}$ converges to y, and we are done. Hence we may assume that $\beta > 0$. By Theorem 1 of §2 there is exactly one instant x_0 on T such that $x_0 < y$ and dur $(x_0, y) = \beta$. Since dur $(x_n, y) \geq \beta =$ dur (x_0, y) we see that $x_n \leq x_0$ for all n. Thus dur $(x_n, y) =$ dur $(x_n, x_0)+$ dur (x_0, y).

It follows that $\beta =$ lim dur $(x_n, y) =$ lim dur $(x_n, x_0)+$ dur (x_0, y). Hence the lim dur (x_n, x_0) exists and is zero. Thus $\{x_n\}$ converges to x_0.

Let T be a time track and let S be a non-empty subset of T. An element z of T is said to be an upper bound for S (resp. lower bound) if $y \leq z$ for all y in $S (z \leq y)$ all y in S. An element z_0 of T is said to be the least upper bound for S if: (i) z_0 is an upper bound for S. (ii) Whenever z is an upper bound for $S, z_0 \leq z$. These ideas are discussed in [3, p. 92] and in [1, p. 31].

Corollary 1. Let T be a time track and let S be a non-empty subset of T. If S has an upper bound on T, then it has a least upper bound on T.

Proof. Let x_0 be an upper bound for S. If x_0 is in S, then it is clearly the least upper bound for this set. Thus we may suppose that x_0 is not in S. Then the set $\{$dur$(s, x_0)|s$ in $S\}$ consists of positive real numbers, and hence has a greatest lower bound γ. It is easy to see that we can find a sequence $\{s_n\}$ of instants in S such that dur $(s_n, x_0) \geq$ dur (s_{n+1}, x_0) for all n, and lim dur $(s_n, x_0) = \gamma$. It is clear that $s_n \leq s_{n+1}$ for all n, and hence that the sequence $\{s_n\}$ has a limit s_0; moreover, $s_0 \leq x_0$. Also note that dur $(s_0, x_0) = \gamma$.

If s is in S, then dur $(s, x_0) \geq \gamma =$ dur (s_0, x_0) and so $s \leq s_0$. Thus s_0 is an upper bound for S. Suppose that t is on T and $t < s_0$. Then, for some n, $t < s_n \leq s_0$. But this says t is not an upper bound for S and so s_0 is the least upper bound for this set.

4. Time: A Non-linear View

We are going to try, in this paragraph, to understand how instants on different time tracks might be related. As a guide to our intuition we shall rely on some simple facts from astronomy. Imagine a star α with time track T_α, and the earth with its time track T_e. We can observe α tonight or tomorrow night or five years from tonight. Using astronomical records we can view α last night or five years ago. So at any instant on the time track of earth we can "interact" with the time track of α, and we can do this in a time symmetric way. This leads to our first definition:

Definition 1. Let T be a time track and let y be an instant that is not on T. We shall say that T is symmetric with respect to y if: (i) There is an instant x on T such that is before y and whenever x on T is before y there is a z on T after y such that dur $(x, y) =$ dur (y, z); (ii) There is an instant z on T that is after y, and whenever z on T is after y there is an x on T that is before y such that dur $(x, y) =$ dur (y, z).

Men first walked on the moon, say, 30 years ago and comet Shoemaker-Levy struck Jupiter, say 3 months ago. Hence men walked on the moon before the comet hit Jupiter. On the other hand, if an eclipse of the sun is to happen in 500 years while an eclipse of the moon will happen in 50 years, then the lunar eclipse will happen before the solar one. These observations motivate our next definition.

Definition 2. Let T be a time track and let y be an instant not on T. We shall say that y distinguishes past instants on T if whenever x_1, x_2 on T are before y, then $x_1 < x_2$ if, and only if, dur $(x_2, y) <$ dur (x_1, y). We shall say that y distinguishes future instants on T if whenever z_1, z_2 on T are after y, then $z_1 < z_2$ if, and only if, dur $(y, z_1) <$ dur (y, z_2).

We immediately have:

Theorem 1. Let T be a time track, y an instant not on T, and suppose that y distinguishes both past and future instants on T. If x, z are on T and if dur $(x, y) =$ dur (y, z), then y is between x and z.

Let us return now to our star α with time track T_α. Suppose that α is five light years from earth. A light signal sent from earth at the instant x on T_e will take five years to arrive at α-at, say, the instant y on T_α. So dur $(x, y) = 5$ years. If the signal is reflected back to earth it will arrive here at

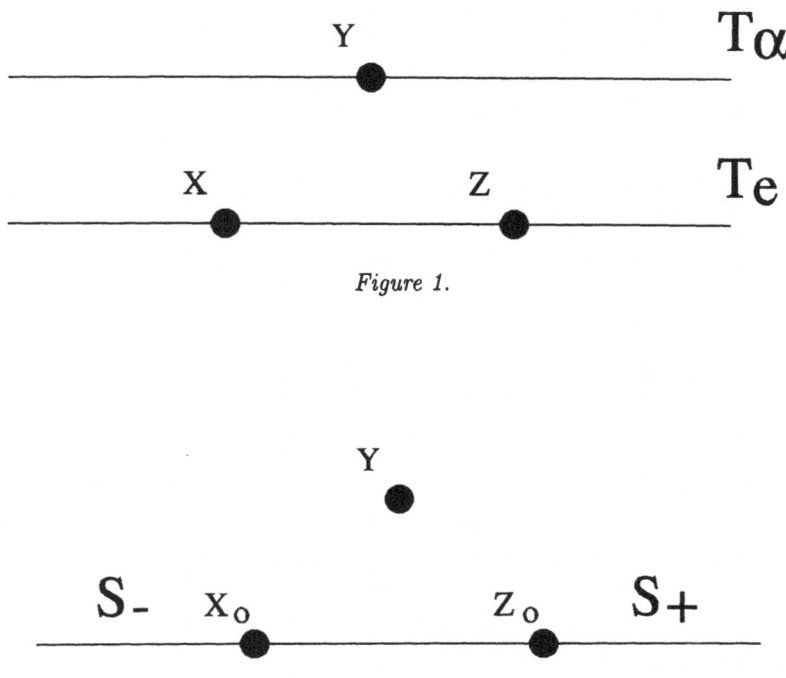

Figure 1.

Figure 2.

an instant z on T_e. Again dur (y, z) 5 years. An observer on α at instant y can have no knowledge of what is happening on earth between the instants x and z. This "gap" is caused by the distance between earth and α, and by the fact that the velocity of light is finite. However we can show that such a temporal gap must exist as a consequence of the relationship between the instant y and the time track T_e. We need make no reference to distance or to velocity. Let us explain what we can show. The proof is given below (see Theorem 2).

Let T be a time track and let y be an instant not on T. Suppose that T is symmetric with respect to y and that y distinguishes both past and future instants on T. Then S_-, the set of instants on T before y (i.e., $S_- = \{x \in T | x < y\}$) is not empty. It has a least upper bound x_0 on T, and x_0 is as close to y as points on T before y can get. More precisely: dur $(x_0, y) = \inf\{\text{dur } (x, y) | x \in S_-\}$.

Similarly, the set S_+ of instants on T after y (i.e., $S_+ = \{z \in T | y < z\}$) is not empty. It has a greatest lower bound z_0 and dur $(y, z_0) = \inf\{\text{dur } (y, z) | z \in S_+\}$.

It also follows that $x_0 < y < z_0$, that dur $(x_0, y) = $ dur (y, z_0) and that any

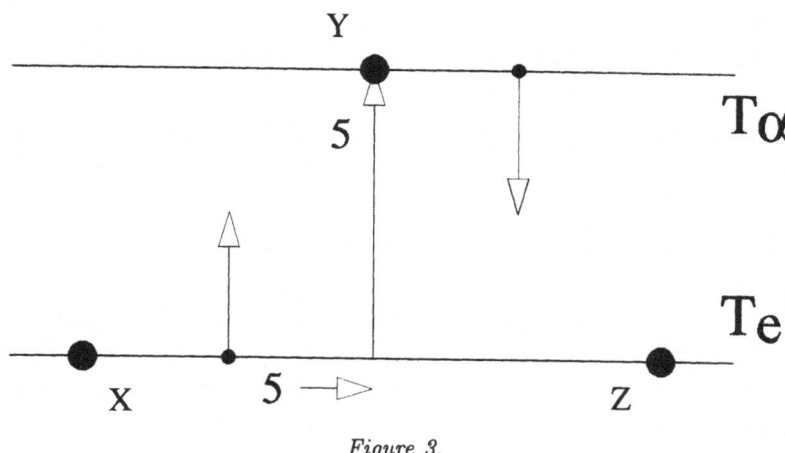

Figure 3.

instant on T between x_0 and z_0 is incomparable to y – so here we have our information gap. We stress again that all of this follows from the temporal relationship between y and T as spelled out in definitions 1 and 2.

In Figure 1 we have dur $(x, y) = 5$ years $=$ dur (y, z). We also know that the duration between the instant we sent the signal, x, to the instant we received its reflection, z, is 10 years. Thus the "triangle" x, y, z in the figure seems to "collapse" to a line. But this "triangle paradox", which seems to be forcing a linear model of time, is based on a mis-interpretation of the diagram. The correct interpretation is given in Figure 3. It is based on the fact that light is traveling from earth to α and back and dur (x, y) is a measure of this travel time.

Referring now to Figure 2, we have dur $(x_0, y) =$ dur (y, z_0) but we cannot say how these quantities relate to dur (x_0, z_0) since we have made no assumptions about the global properties (i.e., the properties "off" a time track) of the function dur. We shall make one such assumption here. It involves sequences on time tracks.

Axiom 5. Let T be a time track and let $\{x_n\}$ be a monotonic sequence of instants on T that is convergent to $x_0 \in T$. Then: (a) If y in I is comparable to every x_n, then y is comparable to x_0 and dur $(x_0, y) = \lim$ dur (x_n, y); (b) If $\{x_n\}$ is increasing and $x_n < y$ for all n, then $x_0 \le y$; (b) If $\{x_n\}$ is decreasing and $x_n > y$ for all n, then $x_0 \ge y$.

Note that when y is on T statement (a) was proved [Theorem 3.2 (b)] and statements (b) and (c) were also proved [Theorem 3.1]. The force of the axiom is that these statements remain true even when the instant y is not on T.

Theorem 2. Let T be a time track and let y be an instant that is not on T. Suppose that T is symmetric with respect to y and that y distinguishes both past and future instants on T. Then the set of instants on T that are before y, call it S_-, is not empty and has a least upper bound x_0 and the set of instants on T that are after y, call it S_+, is not empty and has a greatest lower bound. Moreover, $\operatorname{dur}(x_0, y) = \inf\{\operatorname{dur}(x, y) | x \text{ in } S_-\} = \inf\{\operatorname{dur}(y, z) | z \text{ in } S_+\} = \operatorname{dur}(y, z)$.

Proof. The existence of an instant y not on T is assured by Axiom 4. Since T is symmetric with respect to y the sets S_- and S_+ are non-empty. They are bounded above and below respectively and so x_0 and z_0 exist. Moreover, the sets $\{\operatorname{dur}(x, y) | x \text{ in } S_-\}$ and $\{\operatorname{dur}(y, z) | z \text{ in } S_+\}$ are the same. Consider the first of these. Since it consists of positive, real numbers it has a greatest lower bound β. We can choose $\{x_n\}$, a sequence of instants in S_- such that $\operatorname{dur}(x_n, y) \geq \operatorname{dur}(x_{n+1}, y)$ for all n, and $\lim \operatorname{dur}(x_n, y) = \beta$. But y distinguishes past instants on T and so we must have $x_1 \leq x_2 \leq x_3 \leq \cdots \leq x_n \leq \cdots$. Moreover, every x_n is in S_- and no $x_n \leq x_0$ for all n. It follows from Theorem 3.3 that this sequence has a limit y_1 in T and that $y_1 \leq y$ (Axiom 5). Furthermore, $\beta = \lim \operatorname{dur}(x_n, y)$ must equal $\operatorname{dur}(y_1, y)$ also by Axiom 5. One last observation $y_1 \leq x_0$ by Theorem 3.1.

Now x_0 is the least upper bound for the set S_- and so we can find a sequence $\{x_n'\}$ of instants in S_- that is increasing and converges to x_0; just note that $t_{-\frac{1}{n}}(x_0)$ is not an upper bound for $S_-, n = 1, 2, 3, \cdots$. But $x_n' < y$ for all n so $x_0 < y$. Thus we now have $y_1 \leq x_0 < y$ and since y distinguishes past instants on T, $\operatorname{dur}(x_0, y) \leq \operatorname{dur}(y_1, y)$. Finally, let us show that $\operatorname{dur}(y_1, y) \leq \operatorname{dur}(x_0, y)$. Since $\operatorname{dur}(y_1, y) = \inf\{\operatorname{dur}(x, y) | x \text{ in } S_-\}$ and every x_n' is in S_- we have $\operatorname{dur}(y_1, y) \leq \operatorname{dur}(x_n', y)$ for all n. But since $\lim \operatorname{dur}(x_n', y) = \operatorname{dur}(x_0, y)$ we must also have $\operatorname{dur}(y_1, y) \leq \operatorname{dur}(x_0, y)$. Thus $\operatorname{dur}(y_1, y) = \operatorname{dur}(x_0, y)$, and we are done.

The other half of the theorem is proved similarly.

Corollary 1. The instants x_0 and z_0 on T are not equal.

Proof. We have $x_0 < y$ and $y < z_0$.

Corollary 2. If w is an instant on T and if $x_0 < w < z_0$, then w and y are incomparable.

Proof. If w is comparable to y then either w is in S_-, in which case $w \leq x_0$, or w is in S_+, in which case $z_0 \leq w$. Either of these statements leads to a contradiction.

Let us continue our discussion of the situation described in Theorem 2. Suppose that the instant y is also on a time track, call it T', and suppose that $\operatorname{dur}(x_0, y) < \operatorname{dur}(x_0, z_0)$. We may choose x_1 on T such that $x_0 < x_1$ and $\operatorname{dur}(x_0, x_1) = \operatorname{dur}(x_0, y)$. It follows then, from Corollary 2, that all instants between x_0 and x_1 are incomparable to y. Hence we may define a spliced, or second order, time track $S = S(T, T')$ as follows:

(a) S consists of all instants on T before x_1, and all instants on T' that are at or after y; i.e., $S = \{x \text{ on } T | x < x_1\} \cup \{w \text{ on } T' | y \leq w\}$.

(b) Given two instants s_1, s_2 on S we may order these instants by:

(i) Using the order relation on T when both s_1 and s_2 are on T; (ii) Using the order relation on T' when both s_1 and s_2 are on T'; (iii) Defining s_1 to be earlier than $s_2(s_1 < s_2)$ when s_1 is on T and s_2 is on T'. (c) The duration between s_1, s_2, dur (s_1, s_2), is well-defined in cases (i) and (ii) of (b). In case (iii) we set dur $(s_1, s_2) = $ dur $(s_1, x_1) + $ dur (y, s_2).

It is easy to check that the triple $(S, \leq, \text{dur})$ defined in (a), (b) and (c) has all of the properties of a time track (Definition 2.1).

Note that any object that "lives" on S starts life on T and continues on this time track until the instant x_1. At this instant it "jumps" to T'; specifically, to the instant y. Moreover, dur $(x_0, x_1) = $ dur (x_0, y). This may help explain the redshift variability observed in connection with some remote galaxies [4]. In the standard view any change in the redshift of such a galaxy will take decades, or even centuries, to reach us. However, if a photon leaving such a galaxy remains on the time track of that galaxy, as our model suggests, until it reaches our telescopes, then any change in the redshift of the parent galaxy might be immediately "felt" by the traveling photon-because these two objects, though far apart in space, are still connected in time! So perhaps the extra-galactic redshift is a temporal phenomena of some kind.

Finally, let us mention that in a recent paper on time in quantum mechanics [2, p. 173] a model was introduced that is very much like the following: We have, say, three time tracks T, T' and T'', we have y on T', y'' on T'' and we suppose that x_0 and z_0 on T satisfy the conclusions of Theorem 2 for both y and y''. Furthermore, we suppose that dur $(x_0, y) < $ dur (x_0, z_0) and that dur $(x_0, y'') < $ dur (x_0, z_0). This insures that the spliced time tracks $S(T, T')$ and $S(T, T'')$ are defined. If we imagine that T is the time track of a simple "universe" and an experiment is performed at the instant x_0, then the universe will go to the track $S(T, T')$ or $S(T, T'')$ depending on the outcome of the experiment. Since probabilities can be assigned to the outcomes of the experiment we have probabilities assigned to the possible time tracks. We refer to [2] for more details.

References

1. Royden, H. L.: 1968, *Real Analysis* 2nd Edition, Macmillan Pub. Co., New York
2. Savitt, S. F. (editor): 1995, *Time's Arrows Today*, Cambridge University Press
3. Taylor, A. E.: 1959, *Advanced Calculus*, Ginn and Company
4. Tifft, W. G.: 1991, *ApJ*, **382**, 396

DEFINING SPACETIME

W. M. STUCKEY
Dept. of Physics
Elizabethtown College
Elizabethtown, PA 17022

Abstract. It has become apparent that our intuitive notions of space and time are inadequate for developing a theory of quantum gravity. It is perhaps worthwhile to understand where our macroscopically-developed spatial instinct is implicit in the concept of manifold, and to consider alternative methods for defining (vis-a-vis explicating) space and time. A simple example for generating a *topos* over a fundamental set is provided to illustrate the potential basis of such a definition.

"Many theoretical physicists believe that to progress much further it will be necessary to rethink our understanding of space-time. The 4D manifold structure of general relativity does not seem adequate to describe the kind of processes which are implicated in quantum gravity [1]."

1. Space, Time, and *Topos*

If one accepts as in set theory multiplicity if and only if discernibility (*per* Leibniz's principle of the Identity of Indiscernibles [2], then one may understand space to be that which allows for multiplicity in the face of apparent identity. For example, concerning the points of a spacetime manifold Nerlich writes [3], "This is not to say that the points are different from one another in themselves. They are indistinguishable, each taken by itself." When visualizing a manifold, we commonly consider its elements to be intrinsically indistinguishable. Since we recognize multiplicity in manifold, there must be some implicit method for discrimination. That which we tacitly rely upon to visualize the otherwise indistinguishable elements of a manifold is space – one element is here and another there.

Astrophysics and Space Science **244**:371-374,1996.

In contrast to space, one might describe time as a construct of apparent identity from multiplicity. For example, I identify the computer I'm using now with the computer I used when I began writing this manuscript. There is a discrimination (computer now as opposed to then) which implies multiplicity, yet I understand that I have ownership of but one computer.

In what follows, space and time are considered varieties of *topos* in a preliminary attempt at constructing a definition of spacetime based in set theory. Specifically, a foundation for constructing potential spatio-temporal relationships between proper subsets of any fundamental, well-defined set is introduced. The goal here is to avoid contaminating the definition *a priori* with our macroscopically-developed spatial instinct, as is the case with manifold.

2. Spatiality in Manifold

Our reliance upon space for *visualizing* a manifold is a matter of convenience not necessity, as we may understand our manifold to be constructed from a well-defined set. Nonetheless, our intuitive concept of spatiality is implicit in the construct of manifold. Refer to Wald's definition of an "n-dimensional, C^∞, real manifold" as "a set together with a collection of subsets O_i which satisfy: (1) Each $p \in M$ is a member of at least one O_i, (2) For each i, $\exists$ a one-to-one, onto map F_i: $O_i \rightarrow U_i$ where U_i is an open subset of $\Re^N$, (3) If for any two subsets O_i and O_k we have $O_i \cap O_k \neq \emptyset$, then $F_k \circ F_i^{-1}$ is infinitely continuously differentiable; and the open subsets of $\Re^N$ are expressible as the union of open balls [4]". Items (2) and (3) impose the *topos* implicit in the concept of open balls of $\Re^N$ on each O_i of our manifold.

But the *topos* of an open ball in $\Re^N$ is clearly in accord with our macroscopically-developed spatial instinct. As Finkelstein has stated [5], "The idea of infinitesimal locality presupposes that the world is a manifold." And that the human brain relies subconsciously on this notion of spatiality is evident in part by the continuous, smooth visual perceptions constructed from coarsely granulated, disjoint collections of optic sensations [6].

It is therefore not surprising that manifold would provide an excellent model of spacetime in a classical theory such as general relativity, yet fail to furnish a spatio-temporal platform for quantum gravity. In the following simple example, we attempt to "rethink our understanding of space-time" by resorting to mathematical definition. Specifically, we introduce a basis for the definition of spacetime via the explicit construct of a *topos* over a fundamental set.

3. *Exempli Gratia*

In constructing his quantum network dynamics, Finkelstein eschews the notion of set complaining [7], "Sets do nothing, they simply are." But for the set over which we will define space and time, the elements aren't even are. Indeed, we have no state of being verb to describe this set, because all such descriptions are inherently temporal. In order to properly define time, we must consider the fundamental set nontemporally.

Similarly the elements of this set cannot, without introducing further mathematical structure, be said to exist "here or there." (When dealing with the integers for example, 3 cannot be said *a priori* to lie "between" 2 and 4.) Nor can the elements be said to possess any spatio-temporal extent (as in superstring theory). It may be that given the mathematical structure required to define spacetime, the elements will have spatio-temporal relationships one to another. Or it may turn out that spatio-temporal relationships obtain only between proper subsets of order greater than unity. But for the set alone, the concepts of space and time are meaningless.

It is also true that the elements of the fundamental set are neither kinematic nor dynamic entities such as velocity, momentum, or energy, since these concepts are less fundamental than those of space and time which we seek to define. Rather, the elements of this fundamental set are truly undefinable, since definition implies an explanation in terms of more fundamental concepts. Therefore, the assumption herein is that sets and mappings are meaningful in the absence of space and time.

As an example of how one might attempt to define spacetime via a mathematical structure over a fundamental set, we introduce the following algorithm. Choose a set and investigate those mappings from the set to itself which give rise to an ordered partition of the set. Then construct rules by which these mappings establish spatio-temporal relationships between proper subsets in the cells (subcells). To complete any such effort, one would have to produce dynamics via an appropriate definition of substance.

Consider the case $(1,2,3,4,5,6,7,8,9,10) \longrightarrow (6,6,7,7,8,9,9,10,10,9)$. An ordered partition which arises naturally from this mapping is $[(1,2,3,4,5),$ $(6,7,8), (9,10)]$. The first cell is composed of those elements which do not lie in the range. The second cell contains those elements which are the range of the first cell, etc. To partition a set completely in this fashion, one necessarily considers only those maps for which the range of the last cell is precisely its domain. As an "initial/boundary condition" choose a subcell composition for the first cell. Then, establish a rule in concert with the mapping by which subcells can be identified from cell to adjacent cell. With $(1,2,3)$ and $(4,5)$ as initial subcells, a simple identification would be (6) with $(1,2,3)$, (8) with $(4,5)$, and (7) being shared. To complete the identification of subcells in this fashion, (9) is identified with (6) and (10) with (8), these having no shared elements.

Elements which are shared by two or more subcells in this identification process might provide the basis for defining spatial relationships. And patterns among collections of identified subcells might serve to define units of proper time. Thus, the ordered partition serves as a basis for the arrow of time, but does not suggest a unique global spatio-temporal foliation. Such a formalism defines both apparent identity in the face of multiplicity and a concept of "here and there" once identity is established.

Again, we point out that 3 does not lie *a priori* "between" 2 and 4 in the foundational set of the above example. Indeed, it may be replaced with 300 and the above analysis still obtains. In this regard, there is no special set underlying a fundamental theory of physics. Note also there are many sets and maps which might give rise to equivalent physics. In this regard, the mathematics of fundamental theoretical physics is that of combinatorics as has been suggested by Penrose *inter alios*.

4. Conclusion

While the algorithm *supra* is cursory and debutante, it illustrates the potential for a post modern physics. Accordingly, theoretical physics is the process of defining space, time, and substance in terms of mappings over a well-defined set of undefinable elements. As suggested by Rosen [8] we obtain a mathematical "glimpse of a deeper level of reality than that dealt with by physics" which is "fundamentally and predominantly nonspatial and nontemporal." As a reductionistic enterprise, theoretical physics cannot rightly terminate short of such a Platonic base.

References

1. Gibbs, P.: 1995, *Small Scale Structure of Space-Time: A Bibliographical Review*, hep-th/9506171, p. 2
2. Copi, I.: 1954, *Symbolic Logic*, (The MacMillan Company, New York), p. 149
3. Nerlich, G.: 1994, *The Shape of Space*, 2nd ed, (Cambridge Univ Press, Cambridge), p. 98
4. Wald, R. M.: 1984, *General Relativity*, (University of Chicago Press, Chicago), p. 12
5. Finkelstein, D.: 1991, *The Philosophy of Vacuum* (Clarendon Press, Oxford), eds. Saunders, S. and Brown, H. R., p. 254
6. Ratner, H. (director): 1989, "Sensation and Perception," from *Psychology, The Study of Human Behavior*, produced by Coastline Community College
7. Finkelstein, D. and Hallidy, W. H.: 1991, *Intl. J. Theoretical Physics*, **30**, 481
8. Rosen, J.: 1994, *Physics Essays*, **7**, 335-340

SOME ELEMENTARY GEOMETRIC ASPECTS IN EXTENDING THE DIMENSION OF THE SPACE OF INSTANTS

A. P. PITUCCO
Department of Physics
Pima County Community College West Campus
Tucson, Arizona 85709

Abstract.

A local geometric construction is proposed on the *partially ordered* set of instants $\mathcal{I}$. A totally ordered subset $\mathcal{C}(\mathcal{I}) \subset \mathcal{I}$ is assumed to have 3-dimensional affine coordinate structure, *without* a specified metric, called the τ-space of $\mathcal{C}(\mathcal{I})$. Guided by a strong analogy with analytical mechanics the *T-configuration space* (θ, τ^α), θ a real parameter, is constructed whereupon the usual Hamilton-Jacobi theory establishes a simple geometrical construction, viz., the complete figure from the calculus of variations. The duration function, $\text{dur}:\mathcal{C}(\mathcal{I}) \to \mathbf{R}$ is associated with *temporally equidistant* hypersurfaces through which pass a congruence of extremal curves to the fundamental integral.

1. Introduction

In a recent paper by DeVito, (DeVito, 1995), it has been suggested that time, because of its association with the real line, has escaped a more general local topological inquiry. As such, the usual distinction between space and time is evident with space being endowed with a rich geometric and topologic structure. However, when considering the configuration space of a particle in a dynamical setting, *viz.*, $(t, x^i), i = 1, \ldots, n$, time is treated simply as an additional coordinate of space. In this respect, both space and time have been regarded as geometrical objects of the same kind. As such, a likely extension is to consider time as a multi-dimensional geometric quantity as one does naturally when considering the structure of space alone. This is neither a new idea nor one ignored, since multiple dimensions of time

Astrophysics and Space Science **244**:375-385,1996.

have been discussed previously by Cole (1980), Cole and Starr (1990), Lehto (1990), and Tifft (1996). Although some interesting physical consequences have arisen in particle decay, relativity, and, cosmology this approach has not been problem free. Certainly more questions have been raised than answered as to the physical meaning of extending the dimension of time. Cole introduced a 3-d modification to the temporal component in the usual metric used in general relativity for the static spherically symmetric case. The resulting solution predicted an unsatisfactory Mercury perihelion advance 7/3 times that predicted in usual 4-d GR. (Cole 1980).

In an attempt to address these problems it has been suggested (DeVito 1995) that a mathematical model of time be developed, *ab initio*, without a component of space, while still preserving the usual properties attributed to our familiar concept of, what we now refer to as, *observable time*. To this end DeVito has introduced a *partially ordered set* $\mathcal{I}$ consisting of elements $i, j, k, \ldots$ called *instants* with no immediate or apparent connection to our observable time.

Together with the partial ordering $\leq$ on $\mathcal{I}$, we say that i and j are *comparable* if $i \leq j$ or $j \leq i$ and we let

$$C(\mathcal{I}) = \{(i,j) \in I \times \mathcal{I} | \ i \text{ and } j \text{ are comparable}\}.$$

Also, defined on $C(\mathcal{I})$ is a map to the non-negative reals $\mathbf{R}_+$ called the "duration" function denoted by $\mathrm{dur}(i,j)$ such that:

$$\text{a) } \mathrm{dur}(i,j) = 0 \text{ iff } i = j; \tag{1}$$
$$\text{b) } \mathrm{dur}(i,j) = \mathrm{dur}(j,i) \text{ for any } (i,j) \text{in } C(\mathcal{I}).$$

Remark. i) We note that the duration function, being defined for all comparable i, j in $\mathcal{I}$, is globally defined and that no distinction is made between (i,j) and (j,i).

Essential to this theory is the definition of a *time track* on $\mathcal{I}$:

Definition 1. A non-empty set $C \subseteq \mathcal{I}$ is a *time track* on $\mathcal{I}$ if:

a) $\leq$ is a total ordering on C;
b) If $i, j, k \in \mathcal{I}$ with j between i and k, then $\mathrm{dur}(i,k) = \mathrm{dur}(i,j) + \mathrm{dur}(j,k)$;
c) For any fixed $i \in \mathcal{I}$ and any fixed $\rho \in \mathbf{R}_+$ there are exactly two *distinct* $x, y \in C$ such that $\mathrm{dur}(i,x) = \rho = \mathrm{dur}(i,y)$.

In this paper we adhere closely to this structure, however; in addition to some notational modifications we do impose a significant additional assumption on $\mathcal{I}$, namely, that the set $\mathcal{I}$ may be considered within the context of an n-dimensional space $\mathbf{R}^n$. However, at this early stage it must be emphasized that a metric on this space has not been specified nor should one be presumed.

Earlier attempts at extending such a coordinate structure to time while preserving the standard Euclidean metric have met with considerable difficulty. As mentioned earlier, in the context of general relativity, Cole (1980) introduced a metric for the static spherically symmetric case that included three dimensions of time which was given as

$$ds^2 = -a(r)dr^2 - r^2(d\theta^2 + \sin^2\theta\,d\psi^2) + b(r)[dt^1]^2 + (dt^2)^2 + (dt^3)^2].$$

This, however, resulted in predicting an orbital perihelion shift, for the case of Mercury, over twice that predicted using 4-d GR. This approach immediately suggests two problems, firstly, the incorporation of such a temporal decomposition into the usual space-time manifold structure; and secondly, the assumption that *observable time* can be expressed with the usual Euclidean metric, i.e., $dt^2 = (dt^1)^2 + (dt^2)^2 + (dt^3)^2$.

Thus, our initial assumption, *viz.* to consider $\mathcal{I}$ apart from space is a necessary one, and as such $\mathcal{I}$ does not immediately inherit those properties one normally attributes to space in various physical settings. In this regard we introduce the next section.

2. Mathematical Foundations

I. Preliminaries. We begin our discussion with the following assumptions and definitions. Let the sets $\mathcal{I}$ and $C(\mathcal{I})$ be given and defined as above. These sets are said to consist, respectively, of *points* called *instants* $P, Q, R, . \in \mathcal{I}$, and, *comparable points* $(P, Q) \in C(\mathcal{I})$ with dur : $C(\mathcal{I}) \to \mathbf{R}+$ as defined by (1). We note that, by virtue of equation (1a,b) and definition 1, the duration function does not satisfy the triangle inequality for points R between P and Q and, as such, is not a metric on the set $\mathcal{I}$. Therefore, $\mathcal{I}$ is not *metrizable* by dur(P, Q) and is not endowed with the *usual* topology. Because of this and the desire to keep to a minimum the number of assumptions imposed on the structure of $\mathcal{I}$, we are hesitant, at this early stage, to introduce any other topologies on $\mathcal{I}$. However; it is immediately recognized that such considerations are necessary, in the sequel, if a *meaningful* theory is to be unearthed here, particularly as it relates to physics. Of course the desired approach is to map open sets of $\mathcal{I}$ homeomorphically to open sets of $\mathbf{R}^n$. However, at this early stage we do not assume such a structure, and we note that by virtue of definition (1), a *time track* has the structure of the real line $\mathbf{R}$, and prefer to begin our discussion with a *local theory* based on the following assumptions:

1. A point P in the set of instants may be considered a point in *observable time* with no connection to *observable space*.

2. The set $\mathcal{C}(\mathcal{I})$ has a local three dimensional affine coordinate structure called the τ-*space* which we denote by τ.

3. There exists a 1-1 map by which $P \in \mathcal{C}(\mathcal{I})$ is mapped to $\tau, \phi : \mathcal{C}(\mathcal{I}) \to$ τ:

$$\phi(P) = (\tau^\alpha), \alpha, \beta, \ldots = 1, 2, 3. \tag{2}$$

Definition 2. Any set which consists of the above elements is said to have a τ-*Space Structure.*

To introduce a geometrical picture to the τ-*Space Structure* we introduce a parameter, $\theta \in \mathbf{R}$, and consider the $(\alpha + 1)$-dimensional space $\mathbf{R}^{\alpha+1}$ called the *T-configuration space* denoted as (θ, τ^α).

Remark. No attempt is made here to disguise this as being anything other than analogous to the usual construction in classical analytical mechanics wherein (t, x^α) is considered the configuration space of a particle through which the particles motion is described. In fact, we are strongly guided by such an analogy in our attempts at discovering the physics, if any, of extending the dimension of $\mathcal{I}$. However, we must be reminded that although the construction is similar, there is no space component to the *T-configuration space* and, more importantly, no metric is specified on $\mathbf{R}^{\alpha+1}$ at this early stage.

Guided by this mechanical analogue, we wish to consider a geometric model based upon the Hamilton-Jacobi theory, wherein, consistent with the above construction, a 'complete figure' (Rund 1973) in the *T-configuration space* is introduced. To this end, we require the introduction of a function on $\mathbf{R}^{\alpha+1}$ which is analogous to a Lagrangian, from which is constructed a Hamiltonian in the usual way; however it must be emphasized that an *ad hoc* approach in forming such a construction, by analogy alone, is immediately suspect in establishing a meaningful theory. Thus, we introduce a general theorem (see appendix) (Osgood 1946).

Theorem 1. Let $f(\xi^i), i, j, k, \ldots = 1, \ldots, n$ be an arbitrary C^2 function on $\mathbf{R}^n$ such that

$$\partial(f_{,1}, \ldots f_{,n})/\partial(\xi^1, \ldots \xi^n) \neq 0 \tag{3}$$
$$\text{where } f_{,j} = \partial f/\partial \xi^j \tag{4}$$

Let T denote the transformation

$$y_j = f_{,j} \quad j = 1, \ldots, n \tag{5}$$

now, assuming the summation convention holds, if the function $h(y^j)$ is constructed, as:

$$h(y_j) = \xi^k y_k - f(\xi^j) \tag{6}$$

where $\xi^j = \xi^j(y_k)$ is determined from T^{-1} (by virtue of (3)), then T^{-1} is given by:

$$\xi^j = \partial h/\partial y_j = h_{,j} \tag{7}$$
$$\text{with } \partial(h_1, \ldots, h_n)/\partial(y_1, \ldots, y_n) \neq 0. \tag{8}$$

To summarize we write (6) as:

$$f(\xi^j) + h(y_j) = \xi^k y_k \tag{9}$$

where ξ^j and y_j are given, respectively, by (5) and (7) with their respective Hessian determinates non-zero. As will be immediately evident, a useful corollary (see Appendix) now follows:

Cor. If f and h are defined as in the theorem, and, if each depends on a parameter λ, then

$$\partial f/\partial\lambda + \partial h/\partial\lambda = 0. \tag{10}$$

Remarks: 1. It is immediate from the proof of the corollary that (10) holds in the case of any finite number of parameters, λ^k.

2. The above construction is a general result, and, as such is not necessarily linked to a mechanical system, however; if $f(\lambda^k, \xi^j)$ is associated with a given *Lagrangian* $L(t, x^j, x'^j)$, defined on $\mathbf{R}^{2n+1}$ where $x^j = x^j(t)$, with $x'^j \equiv dx^j/dt = \xi^j$ with t and x^j regarded as parameters, then theorem 1 and its corollary prove Hamilton's Equations when the curves x^j satisfy the Euler-Lagrange equation $E_j(L) = 0$, $d/dt\{\partial L/\partial x'^j\} = \partial L/\partial x^j$.

For, suppose we set $y_j \equiv p_j = \partial L/\partial x'$, as in (5), and construct the function $h(y_j)$ according to the prescription given by (6) we then obtain, noting condition (3),

$$h(y_j) \equiv H(t, x^j, p_j) = -L(t, x^j, x'^j(p_k)) + p_j x'^j(p_k),$$

The desired result now follows, by using (7) with $\xi_j = x'^j$, followed by (10), with $E_j(L) = 0$:

$$x'^j = dx^j/dt = \partial H/\partial p_j$$
$$dp_j/dt = \partial L/\partial x^j = -\partial H/\partial x^j.$$

In a similar fashion theorem 1 and its corollary may be used to prove the converse.

3. We note that the underlying theory lies solely within the context of the Calculus of Variations and that the functions $x^j(t)$ are thus regarded as defining a curve γ connecting two *fixed* points p and q in the configuration space, $\mathbf{R}^{n+1}$, of a particle in the variables (t, x^j, x'^j):

$$\gamma : x^j = x^j(t), j \quad j = 1, \ldots, n. \tag{11}$$

Also, the n functions $x^j = x^j(t)$, possessing derivatives,

$$x'^j = dx^j/dt \tag{12}$$

satisfying $E_j(L) = 0$ are such that, for a given C^2 Lagragian L, the fundamental integral I, taken between p and q,

$$I = \int_p^q L(t, x^j, x'^j) dt \tag{13}$$

assumes an extreme value as compared with other functions $\bar{x}^j = \bar{x}^j(t)$ which coincide at the endpoints.

II. The T-Configuration Space. The point of view we wish to adopt here is one motivated by the above remarks and, as such, we wish to regard the *T-configuration space*, (θ, τ^α), (or *T-Space*) in much the same way. To this end, we shall regard θ as a single parameter and write $\tau^\alpha = \tau^\alpha(\theta)$ as defining a curve γ:

$$\gamma : \tau^\alpha = \tau^\alpha(\theta) \tag{14}$$

in the $\mathbf{R}^{\alpha+1}$ *T-configuration space* of the variables (θ, τ^α).

It is recognized that θ in this setting is simply being regarded in the same manner in which "time", t, has been regarded previously. However; with no physical significance assigned to θ, there is a fundamental difference. It is introduced here, in this parametrical setting, in the desire to extract the physical significance (if any) of multi-dimensional time. Thus, let the two *fixed* points p, q with respective coordinates $(\theta_1, \tau_1^\alpha)$, $(\theta_2, \tau_2^\alpha)$ be the respective *fixed* endpoints on γ. Also, the quantities $\tau'^\alpha \equiv d\tau^\alpha/d\theta$ are interpreted as the components of the *tangent vector* $(1, \tau'^\alpha)$ of γ. Let us now suppose that we are given a C^2 function, $\mathcal{L}(\theta, \tau^\alpha, \tau'^\alpha)$, in the $2\alpha + 1$ arguments $(\theta, \tau^\alpha, \tau'^\alpha)$, which we identify with the function $f(\xi^j)$ in theorem 1, with θ and τ^α regarded as parameters.

We further assume that it satisfies condition (3), viz.,

$$\partial(\mathcal{L}_{,1}, \ldots, \mathcal{L}_{,n})/\partial(\tau'^1, \ldots \tau'^n) \equiv \det(\partial^2 \mathcal{L}/\partial\tau'^i \partial\tau'^j) \neq 0. \tag{15}$$

Thus, in accordance with the prescription of the theorem, with $\xi^\alpha \equiv \tau'^\alpha$, and $y_\alpha \equiv p_\alpha$ we define the transformation T as:

$$p_\alpha = \partial\mathcal{L}/\partial\tau'^\alpha \tag{16}$$

so that T^{-1}, by (7), becomes

$$\tau'^\alpha = \partial\mathcal{H}/\partial p_\alpha \tag{17}$$

where, because of (6), $h(y_j) \equiv \mathcal{H}(p_\alpha) = \tau'^\beta p_\beta - \mathcal{L}(\tau'^\alpha)$, which we express as

$$\mathcal{L}(\tau'^\alpha) + \mathcal{H}(p_\alpha) = \tau'^\beta p_\beta. \tag{18}$$

Remarks: 1. By virtue of (15), we can solve for the τ'^j as functions of $(\theta, \tau^\alpha, p_\alpha)$, and write

$$\tau'^\alpha = \varphi(\theta, \tau^\alpha, p_\alpha). \tag{19}$$

Thus, we now express $\mathcal{H}$ in all of its arguments

$$\mathcal{H}(\theta, \tau^\alpha, p_\alpha) = \tau'^\beta p_\beta - \mathcal{L}(\theta, \tau^\alpha, \varphi(\theta, \tau^\alpha, p_\alpha)), \tag{20}$$

and by (8) we have

$$\partial(\mathcal{H}_1, \ldots, \mathcal{H}_n)/\partial(p_1, \ldots, p_n) = \det(\partial^2 \mathcal{H}/\partial p_\alpha \partial p_\beta) \neq 0. \tag{21}$$

2. In our continued attempt to maintain a strong analogy with analytical mechanics we call $\mathcal{L}(\theta, \tau^\alpha, \tau'^\alpha)$ the *T-Space Lagrangian*, and $\mathcal{H}(\theta, \tau^\alpha, p_\alpha)$ the associated *T-Space Hamiltonian*, written in terms of p_α, which we call the *generalized T-Space momentum*. (These definitions are for identification purposes only and obviously no physical meaning is implied nor should be inferred here.) Also, we denote the *Euler-Lagrange* equation on *T*-Space as

$$\mathcal{E}_\alpha(\mathcal{L}) \equiv d/d\theta\{\partial \mathcal{L}/\partial \tau'^\alpha\} - \partial \mathcal{L}/\partial \tau^\alpha \tag{22}$$

so that $\mathcal{E}_\alpha(\mathcal{L}) = 0$ constitutes the necessary condition to be satisfied by the curves $\tau^\alpha = \tau^\alpha(\theta)$ in order that the fundamental integral

$$I = \int_p^q \mathcal{L}(\theta, \tau^j, \tau'^j) d\theta \tag{23}$$

be extremalized.

3. The function $\mathcal{H}$, in terms of $(\theta, \tau^\alpha, p_\alpha)$, is thus said to be derived from the function $\mathcal{L}$, in terms of $(\theta, \tau^\alpha, \tau'^\alpha)$, by virtue of a *T-Legendre transformation* of these variables.

We may summarize the results of theorem 1, viz., equations (7) and (10) taken with (19), in the form of the following lemmas:

Lemma 1. The *T*-Hamiltonian function is of class C^2 and satisfies the identities:

$$\tau'^\alpha = \varphi(\theta, \tau^\beta, p_\beta) = \partial \mathcal{H}(\theta, \tau^\beta, p_\beta)/\partial p_\alpha \tag{24}$$

$$\partial \mathcal{H}/\partial \tau^\alpha = -\partial \mathcal{L}/\partial \tau^\alpha \tag{25}$$

$$\partial \mathcal{H}/\partial \theta^\alpha = -\partial \mathcal{L}/\partial \theta^\alpha \tag{26}$$

Lemma 2. If $\gamma : \tau^\alpha = \tau^\alpha(\theta)$ in T-Space satisfies $\mathcal{E}_\alpha(\mathcal{L}) = \prime$ then

$$d\tau^\alpha/d\theta = \partial \mathcal{H}/\partial p_\alpha \tag{27}$$

$$dp_\alpha/d\theta = -\partial \mathcal{H}/\partial \tau^\alpha. \tag{28}$$

III. The Hamilton-Jacobi Equation on T-Space. Let us next consider an *arbitrary* one-parameter family of hypersurfaces on the T-Space of the variables (θ, τ^α) in terms of the parameter χ characterized by the C^2 function $S(\theta, \tau^\alpha)$

$$S(\theta, \tau^\alpha) = \chi. \tag{29}$$

We assume this family to cover a region $\mathcal{R}$ of T simply, thus to each point **p** in γ there is a unique χ belonging to the family. Let γ be a C^2 curve in T

$$\tau^\alpha = \tau^\alpha(\theta) \tag{30}$$

which intersects the family (29), nowhere tangentially to any one of its members, and is such that the components of the tangent vector τ'^α are so as to generate a direction to *minimize* the fundamental integral (23). We call this direction the *temporal gradient* which is given, in general, by the condition (Rund 1973, pg 20)

$$\partial \mathcal{L}/\partial \tau'^\alpha = (\mathcal{L}/\chi')\partial S/\partial \tau^\alpha \tag{31}$$
$$\text{where } \chi' = d\chi/d\theta$$

The *temporal gradient* direction together with (31) and the definition of the T-momentum (16) immediately suggest that a condition be placed on the selection of hypersurfaces, (29). Let the family of hypersurfaces which satisfy the condition

$$\mathcal{L} = \chi' \tag{32}$$

be referred to as *temporally equidistant with respect to the T-Lagrangian* $\mathcal{L}(\theta, \tau^\alpha, \tau'^\alpha)$. So that under this condition we have

$$p_\alpha = \partial \mathcal{L}/\partial \tau'^\alpha = \partial S/\partial \tau^\alpha \tag{33}$$

from which we have by (19), $\tau'^\alpha = \varphi(\theta, \tau^\alpha, p_\alpha)$. Thus by virtue of (33) and (29) we obtain a set of first order o.d.e.'s

$$\tau'^\alpha = F^\alpha(\theta, \tau^\beta) \tag{34}$$

the solutions of which yield a 3-parameter family of curves called the T-congruence of curves belonging to the family of hypersurfaces $S(\theta, \tau^\alpha) = \chi$. These results may now be used to establish the next lemma (Rund 1973):

Lemma 3. Given **p** an arbitrary point on a given hypersurface $S(\theta, \tau) = \chi_1$ with **q** a member of the *T-congruence*, belonging to the family, which intersects any other member of the family with parameter value χ_2, then

$$I = \int_{\mathbf{p}}^{\mathbf{q}} \mathcal{L}(\theta, \tau^\alpha, \tau'^\alpha) d\theta = \chi_1 - \chi_2 \tag{35}$$

and, is independent of the position of the *initial point* **p** on the first hypersurface.

Remarks. 1. In a sense, we may say that two temporally equidistant hypersurfaces, characterized by their respective parameters χ_1 and χ_2, cut off "equal parts" from every member of the congruence belonging to the family.

2. We are thus motivated to establish a relationship between such hypersurfaces and the *duration function*, dur(P, Q), on the Set of Instants $\mathcal{I}$, wherein, for I given by (35):

$$\text{dur}(P, Q) \simeq I. \tag{36}$$

3. Before establishing such a relationship we must specify a kind of *reality condition* that we impose on the nature of the curves characterized by (30) that is consistent with "observable time". To this end we recognize that to each point **p** in γ there is associated a unique χ which generates the hypersurface (29), and as such, χ may be regarded as a monotonic function of θ along γ to which we impose this *reality condition*, viz.,

$$\chi' = d\chi/d\theta > 0. \tag{37}$$

We state the following theorem (Rund 1973) which establishes the necessary and sufficient conditions which must be satisfied by a family of hypersurfaces in order that they be *temporally equidistant with respect to the T-Lagrangian* $\mathcal{L}(\theta, \tau^\alpha, \tau'^\alpha)$.

Theorem 2. A family of hypersurfaces $S(\theta, \tau^\alpha) = \chi$ is *temporally equidistant* with respect to the *T-Lagrangian* $\mathcal{L}(\theta, \tau^\alpha, \tau'^\alpha)$ if, and only if, $S(\theta, \tau^\alpha)$ is a solution of the Hamilton-Jacobi equation,

$$\partial S/\partial\theta + \mathcal{H}(\theta, \tau^\alpha, \partial S/\partial\tau^\alpha) = 0 \tag{38}$$

$$\text{with } p_\alpha = \partial S/\partial\tau^\alpha \tag{39}$$

where $\mathcal{H}(\theta, \tau^\alpha, p_\alpha)$, given by (20), is the *T-Hamiltonian* associated with the given *T-Lagrangian*.

Remark. This construction, namely, establishing the family of *temporally equidistant* hypersurfaces together with the congruence of curves belonging to it, is tantamount to constructing the *complete figure of the problem* in the calculus of variations.

IV. The duration function on T-Space. Let us suppose that we are given a family of hypersurfaces $S(\theta, \tau^\alpha) = \chi$ on T-Space which are solutions to (38) with (39) and are thus *temporally equidistant*. This suggests, according to remark 2 above, that we establish a firm relationship between $\mathrm{dur}(P, Q)$ on the Set of Instants $\mathcal{I}$ and $I = \int_p^q \mathcal{L}(\theta, \tau^\alpha, \tau'^\alpha)d\theta = \chi_2 - \chi_1$ on the T-Space. To this end we are motivated by the desire to maintain a clear connection between a *time track* on $\mathcal{I}$ and curves in T-Space, for, the *duration function*, as introduced by DeVito, adheres to properties common to "observable time". By constructing the *complete figure of the problem* we are guided by the geometric simplicity of *temporally equidistant hypersurfaces*. That is, an "observable time" difference (ΔT) should not only be equivalent to all observers on a *time track* on $\mathcal{I}$, but should also be equivalent when considered with respect to the T-*Space* of that *time track*. In this regard let us now require that the real number $|\chi_2 - \chi_1|$ in (38) be identified with the real number defined by $\mathrm{dur}(P, Q)$ in (1) for $P, Q \in \mathcal{C}(\mathcal{I})$. We refer to this identification as the *local temporal equivalent condition*. Thus let fixed points $P, Q \in \mathcal{C}(\mathcal{I})$ with $\phi(P) = (\tau_1^\alpha)$ and $\phi(Q) = (\tau_2^\alpha)$ be identified with the endpoints $\mathbf{p}, \mathbf{q}$ of γ a member of the T-congruence with respective coordinates $(\theta_1, \tau_1^\alpha)$ and $(\theta_2, \tau_2^\alpha)$ which determine the members χ_1 and χ_2 of the family of hypersurfaces through which γ passes and by the *local temporal equivalence condition* write

$$\mathrm{dur}(P, Q) = |\rho| = |\chi_2 - \chi_1| \tag{40}$$

which immediately satisfy (1a,b) for $P, Q \in \mathcal{C}(\mathcal{I})$.

Lemma 3. The *local temporal equivalence condition* constitutes a duration function restricted to $\mathcal{C}(\mathcal{I})$

APPENDIX

1. The proof of the theorem follows simply by partial differentiation of (6) with respect to y_j followed by substitution of (5), noting that $\xi^j = \xi^j(y_k)$ by virtue of (3):

$$\partial h/\partial y_j = \xi^k \delta_k^j + y_k \partial \xi^k/\partial y_j - (\partial f/\partial \xi^k)(\partial \xi^k/\partial y_j) \; \square$$

Also, since T followed by T^{-1} is the identity I we have

$$\partial(y,_1, \ldots, y,_n)/\partial(\xi_1, \ldots, \xi_n) \cdot \partial(\xi,_1, \ldots, \xi,_n)/\partial(y_1, \ldots, y_n) = 1$$

which, by (5) and (7), yields (8).

2. The corollary follows immediately with (9) written in terms of λ as $f(\xi^j, \lambda) + h(y^j, \lambda) = \xi^k y_k$ regarding (ξ^j, λ) as independent variables then

$\partial f/\partial\lambda + \partial h/\partial\lambda + (\partial h/\partial y^j)(\partial y^j/\partial\lambda) = \xi^k(\partial y_k/\partial\lambda)$ so that from (7) the result now follows.

References

Cole, E. A. B.: 1980, *J. Phys. A: Math*, **13**, 109-115

Cole, E. A. B. and Starr, I. M.: 1990, *Nouvo Cim. B*, **105**, 1091-1102

DeVito, C.: 1995, *"First Steps Towards a Mathematical Theory of Time*, (unpublished)

Lehto, A.: 1990, *Chinese Journal of Physics*, **28**, 215-235

Osgood, W. F.: 1946, *Mechanics*, McMillan, New York

Rund, H.: 1973, *The Hamilton-Jacobi Theory in the Calculus of Variations*, Robert E. Kreiger Publishing Co., Huntington, New York

Tifft, W. G.: 1996, *ApJ*, **468**, 491.

FISHER INFORMATION AS A MEASURE OF TIME

B. R. FRIEDEN
Optical Sciences Center
University of Arizona
Tucson, Arizona 85721-0094

Abstract.
Fisher information I is a classical concept that originates in estimation theory. Through the Cramer-Rao inequality, it defines the smallest possible error in the estimation of a parameter in the presence of noise obeying a given probability law. More recently, Fisher information has been incorporated within a variational principle for forming the laws of physics (Schrödinger wave equation, Dirac equation, etc.). The premise is that $dI/dt \leq 0$, with t the time, so that, at equilibrium, $I = \min$. The premise has recently been proven for any process obeying a Fokker-Planck differential equation. Hence, Fisher information provides a new measure of the passage of time. All errors of estimation increase, on average, with time.

1. "Smart" Measurements

Consider the following problem:

Data

$$\mathbf{y} = y_1, ..., y_N \qquad (1)$$

obey

$$y_n = \theta + x_n,$$

parameter θ to be found. An unbiased estimate $\hat{\theta} \equiv \hat{\theta}(\mathbf{y})$ is formed. It has a mean-squared error

$$e_\theta^2 = < [\theta - \hat{\theta}(\mathbf{y})]^2 > . \qquad (2)$$

Astrophysics and Space Science **244**: 387-391,1996.

The error obeys the Cramer-Rao relation

$$e_\theta{}^2 I(t) \geq 1, \tag{3}$$

where $I(t)$ is the *Fisher* information

$$I(t) = \int d^N y \frac{[\partial p/\partial \theta]^2}{p}, \ p \equiv p(\mathbf{y} \mid \theta, t). \tag{4}$$

[Note: When θ is particle position, Cramer-Rao relation (3) is equivalent to the Heisenberg uncertainty principle [1].]

2. Fisher I as a Cross Entropy

Alternatively, I is a cross entropy,

$$I(t) = -2 \lim_{\Delta\theta \to 0} \Delta\theta^{-2} \int d^N y \ p \ ln(p_{\Delta\theta}/p) \tag{5}$$
$$p_{\Delta\theta} \equiv p(\mathbf{y} \mid \theta + \Delta\theta, t), \ \ p \equiv p(\mathbf{y} \mid \theta, t).$$

(This may be verified by two uses of l'Hôpital's rule.)

Suppose that both p and $p_{\Delta\theta}$ obey (a) zero memory and (b) time reversal. Then [2] they obey a *Fokker-Planck* equation

$$\partial p/\partial t = \mathcal{L}p \tag{6}$$

$$\mathcal{L} \equiv -\partial/\partial y_i D_i(\mathbf{y}) + \partial^2/\partial y_i \partial y_j D_{ij}(\mathbf{y})$$

for arbitrary $D_i(\mathbf{y})$ and $D_{ij}(\mathbf{y})$. It follows that, analogous with the Boltzmann H-theorem [2], information I obeys an "I-theorem"

$$dI/dt \leq 0. \tag{7}$$

As an example look at Brownian motion. Here

$$y = \ \theta + x, \text{ so that} \tag{8}$$
$$p(y \mid \theta) = p(y - \theta).$$

Then

$$I = \int dx p'^2(x)/p(x). \tag{9}$$

Since $p(x) = N(0, \sigma^2)$ with $\sigma^2 = Dt$, this gives

$$I = 1/\sigma^2 = 1/Dt, \tag{10}$$

so that

$$dI/dt = -1/Dt^2 \leq 0 \tag{11}$$

as was predicted.

3. Estimates and Equilibrium Distributions

The foregoing actually interprets time as a measure of the ability to estimate. This follows because, by Eq. (3), the minimum possible mean square error obeys

$$e_{min}^2(t) = 1/I(t). \tag{12}$$

Then

$$de_{min}^2/dt = (-1/I^2)dI/dt \tag{13}$$

or

$$de_{min}^2/dt \geq 0. \tag{14}$$

Example: In the Brownian case, by Eqs. (10) and (13),

$$de_{min}^2/dt = -(Dt)^2(-1/Dt^2) \tag{15}$$
$$= D \ (Const.)$$

Here the minimized error increases in time at a *constant* rate.
The I-theorem (7) implies that

$$\lim_{t \to \infty} I(t) = min, \text{or } \delta I = 0 \tag{16}$$

at equilibrium. The latter suggests a variational principle for finding the underlying probability density function (PDF) $p(x)$. Appropriate constraints are suggested by positing information flow δJ from the measured phenomenon into the data, obeying a conservation law

$$\delta J = \delta I, \text{ or } \delta(I - J) = 0. \tag{17}$$

This is called the principle of extreme physical information (EPI). It parallels the second law of statistical mechanics in predicting increasing disorder for a system. However, in that $p(x)$ is of a general nature, EPI embraces all physical phenomena (not just statistical mechanics).

The following physical laws $p(x)$, without the time, have been derived using the EPI principle:

(1) The Schroedinger wave equation [3]
(2) The Maxwell-Boltzmann and Boltzmann distributions [4]
(3) The Dirac equation, Klein-Gordon equation [5]
(4) The $1/f$ power spectral noise law [6]

4. Covariant EPI Derivations (with the time)

Now regard time as a fourth measured coordinate (time is now treated as a random variable like x, y, z). Suppose that N independent 4-measurements of (now) N generally different parameters $\theta_n, n = 1, ..., N$ are made. Then the Fisher information is [7]

$$I = 4 \int d^4x q_{n,\lambda} \, q_n{}^{,\lambda} \tag{18}$$

where

$$p_n \equiv p_n(\mathbf{x}) \equiv q_n^2(\mathbf{x}) \tag{19}$$

is the PDF for 4-noise $\mathbf{x}$ in measurement number n. The $q_n(\mathbf{x})$ are defined as real probability amplitudes. Then EPI derives, *with* the time, the following physical laws:

(5) The Dirac equation, Klein-Gordon equation [8]
(6) The *constancy* of physical constants c, $\hbar$, e [9]
(7) The Einstein field equations, Planck length [10]

5. EPI as a Physical Process

The variation δI essential to EPI Eq. (17) is due to variations δp_n in the PDFs or equivalently variations δq_n in the probability amplitudes defined in Eq. (19). When measurements $\mathbf{y}$ are made, the variations δq_n arise as quantum perturbations caused by the measurement process [11].

We conclude that when measurements $\mathbf{y}$ are made, EPI results as a physical process which culminates in the physical law governing the amplitude functions $q_n(\mathbf{x})$. Interestingly, each measurement y_n defines a corresponding amplitude $q_n(\mathbf{x})$. Hence, each physical law $q_n(\mathbf{x})$ arises 1:1 as a reaction to measurement.

References

1. Frieden, B. R.: 1992, *Phys. Lett. A*, **169**, 123
2. Risken, H.: 1984, *The Fokker-Planck Equation*, (Springer-Verlag, New York), p. 133
3. Frieden, B. R.: 1989, *Am. J. Phys.*, **57**, 1004
4. Frieden, B. R.: 1990, *Phys. Rev. A*, **41**, 4265
5. Frieden, B. R.: 1993, *Phys. Rev. A*, **198**, 262
6. Frieden, B. R.: 1994, *Phys. Rev. E*, **49**, 2644
7. Frieden, B. R. and Cocke, W. J.: 1996, *Phys. Rev. E*, **54**, 257
8. Frieden, B. R.: 1995, *Adv. in Imaging and Electron Phys.*, **90** (Acad. Press, Orlando), p. 123
9. Frieden, B. R. and Soffer, B. H.: 1995, *Phys. Rev. E*, **52**, 2274
10. Cocke, W. J. and Frieden, B. R.: 1996, in *Classical and Quantum Gravity*, under review
11. Frieden, B. R. and Soffer, B. H.: 1996, *Phys. Rev. E*, under review

LIST OF FORTHCOMING PAPERS

P.G. Kazantzis: The Structure of Periodic Solutions in the Restricted Problem of the Three Bodies. II. Sun-Jupiter Case (Received 4 July, 1996; accepted 10 January, 1997)

M.K. Mak, T. Harko and P.C.W. Fung: Viscous Fluid Cosmological Models in a Bianchi Type I Universe (Received 22 October, 1996; accepted 4 February, 1997)

Ng. Ibohal: On the relationship between Killing-Yano Tensors and Electromagnetic Fields on Curved Spaces (Received 25 March, 1996; accepted 10 February, 1997)

D.M.Z. Jassur: Close Binary System Go Cygnus: Standard UBV Observations and Light Curve Analysis (Received 14 January, 1997; accepted 19 February, 1997)

N.A. Silant'ev: Methods of Calculation of Turbulent Diffusivities (Received 6 August, 1996; accepted 19 February, 1997)

Yu.N. Gnedin: Resonance Magnetic Conversion of Photons into Massless Axions and Striking Feature in Quasar Polarized Light (Received and accepted 4 March, 1997)

I.L. Rosental and I.V. Belousova: Origin of Cosmic Gamma-Bursts (Received and accepted 4 March, 1997)

C.B. Kilinç and I. Yavuz: Cylindrically-symmetric and Inhomogeneous Cosmological Models with Viscous Fluid (Received 27 June, 1996; accepted 14 March, 1997)

THE KLUWER LATEX STYLE FILE

Kluwer Academic Publishers has developed a special style file for authors who want to submit LaTeX articles. KLUWER.STY is a general LaTeX style file which is used for all Kluwer journals, irrespective of the publication's size or layout. (The specific journal characteristics are added later during the production process.) Authors are kindly requested always to use KLUWER.STY when creating a LaTeX article for a Kluwer journal.

Instruction File
Although KLUWER.STY is very similar to the ARTICLE.STY and uses many of the standard LaTeX commands, there are some differences. These are explained in the accompanying instruction file - KAPINS[number].TEX

Getting the Kluwer Style File

KLUWER.STY is offered at a number of servers around the world. Unfortunately, those are unauthorized copies and authors are strongly advised not to use them. Kluwer can only guarantee the integrity of files obtained directly from Kluwer.

WWW-site, E-mail or Air Mail
Authors can obtain KLUWER.STY and the instruction file from Kluwer Academic Publishers' information system KAPIS. This free service is available at:

WWW.WKAP.NL

The stylefile may be requested from:

Kluwer Academic Publishers, Editorial Department,
P.O.Box 990, 3300 AZ Dordrecht, The Netherlands.
Telephone: (0)78-6392392, Fax: (0)78-6392555
E-mail: EDITDEPT@WKAP.NL

The files can be sent either by e-mail or on diskette. Don't forget to mention the journal's name, your e-mail number, and postal address.

Astrophysics and Space Science **244**: 395–396, 1997.

Submitting Manuscripts

Please send your completed LaTeX article on diskette, together with the appropriate number of hard copies to the address listed in the *Instructions to Authors* of your journal.

Via E-mail
Experience has shown that sending articles via e-mail can sometimes result in lacunas appearing in the article. That's why Kluwer prefers to receive LaTeX articles on diskettes, accompanied by the hard copies. If you must send the LaTeX article via e-mail, don't forget to send the requisite number of hard copies by air mail.

Questions

Should you have any questions or encounter problems using KLUWER.STY, please contact Kluwer Academic Publishers for assistance.

The manufacturer's authorised representative in the EU is Springer
Nature Customer Service Centre GmbH, Europaplatz 3, 69115 Heidelberg,
Germany. If you have any concerns regarding our products, please
contact ProductSafety@springernature.com

Printed and bound by CPI Group (UK) Ltd, Croydon, CR0 4YY
29/04/2026
02099472-0008